Functional Equations – Results and Advances

Advances in Mathematics

VOLUME 3

The titles published in this series are listed at the end of this volume.

Functional Equations – Results and Advances

edited by

ZOLTÁN DARÓCZY

Institute of Mathematics and Informatics,
University of Debrecen, Hungary

and

ZSOLT PÁLES

Institute of Mathematics and Informatics,
University of Debrecen, Hungary

KLUWER ACADEMIC PUBLISHERS

DORDRECHT / BOSTON / LONDON

A C.I.P. Catalogue record for this book is available from the Library of Congress.

ISBN 978-1-4419-5210-3

Published by Kluwer Academic Publishers,
P.O. Box 17, 3300 AA Dordrecht, The Netherlands.

Sold and distributed in North, Central and South America
by Kluwer Academic Publishers,
101 Philip Drive, Norwell, MA 02061, U.S.A.

In all other countries, sold and distributed
by Kluwer Academic Publishers,
P.O. Box 322, 3300 AH Dordrecht, The Netherlands.

Printed on acid-free paper

This book is dedicated
to the Millennium of
The Hungarian State

Contents

Preface

The theory of functional equations has been developed in a rapid and productive way in the second half of the Twentieth Century. First of all, this is due to the fact that the mathematical applications raised the investigations of newer and newer types of functional equations. At the same time, the selfdevelopment of this theory was also very fruitful. This can be followed in many monographs that treat and discuss the various methods and approaches.

These developments were also essentially influenced by a number journals, for instance, by the *Publicationes Mathematicae Debrecen* (founded in 1953) and by the *Aequationes Mathematicae* (founded in 1968), because these journals published papers from the field of functional equations readily and frequently. The latter journal also publishes the yearly report of the International Symposia on Functional Equations and a comprehensive bibliography of the most recent papers. At the same time, there are periodically and traditionally organized conferences in Poland and in Hungary devoted to functional equations and inequalities.

In 2000, the 38th International Symposium on Functional Equations was organized by the Institute of Mathematics and Informatics of the University of Debrecen in Noszvaj, Hungary. The report about this meeting can be found in *Aequationes Math.* **61** (2001), 281–320.

Encouraged by the *Kluwer Academic Publishers*, and personally by *John Martindale*, the organizers of the conference decided to invite the leading experts of functional equations to contribute by original papers to a volume dedicated to the memory of the last millennium. The papers were submitted by the end of the summer of 2000 and then refereed by two independent referees The material accepted finally for publication reflects very well the complexity and applicability of the most active

ix

fields in the theory of functional equations. Now we briefly mention the different subjects and areas considered in this volume together with the names of the contributing authors.

- *Classical functional equations and inequalities.* (R. Badora–R. Ger, K. Baron, B. Choczewski, R. Ger–K. Nikodem, A. Járai–W. Sander, Lajkó–Zs. Páles)

- *Stability of functional equations.* (Z. Boros, A. Gilányi, Z. Moszner, Jacek Tabor–József Tabor)

- *Functional equations in one variable and iteration theory.* (K. Ciepliński–M. C. Zdun, R. Girgensohn, H. H. Kairies)

- *Composite functional equations and the theory of means.* (Z. Daróczy–Zs. Páles, G. Hajdu, L. Losonczi, J. Matkowski, M. Sablik)

- *Functional equations on algebraic structures.* (T. M. K. Davison, F. Skof, L. Székelyhidi, Jacek Tabor)

- *Functional equations in functional analysis,* (J. Baker, B. Ebanks, L. Molnár)

- *Bisymmetry and associativity type equations on quasigroups.* (C. Alsina–E. Trillas, A. Krapež, M. Taylor)

The editors of this volume are convinced that the results and methods represented in this book give a representative crossection of what is recently happening in the field of functional equations.

Finally we wish to thank *Zsolt Szálkai* who helped us in the editorial work mainly by typesetting (LATEXing) and unifying the contributed papers.

Debrecen, June 29, 2001.

THE EDITORS

I
CLASSICAL
FUNCTIONAL EQUATIONS
AND INEQUALITIES

ON SOME TRIGONOMETRIC FUNCTIONAL INEQUALITIES

Roman Badora and Roman Ger

Institute of Mathematics, Silesian University,
ul. Bankowa 14, PL-40-007, Katowice, Poland
robadora@ux2.math.us.edu.pl, romanger@us.edu.pl

Abstract   We deal with d'Alembert's and Wilson's differences

$$f(x+y) + f(x-y) - 2f(x)f(y)$$

and

$$f(x)f(y) - f\left(\frac{x+y}{2}\right)^2 + f\left(\frac{x-y}{2}\right)^2,$$

respectively, assuming that their absolute values (or norms) are majorized by some function in a single variable. The superstability type results obtained are then used to characterize the functions

$$x \longmapsto \cos \alpha x \qquad and \qquad x \longmapsto \sin \alpha x$$

of real variable x with a complex but not real coefficient α.

Keywords: stability, d'Alembert's equation, Wilson's equation, functional inequality, vector-valued solution

Mathematics Subject Classification (2000): 39B82, 39B62, 39B52

1.   INTRODUCTION

The Hyers-Ulam stability of the basic trigonometric functional equations

$$f(x+y) + f(x-y) = 2f(x)f(y) \tag{A}$$

of d'Alembert, and

$$f(x+y)f(x-y) = f(x)^2 - f(y)^2 \tag{W}$$

3

Z. Daróczy and Z. Páles (eds.),
Functional Equations - Results and Advances, 3–15.
© 2002 Kluwer Academic Publishers.

of Wilson, were established by J. A. Baker [1] and P. W. Cholewa [2], respectively. As a matter of fact, they have observed the so called *superstability* phenomenon stating that whenever d'Alembert's difference

$$A_f(x, y) := f(x + y) + f(x - y) - 2f(x)f(y)$$

(resp. Wilson's difference

$$W_f(x, y) := f(x + y)f(x - y) - f(x)^2 + f(y)^2 \,)$$

remains bounded in absolute value then f itself being unbounded has to satisfy equation (A) (resp. (W)). Related results were also obtained by L. Székelyhidi [4], [5] and by W. Förg-Rob and J. Schwaiger [3].

Note that, in the case where the division by 2 is uniquely performable in the domain of f, the inequality $|W_f| \leq \varepsilon$ is equivalent to $|\tilde{W}_f| \leq \varepsilon$, where

$$\tilde{W}_f(x, y) := f(x)f(y) - f\left(\frac{x + y}{2}\right)^2 + f\left(\frac{x - y}{2}\right)^2.$$

In order to achieve a greater uniformity of the presentation, while considering Wilson's differences we shall confine ourselves to the operator $\tilde{W}_f$.

The aim of the present paper is to examine the stability (superstability) of d'Alembert's and Wilson's equations by admitting not necessarily constant bounds. More precisely, we deal with functional inequalities of the type

$$|T_f(x, y)| \leq \varphi(y) \qquad \text{or} \qquad |T_f(x, y)| \leq \varphi(x),$$

where $T_f \in \{A_f, \tilde{W}_f\}$. With no regularity assumptions whatsoever upon f we shall show that the only unbounded solutions f of these stability type functional inequalities are just the solutions of the corresponding functional equation. After establishing superstability results of that kind, we present also some new characterizations of the classical trigonometric functions mapping the real line $\mathbb{R}$ into the complex plane $\mathbb{C}$.

2.   GENERALIZED STABILITY OF D'ALEMBERT'S EQUATION

We begin with two stability results concerning complex valued mappings.

Theorem 1. Let $(G, +)$ be an Abelian group and let $f : G \to \mathbb{C}$ and $\varphi : G \to \mathbb{R}$ satisfy the inequality

$$|f(x + y) + f(x - y) - 2f(x)f(y)| \leq \varphi(y) \qquad \text{for all} \quad x, y \in G. \quad (1)$$

Then either f is bounded or

$$f(x+y) + f(x-y) = 2f(x)f(y) \qquad \text{for all} \quad x, y \in G.$$

Proof. Assume that $f : G \to \mathbb{C}$ yields an unbounded solution of inequality (1). Thus, there exists a sequence $(z_n)_{n \in \mathbb{N}}$ of elements from G such that

$$0 \neq |f(z_n)| \longrightarrow \infty \quad \text{as} \quad n \longrightarrow \infty. \tag{2}$$

On setting $x = z_n$ and $y = x$ in (1), we get

$$|f(z_n + x) + f(z_n - x) - 2f(z_n)f(x)| \leq \varphi(x)$$

for all $x \in G$ and $n \in \mathbb{N}$, whence

$$\left| \frac{f(z_n + x) + f(z_n - x)}{2f(z_n)} - f(x) \right| \leq \frac{\varphi(x)}{2|f(z_n)|}$$

for all $x \in G$ and $n \in \mathbb{N}$. Now, passing to the limit as $n \longrightarrow \infty$ and taking (2) into account, we infer that

$$f(x) = \lim_{n \longrightarrow \infty} \frac{f(z_n + x) + f(z_n - x)}{2f(z_n)} \qquad \text{for all} \quad x \in G. \tag{3}$$

On the other hand, by means of (1), for all $x, y \in G$ and all $n \in \mathbb{N}$, one has

$$
\begin{aligned}
2\varphi(y) \geq &\; |f(z_n + x + y) + f(z_n + x - y) - 2f(z_n + x)f(y)| \\
&+ |f(z_n - x + y) + f(z_n - x - y) - 2f(z_n - x)f(y)| \\
\geq &\; |f(z_n + x + y) + f(z_n - (x + y)) \\
&+ f(z_n + x - y) + f(z_n - (x - y)) \\
&- 2[f(z_n + x) + f(z_n - x)]f(y)|,
\end{aligned}
$$

or, equivalently,

$$
\begin{aligned}
\frac{\varphi(y)}{|f(z_n)|} \geq &\; \left| \frac{f(z_n + x + y) + f(z_n - (x + y))}{2f(z_n)} \right. \\
&+ \frac{f(z_n + x - y) + f(z_n - (x - y))}{2f(z_n)} \\
&\left. - 2 \frac{f(z_n + x) + f(z_n - x)}{2f(z_n)} f(y) \right|,
\end{aligned}
$$

whence, by passing to the limit as $n \longrightarrow \infty$, we arrive at

$$0 \geq |f(x+y) + f(x-y) - 2f(x)f(y)| \qquad \text{for all} \quad x, y \in G,$$

because of (2) and (3). Thus the proof has been completed.   □

Theorem 2. *Let $(G, +)$ be an Abelian group and let $f : G \to \mathbb{C}$ and $\varphi : G \to \mathbb{R}$ satisfy the inequality*

$$|f(x + y) + f(x - y) - 2f(x)f(y)| \le \varphi(x) \qquad \text{for all} \quad x, y \in G. \quad (4)$$

Then either f is bounded or

$$f(x + y) + f(x - y) = 2f(x)f(y) \qquad \text{for all} \quad x, y \in G.$$

Proof. Assume that $f : G \to \mathbb{C}$ yields an unbounded solution of inequality (4). Thus, there exists a sequence $(z_n)_{n \in \mathbb{N}}$ of elements from G such that relationship (2) holds true. Putting $y = z_n$ in (4), dividing both sides by $2|f(z_n)|$ and passing to the limit as $n \longrightarrow \infty$ we obtain that

$$f(x) = \lim_{n \to \infty} \frac{f(x + z_n) + f(x - z_n)}{2f(z_n)} \qquad \text{for all} \quad x \in G.$$

The rest of the proof runs along the same lines as before.   □

As we shall see later on, the latter statement is no longer valid while dealing with the corresponding two cases regarding Wilson's equation. But first, to complete this section, we prove some vector-valued analogues of the two results just obtained.

Theorem 3. *Let $(G, +)$ be an Abelian group and let $(\mathcal{A}, \| \cdot \|)$ be a semisimple commutative Banach algebra. Assume that $f : G \to \mathcal{A}$ and $\varphi : G \to \mathbb{R}$ satisfy anyone of the inequalities*

$$\|f(x + y) + f(x - y) - 2f(x)f(y)\| \le \varphi(y) \qquad \text{for all} \quad x, y \in G, \quad (5)$$

or

$$\|f(x + y) + f(x - y) - 2f(x)f(y)\| \le \varphi(x) \qquad \text{for all} \quad x, y \in G. \quad (6)$$

Then

$$f(x + y) + f(x - y) = 2f(x)f(y) \qquad \text{for all} \quad x, y \in G,$$

provided that for an arbitrary linear multiplicative functional $x^* \in \mathcal{A}^*$ the superposition $x^* \circ f$ fails to be bounded.

Proof. Assume (5) and fix arbitrarily a linear multiplicative functional $x^* \in \mathcal{A}^*$. As well known we have $\|x^*\| = 1$ whence, for every $x, y \in G$, we have

$$\begin{aligned}
\varphi(y) &\ge \|f(x + y) + f(x - y) - 2f(x)f(y)\| \\
&= \sup_{\|y^*\|=1} |y^* \left(f(x + y) + f(x - y) - 2f(x)f(y) \right)| \\
&\ge |x^*(f(x + y)) + x^*(f(x - y)) - 2x^*(f(x))x^*(f(y))|,
\end{aligned}$$

which states that the superposition $x^* \circ f$ yields a solution of inequality (1). Since, by assumption, that superposition is unbounded an appeal to Theorem 1 shows that the function $x^* \circ f$ solves d'Alembert's equation (A). In other words, bearing the linear multiplicativity of x^* in mind, for all $x, y \in G$, d'Alembert's difference $A_f(x, y)$ falls into the kernel of x^*. Therefore, in view of the unrestricted choice of x^*, we infer that

$$A_f(x, y) \in \bigcap \{\, \ker x^* \,:\, x^* \text{ is a multiplicative member of } \mathcal{A}^* \,\}$$

for all $x, y \in G$. Since the algebra $\mathcal{A}$ has been assumed to be semisimple, the last term of the above formula coincides with the singleton $\{0\}$, i.e.

$$f(x + y) + f(x - y) - 2f(x)f(y) = 0 \qquad \text{for all} \quad x, y \in G,$$

as claimed.

If (6) is assumed instead of (5), we proceed in the same way. This completes the proof. $\qquad\qquad\qquad\qquad\qquad\qquad\qquad\qquad\qquad\qquad\qquad\quad$ $\square$

Remark 1. It would be desirable to get rid of somewhat awkward hypothesis that all the superpositions $x^* \circ f$ are unbounded, assuming simply that f itself fails to be bounded. However, as shown by the counterexample below, the assumption that f yields an unbounded solution of inequality (5) (resp. (6)) is insufficient to force f to satisfy equation (A). In fact, consider a Banach algebra $\mathcal{A}$ of all diagonal 2×2-matrices with complex entries and a function $f : \mathbb{R} \to \mathcal{A}$ given by the formula

$$f(x) := \begin{bmatrix} \cos \alpha x & 0 \\ 0 & b \end{bmatrix} \qquad (x \in \mathbb{R}),$$

where $\alpha \in \mathbb{C} \setminus \mathbb{R}$ and $b \in \mathbb{C} \setminus \{0, 1\}$ are arbitrarily fixed. Then f yields an unbounded solution to both the inequalities (5) and (6) with an arbitrary majorizing function $\varphi : \mathbb{R} \to \mathbb{R}$ enjoying the property $\inf\{\varphi(t) : t \in \mathbb{R}\} \geq 2|b| \cdot |1 - b|$. Nevertheless, f fails to satisfy d'Alembert's equation (A).

What is missing? The superposition of f with the linear multiplicative functional $x^* : \mathcal{A} \to \mathbb{C}$ given by

$$x^* \left(\begin{bmatrix} a & 0 \\ 0 & b \end{bmatrix} \right) := b \qquad \left(\begin{bmatrix} a & 0 \\ 0 & b \end{bmatrix} \in \mathcal{A} \right),$$

is constant and hence bounded.

3.   GENERALIZED STABILITY OF WILSON'S EQUATION

Following the scheme applied in the previous section we proceed with two stability results concerning complex valued mappings.

Theorem 4. *Let $(G,+)$ be a uniquely 2-divisible Abelian group and let $f : G \to \mathbb{C}$ and $\varphi : G \to \mathbb{R}$ satisfy the inequality*

$$\left| f(x)f(y) - f\left(\frac{x+y}{2}\right)^2 + f\left(\frac{x-y}{2}\right)^2 \right| \leq \varphi(y) \qquad \text{for all} \quad x,y \in G. \tag{7}$$

Then either f is bounded or

$$f(x)f(y) = f\left(\frac{x+y}{2}\right)^2 - f\left(\frac{x-y}{2}\right)^2 \qquad \text{for all} \quad x,y \in G.$$

Proof. Assume that $f : G \to \mathbb{C}$ yields an unbounded solution of inequality (7). Thus, there exists a sequence $(z_n)_{n \in \mathbb{N}}$ of elements from G such that

$$0 \neq |f(2z_n)| \longrightarrow \infty \quad \text{as} \quad n \longrightarrow \infty. \tag{8}$$

Inequality (7) may equivalently be written as

$$\left| f(2x)f(2y) - f(x+y)^2 + f(x-y)^2 \right| \leq \varphi(2y) \qquad \text{for all} \quad x,y \in G. \tag{9}$$

In particular, (9) implies that

$$\left| f(2y) - \frac{f(z_n+y)^2 - f(z_n-y)^2}{f(2z_n)} \right| \leq \frac{\varphi(2y)}{|f(2z_n)|} \qquad \text{for all} \quad y \in G, n \in \mathbb{N},$$

which jointly with (8) gives

$$f(2y) = \lim_{n \to \infty} \frac{f(z_n+y)^2 - f(z_n-y)^2}{f(2z_n)} \qquad \text{for all} \quad y \in G. \tag{10}$$

Now, with the aid of (7), we arrive at

$$2\varphi(y) \geq \left| f(2z_n + x)f(y) - f\left(z_n + \frac{x+y}{2}\right)^2 + f\left(z_n + \frac{x-y}{2}\right)^2 \right|$$

$$+ \left| f(2z_n - x)f(y) - f\left(z_n - \frac{x-y}{2}\right)^2 + f\left(z_n - \frac{x+y}{2}\right)^2 \right|$$

$$\geq \left| [f(2z_n + x) + f(2z_n - x)]f(y) \right.$$

$$- \left[f\left(z_n + \frac{x+y}{2}\right)^2 - f\left(z_n - \frac{x+y}{2}\right)^2 \right]$$

$$\left. + \left[f\left(z_n + \frac{x-y}{2}\right)^2 - f\left(z_n - \frac{x-y}{2}\right)^2 \right] \right|$$

for all $x, y \in G$ and every $n \in \mathbb{N}$. Consequently,

$$\frac{2\varphi(y)}{|f(2z_n)|} \geq \left| \frac{f(2z_n + x) + f(2z_n - x)}{f(2z_n)} f(y) \right.$$
$$- \frac{f\left(z_n + \frac{x+y}{2}\right)^2 - f\left(z_n - \frac{x+y}{2}\right)^2}{f(2z_n)}$$
$$\left. + \frac{f\left(z_n + \frac{x-y}{2}\right)^2 - f\left(z_n - \frac{x-y}{2}\right)^2}{f(2z_n)} \right|$$

for all $x, y \in G$ and every $n \in \mathbb{N}$. Passing here to the limit as $n \longrightarrow \infty$ with the use of (8) and (10) (note that f must not vanish identically), we conclude that, for every $x \in G$, there exists the limit

$$g(x) := \lim_{n \to \infty} \frac{f(2z_n + x) + f(2z_n - x)}{f(2z_n)}.$$

Moreover, the function $g : G \to \mathbb{C}$ obtained in that way has to satisfy the equation

$$g(x)f(y) = f(x+y) - f(x-y) \qquad \text{for all} \quad x, y \in G. \qquad (11)$$

Directly from the definition of g, we get the equality $g(0) = 2$ which jointly with (11) implies that f has to be odd. Keeping this in mind, by means of (11), we infer that the equalities

$$\begin{aligned}
f(x+y)^2 - f(x-y)^2 &= (f(x+y) + f(x-y))\,(f(x+y) - f(x-y)) \\
&= (f(x+y) + f(x-y))\,g(x)f(y) \\
&= (g(x)f(x+y) + g(x)f(x-y))\,f(y) \\
&= (f(2x+y) - f(-y) + f(2x-y) - f(y))\,f(y) \\
&= (f(2x+y) + f(2x-y))\,f(y) \\
&= (f(y+2x) - f(y-2x))\,f(y) \\
&= g(y)f(2x)f(y)
\end{aligned}$$

hold true for all $x, y \in G$. Since the oddness forces f to vanish at 0, putting $y = x$ in (11) we conclude that

$$f(2y) = f(y)g(y) \qquad \text{for all} \quad y \in G.$$

This, in turn, leads to the equation

$$f(x+y)^2 - f(x-y)^2 = f(2x)f(2y)$$

valid for all $x, y \in G$ which, in the light of the unique 2-divisibility of G, states nothing else but (W) and completes the proof. $\qquad \square$

Theorem 5. Let $(G, +)$ be a uniquely 2-divisible Abelian group and let $f : G \to \mathbb{C}$ and $\varphi : G \to \mathbb{R}$ satisfy the inequality

$$\left| f(x)f(y) - f\left(\frac{x+y}{2}\right)^2 + f\left(\frac{x-y}{2}\right)^2 \right| \leq \varphi(x) \qquad \text{for all} \quad x, y \in G.$$
(12)

Then either f is bounded or

$$f(x)f(y) = f\left(\frac{x+y}{2}\right)^2 - f\left(\frac{x-y}{2}\right)^2 \qquad \text{for all} \quad x, y \in G.$$

Proof. Assume that $f : G \to \mathbb{C}$ yields an unbounded solution of inequality (12) and let $(z_n)_{n \in \mathbb{N}}$ stand for a sequence of elements from G such that relationship (8) holds true. Inequality (12) may equivalently be written as

$$\left| f(2x)f(2y) - f(x+y)^2 + f(x-y)^2 \right| \leq \varphi(2x) \qquad \text{for all} \quad x, y \in G,$$
(13)

whence, in particular, we get

$$\left| f(2x) - \frac{f(x+z_n)^2 - f(x-z_n)^2}{f(2z_n)} \right| \leq \frac{\varphi(2x)}{|f(2z_n)|}$$

for all $x \in G$ and all $n \in \mathbb{N}$. Jointly with (8) this implies that

$$f(2x) = \lim_{n \to \infty} \frac{f(x+z_n)^2 - f(x-z_n)^2}{f(2z_n)} \qquad \text{for all} \quad x \in G.$$

An obvious slight change in the proof steps applied after formula (10) allows one to state the existence of a limit

$$\tilde{g}(y) := \lim_{n \to \infty} \frac{f(y+2z_n) + f(-y+2z_n)}{f(2z_n)}.$$

for every y from G. Moreover, the function $\tilde{g} : G \to \mathbb{C}$ obtained in that way has to satisfy the equation

$$f(x)\tilde{g}(y) = f(x+y) + f(x-y) \qquad \text{for all} \quad x, y \in G.$$

Directly from the definition of $\tilde{g}$, we see that it yields an even function. Clearly, so is also the function $g := \frac{1}{2}\tilde{g}$. Moreover, $g(0) = \frac{1}{2}\tilde{g}(0) = 1$ and

$$f(x+y) + f(x-y) = 2f(x)g(y) \qquad \text{for all} \quad x, y \in G. \qquad (14)$$

We are going to show that

$$f(0) = 0. \tag{15}$$

Suppose that this is not the case. Then in what follows, without loss of generality, we may assume that $f(0) = 1$ (replacing, if necessary the function f by $f/f(0)$) and φ by $\varphi/f(0)$). Plainly, equation (14) applied for $x = 0$ says now that g is nothing else but the even part of f and equation (14) itself assumes the form

$$f(x+y) + f(x-y) = f(x)(f(y) + f(-y)) \qquad \text{for all} \quad x, y \in G.$$

Now, replacing y by $-y$ in (12) and adding the resulting inequality to (12) side by side, we conclude that

$$|f(x)f(y) + f(x)f(-y)| \le 2\varphi(x) \qquad (x, y \in G),$$

whence, by setting here $x = 0$, we obtain that $|f(y) + f(-y)| \le 2\varphi(0)$ for all $y \in G$. In other words, we have

$$|g(y)| \le \varphi(0) \qquad \text{for all} \quad y \in G. \tag{16}$$

Let h denote the odd part of f. Then we have $f = g + h$ as well as

$$
\begin{aligned}
h(x)g(y) - g(x)h(y) &= (f(x) - g(x))g(y) - g(x)(f(y) - g(y)) \\
&= f(x)g(y) - g(x)f(y) \\
&= \frac{1}{2}[f(x+y) + f(x-y)] \\
&\quad - \frac{1}{2}[f(x+y) + f(y-x)] \\
&= h(x-y)
\end{aligned}
$$

for all $x, y \in G$ because of (14) and the fact that h is just the odd part of f. Replacing here y by $-y$ and making use of the evenness of g as well as the oddness of h, we arrive at

$$h(x)g(y) + g(x)h(y) = h(x+y) \qquad (x, y \in G),$$

which implies the dupplication formula for h:

$$h(2x) = 2g(x)h(x)$$

valid for all $x \in G$. Applying (13) for $x = 0$, we get the inequality

$$|f(2y) - f(y)^2 + f(-y)^2| \le \varphi(0) \qquad (y \in G),$$

whence, in view of the representation $f = g + h$ and the dupplication formula just obtained, one has

$$|g(2y) - 2g(y)h(y)| \leq \varphi(0) \qquad (y \in G).$$

In particular,

$$|2g(y)h(y)| \leq |g(2y)| + \varphi(0) \leq 2\varphi(0)$$

for every $y \in G$ because of (16).

Now, on setting $x = y$ in (14), we deduce that

$$
\begin{aligned}
|f(2y) + 1| &= |2f(y)g(y)| = |2(g(y) + h(y))g(y)| \\
&\leq 2|g(y)|^2 + |2g(y)h(y)| \\
&\leq 2\varphi(0)^2 + 2\varphi(0)
\end{aligned}
$$

for all $y \in G$ stating, by means of the 2-divisibility of G, that f itself is globally bounded - a contradiction. Thus, equality (15) has been proved.

Equation (14) applied for $x = 0$ implies now that the even part of f vanishes identically; in other words, f has to be odd.

Applying (14) once again and the oddness of f just derived, we obtain the equalities

$$
\begin{aligned}
f(x+y)^2 - f(x-y)^2 &= (f(x+y) - f(x-y))\,(f(x+y) + f(x-y)) \\
&= (f(x+y) - f(x-y)) \cdot 2f(x)g(y) \\
&= (2f(x+y)g(y) - 2f(x-y)g(y))\,f(x) \\
&= (f(x+2y) + f(x) - f(x-2y) - f(x))\,f(x) \\
&= (f(2y+x) + f(2y-x))\,f(x) \\
&= 2f(2y)g(x)f(x)
\end{aligned}
$$

that hold true for all $x, y \in G$. In view of (15), setting $y = x$ in (14), we conclude that

$$f(2x) = 2f(x)g(x) \qquad \text{for all} \quad x \in G.$$

This, in turn, leads to the equation

$$f(x+y)^2 - f(x-y)^2 = f(2x)f(2y)$$

valid for all $x, y \in G$ which was to be shown. $\qquad \square$

Remark 2. Actually, the proof presented above might be shortened considerably, since once we were able to derive the oddness of f it would suffice to interchange x and y in (12) to get (7) and to apply Theorem 4.

Nevertheless, we think that is of some interest to show that the proof presented might be finished without referring to Theorem 4.

Our next goal is now to establish a vector analogue of the last two results just obtained. Having disposed of Theorems 4 and 5, with the aid of the method presented in the proof of Theorem 3, it is fairly easy to verify the following:

Theorem 6. *Let* $(G, +)$ *be an Abelian group and let* $(\mathcal{A}, \| \cdot \|)$ *be a semisimple commutative Banach algebra. Assume that* $f : G \to \mathcal{A}$ *and* $\varphi : G \to \mathbb{R}$ *satisfy anyone of the inequalities*

$$\left\| f(x)f(y) - f\left(\frac{x+y}{2}\right)^2 + f\left(\frac{x-y}{2}\right)^2 \right\| \leq \varphi(y) \qquad \text{for all} \quad x, y \in G,$$

or

$$\left\| f(x)f(y) - f\left(\frac{x+y}{2}\right)^2 + f\left(\frac{x-y}{2}\right)^2 \right\| \leq \varphi(x) \qquad \text{for all} \quad x, y \in G.$$

Then

$$f(x)f(y) = f\left(\frac{x+y}{2}\right)^2 - f\left(\frac{x-y}{2}\right)^2 \qquad \text{for all} \quad x, y \in G,$$

provided that for an arbitrary linear multiplicative functional $x^* \in \mathcal{A}^*$ *the superposition* $x^* \circ f$ *fails to be bounded.*

4. CHARACTERIZATIONS OF TRIGONOMETRIC FUNCTIONS BY FUNCTIONAL INEQUALITIES

We terminate this paper showing how to apply Theorems 1 and 2 (resp. Theorems 4 and 5) in order to characterize the cosine (resp. sine) function in terms of functional inequalities. Unlike the previous sections, some mild regularity upon the "unknown" function will occur in the statements below.

Corollary 1. *A function* $f : \mathbb{R} \to \mathbb{C}$ *is of the form*

$$f(x) = \cos \alpha x \qquad (x \in \mathbb{R}),$$

with some constant $\alpha \in \mathbb{C} \setminus \mathbb{R}$, *if and only if*

(i) f *is Lebesgue measurable;*

(ii) f *is unbounded;*

(iii) *there exists a function* $\varphi : \mathbb{R} \to \mathbb{R}$ *such that*

$$|f(x+y) + f(x-y) - 2f(x)f(y)| \leq \varphi(y) \qquad \text{for all} \quad x, y \in \mathbb{R},$$

or

$$|f(x+y) + f(x-y) - 2f(x)f(y)| \leq \varphi(x) \qquad \text{for all} \quad x, y \in \mathbb{R}.$$

Proof. Merely the "if" part requires a motivation. Conditions (iii) and (ii) jointly with Theorem 1 or Theorem 2 force the function f in question to satisfy d'Alembert's equation (A). It is well known that a Lebesgue measurable solution $f : \mathbb{R} \to \mathbb{C}$ of the d'Alembert's equation (see (i)) has to have the form $f(x) = \cos \alpha x, x \in \mathbb{R}$, where α is a complex constant. Clearly, a function of that form is unbounded if and only if the constant α falls into the set $\mathbb{C} \setminus \mathbb{R}$. This ends the proof.   □

Corollary 2. *A function* $f : \mathbb{R} \to \mathbb{C}$ *is of the form*

$$f(x) = \beta \sin \alpha x \qquad (x \in \mathbb{R}),$$

with some constants $\alpha \in \mathbb{C} \setminus \mathbb{R}$ *and* $\beta \in \mathbb{C} \setminus \{0\}$, *if and only if*

(j) f *is Lebesgue measurable;*

(jj) f *is unbounded;*

(jjj) *the function*

$$\mathbb{R} \setminus \{0\} \ni x \longmapsto \frac{f(x)}{x} \in \mathbb{C}$$

is nonconstant;

(jv) *there exists a function* $\varphi : \mathbb{R} \to \mathbb{R}$ *such that*

$$\left| f(x)f(y) - f\left(\frac{x+y}{2}\right)^2 + f\left(\frac{x-y}{2}\right)^2 \right| \leq \varphi(y)$$

$$\text{for all} \quad x, y \in \mathbb{R},$$

or

$$\left| f(x)f(y) - f\left(\frac{x+y}{2}\right)^2 + f\left(\frac{x-y}{2}\right)^2 \right| \leq \varphi(x)$$

$$\text{for all} \quad x, y \in \mathbb{R}.$$

Proof. Conditions (jv) and (jj) jointly with Theorem 4 or Theorem 5 force the function f in question to satisfy Wilson's equation (W). It is well known that a Lebesgue measurable solution $f : \mathbb{R} \to \mathbb{C}$ of the

Wilson's equation (see (j)) is either linear (excluded because of (jjj)) or has to have the form $f(x) = \beta \sin \alpha x$, $x \in \mathbb{R}$, where α and β are complex constants. Clearly, a function of that form is unbounded if and only if $\beta \neq 0$ and α falls into the set $\mathbb{C} \setminus \mathbb{R}$. This finishes the proof because the converse implication is trivial. $\qquad\square$

Remark 3. The characterization results just derived may easily be extended to the case where the real line $\mathbb{R}$ is replaced by an abstract locally compact Abelian group $(G,+)$ (possibly with a uniquely perfomable division by 2) and the measurability understood in the sense of Haar. The roles of cosine and sine functions would then be played by mappings of the form

$$G \ni x \longmapsto \frac{1}{2}(m(x) + m(-x)) \in \mathbb{C}$$

(resp.

$$G \ni x \longmapsto \beta(m(x) - m(-x)) \in \mathbb{C})$$

where m stands for a Haar measurable (and hence continuous) exponential function mapping G into $\mathbb{C}$.

References

[1] J. A. Baker, *The stability of the cosine equation*, Proc. Amer. Math. Soc. **80** (1980), 411–416.

[2] P. W. Cholewa, *The stability of the sine equation*, Proc. Amer. Math. Soc. **88** (1983), 631–634.

[3] W. Förg-Rob and J. Schwaiger, *On the stability of some functional equations for generalized hyperbolic functions and for the generalized cosine equation*, Results in Math. **26** (1994), 274–280.

[4] L. Székelyhidi, *The stability of D'Alembert type functional equations*, Acta Sci. Math. (Szeged) **44** (1982), 313–320.

[5] L. Székelyhidi, *The stability of the sine and cosine functional equation*, Proc. Amer. Math. Soc. **110** (1990), 109–115.

ON THE CONTINUITY
OF ADDITIVE–LIKE FUNCTIONS
AND JENSEN CONVEX FUNCTIONS
WHICH ARE BOREL ON A SPHERE

Karol Baron

Instytut Matematyki, Uniwersytet Śląski,

PL–40-007 Katowice, Poland

baron@uranos.cto.us.edu.pl

Abstract   Assume X is a separable Banach space with $\dim X \geq 2$.

1. If T is a topological space with a countable base, $F : T \to T$ is a Borel function such that $F(\cdot, t)$ is continuous for every $t \in T$, $f : X \to T$ satisfies

$$f(x + y) = F(f(x), f(y))$$

for all $x, y \in X$, and the restriction of f to a sphere is a Borel function, then f is continuous.

2. If D is an open and convex subset of X, $f : D \to \mathbb{R}$ is Jensen convex and the restriction of f to a sphere contained in D is a Borel function, then f is continuous.

Keywords: additive–like function, Jensen convex function, Borel function, selector having the Baire property

Mathematics Subject Classification (2000): 39B52, 39B62, 26B25

Following F. Zorzitto [10], W. Ring, P. Schöpf and J. Schwaiger [9] and R. Ger [2], we consider the Cauchy–type equation

$$f(x + y) = F(f(x), f(y)) \tag{1}$$

and the Jensen inequality

$$f\left(\frac{x + y}{2}\right) \leq \frac{f(x) + f(y)}{2} \tag{2}$$

17

Z. Daróczy and Z. Páles (eds.),

Functional Equations - Results and Advances, 17–20.

© 2002 *Kluwer Academic Publishers.*

and the problem of continuity of their solutions which are regular on a sphere. Instead of continuity assumed in the above mentioned papers we suppose the Borel measurability but we need additionally countable bases in the spaces considered and completeness of the domain. Our results read.

Theorem 1. *Assume X is a separable Banach space with $\dim X \geq 2$, T is a topological space with a countable base and $F : T \times T \to T$ is a Borel function such that $F(\cdot, t)$ is continuous for every $t \in T$.*

If $f : X \to T$ is a solution of (1) *and the restriction of f to a sphere is a Borel function, then f is continuous.*

Theorem 2. *Assume X is a separable Banach space with $\dim X \geq 2$ and D is its open and convex subset.*

If $f : D \to \mathbb{R}$ is a solution of (2) *and the restriction of f to a sphere contained in D is a Borel function, then f is continuous.*

Denote

$$B = \{x \in X : \; \|x\| < 1\}, \qquad S = \{x \in X : \; \|x\| = 1\}.$$

We start with the following lemma (cf. [9; Lemma 2]).

Lemma. *If X is a separable Banach space with $\dim X \geq 2$, then there exists a function $\varphi : B \to X$ having the Baire property and such that*

$$\varphi(x), \quad x - \varphi(x) \in S \qquad \text{for} \quad x \in B. \tag{3}$$

Proof. Define $\Phi(x) = S \cap (S + x)$ for $x \in B$. Obviously, for every $x \in B$, the set $\Phi(x)$ is a closed subset of X and (see [1; p. 319]) it is also nonempty. Moreover,

$$\{x \in B : \; \Phi(x) \cap U \neq \emptyset\} = B \cap ((S \cap U) + S) \qquad \text{for} \quad U \subset X.$$

Assume now that $U \subset X$ is open. Then $(S \cap U) + S$ is the continuous image of the Borel subset $(S \cap U) \times S$ of $X \times X$ and so (see [5; pp. 434–437] or [4; pp. 38–44]) it has the Baire property. Applying the theorem of K. Kuratowski and C. Ryll–Nardzewski [6; p. 398], [5; p. 460] we obtain a Baire–measurable selector $\varphi : B \to X$ of Φ. □

Proof of Theorem 1. Let $f : X \to T$ be a solution of (1) such that its restriction to a sphere $rS + x_0$ is a Borel function. Define functions $g, h : X \to T$ by

$$g(x) = f(rx + x_0), \qquad h(x) = f(rx + 2x_0).$$

Then $g|_S$ is a Borel function and

$$h(x + y) = F(g(x), g(y)) \quad \text{for} \quad x, y \in X.$$

Let $\varphi : B \to X$ be a function having the Baire property and satisfying (3). Then

$$h(x) = F(g(\varphi(x)), g(x - \varphi(x))) \quad \text{for} \quad x \in B$$

which shows that $h|_B$ has the Baire property. Consequently, f restricted to the ball $rB + 2x_0$ has the Baire property. Applying [3; Theorem 5.B3] by K-G. Grosse–Erdmann, we obtain the continuity of f. $\quad\Box$

Proof of Theorem 2. Replacing f by g defined (on $\frac{1}{r}(D - x_0)$) as in the proof of Theorem 1, we may (and we do) assume that $f|_S$ is a Borel function. Let $\varphi : B \to X$ be a function having the Baire property and satisfying (3). Then

$$f\left(\frac{x}{2}\right) \le \frac{f(\varphi(x)) + f(x - \varphi(x))}{2} \quad \text{for} \quad x \in B$$

which shows that f restricted to the ball $\frac{1}{2}B$ is majorized by a function having the Baire property. Applying the theorem of Z. Ciesielski and W. Orlicz (see [8; 7.21] and the proof of Theorem 1 on p. 218 of [4]; cf. also [7; Theorem 4] by M.R. Mehdi) we obtain the continuity of f. $\quad\Box$

Remark (added in proof). Theorems 1 and 2 also are valid for non separable Banach spaces X. In fact, according to an observation of Peter Volkmann, the general case can easily be reduced to the separable one: it is sufficient to argue by contradiction.

Acknowledgments. The paper was inspired by a lecture of Professor Roman Ger presenting [2] on the Silesian University Seminar on Functional Equations. The research was supported by the Silesian University Mathematics Department (Iterative Functional Equations and Real Analysis program).

References

[1] R. Ger, *Some remarks on quadratic functionals*, Glasnik Mat. **23** (1988), 315–330.

[2] R. Ger, *Regularity behaviour of functions admitting addition formulae.*

[3] K.-G. Grosse–Erdmann, *Regularity properties of functional equations and inequalities*, Aequationes Math. **37** (1989), 233–251.

[4] M. Kuczma, *An Introduction to the Theory of Functional Equations and Inequalities*, Państwowe Wydawnictwo Naukowe & Uniwersytet Śląski, Warszawa–Kraków–Katowice, 1985.

[5] K. Kuratowski and A. Mostowski, *Set Theory*, Państwowe Wydawnictwo Naukowe & North–Holland, 1976.

[6] K. Kuratowski and C. Ryll–Nardzewski, *A general theorem on selectors*, Bull. Acad. Polon. Sci. Sér. Sci. Math. Astronom. Phys. **13** (1965), 397–403.

[7] M. R. Mehdi, *On convex functions*, J. London Math. Soc. **39** (1964), 321–326.

[8] W. Orlicz and Z. Ciesielski, *Some remarks on the convergence of functionals on bases*, Studia Math. **16** (1958), 335–352.

[9] W. Ring, P. Schöpf, and J. Schwaiger, *On functions continuous on certain classes of 'thin' sets*, Publ. Math. Debrecen **51** (1997), 205–224.

[10] F. Zorzitto, *Homogeneous summands of exponentials*, Publ. Math. Debrecen **45** (1994), 177–182.

NOTE ON A FUNCTIONAL–DIFFERENTIAL INEQUALITY

Bogdan Choczewski

Institute of Mathematics, Pedagogical University,
Podchorgżych 2, 30-084 Kraków, Poland
and
Faculty of Applied Mathematics, University of Mining and Metallurgy,
Mickiewicza 30, 30-059 Kraków, Poland
e-mail: smchocze@cyf-kr.edu.pl

Abstract   The inequality

$$f(t) - f(s) - f'(s)(t - s) \geq f(t - s) \qquad (t, s \in \mathbb{R}),$$

is considered. The result obtained gives a partial answer to a question posed by S. Rolewicz in his paper [2].

Keywords: S. Rolewicz's problem, functional-differential inequality

Mathematics Subject Classification (2000): 26D10

In his study of strong subdifferentials in the theory of monotone multifunctions (cf. [1] and [2]), S. Rolewicz was led to the following

Problem. *Assume that (H) $f : \mathbb{R} \to \mathbb{R}$ is a nonnegative and even function. Is it true that if f is differentiable in $\mathbb{R}$ and satisfies the inequality*

$$f(t) - f(s) - f'(s)(t - s) \geq f(t - s) \qquad (t, s \in \mathbb{R}), \qquad (P)$$

then necessarily

$$f(t) = Kt^2 \qquad (t \in \mathbb{R}), \qquad (Q)$$

where $K \geq 0$ is a constant?

Z. Daróczy and Z. Páles (eds.),
Functional Equations - Results and Advances, 21–24.
© 2002 *Kluwer Academic Publishers.*

The aim of this note is to prove the following

Theorem. *Assume (H) to hold. If f is twice differentiable in an interval $[0,a), a > 0$, and*

$$f''(0) = 2K := 2a^2 f(a), \qquad (*)$$

then (P) is satisfied if and only if f is given by (Q).

Proof. Obviously the function f given by (Q) satisfies (P). Let f be a solution to (P) obeying our assumptions. Since f is even, it is enough to establish (Q) in $[0, +\infty)$.

1. Put $t = s$ in (P) to find $0 \geq f(0)$, whence

$$f(0) = 0. \qquad (1)$$

Further, by dividing (P) by $h := t - s > 0$, resp. < 0, we get

$$\frac{f(s+h) - f(s)}{h} - f'(s) \geq \frac{f(h)}{h},$$

resp. with the inequality sign reversed. Consequently,

$$\lim_{h \to 0+} \frac{f(h)}{h} \leq 0, \qquad \lim_{h \to 0-} \frac{f(h)}{h} \geq 0,$$

whence

$$f'(0) = 0. \qquad (2)$$

2. Exchange t with s in (P):

$$f(s) - f(t) - f'(t)(s - t) \geq f(t - s) \qquad (t, s \in \mathbb{R}) \qquad (3)$$

(as f is even) and add the corresponding sides of (3) and (P) to arrive at:

$$(f'(t) - f'(s))(t - s) \geq 2f(t - s). \qquad (4)$$

When $s = 0$, thanks to (2), we have from (4)

$$tf'(t) \geq 2f(t) \qquad (t \in \mathbb{R}). \qquad (5)$$

It follows that $f'(t) \geq 0$ for $t \geq 0$. We claim that f is either the zero function or it is positive on $(0, +\infty)$. For, put $-s$ in place of s in (P). Since f' is odd, we have

$$f(t) - f(s) + f'(s)(t + s) \geq f(t + s) \qquad (t, s \in \mathbb{R})$$

and, on taking $s = t$ here,

$$2tf'(t) \geq f(2t) \qquad (t \in \mathbb{R}). \tag{6}$$

If $f(u) = 0$ for a $u > 0$, then $f(t) = 0$ for $t \in [0, u]$. Put $t = u$ in (6) to get $f(2u) = 0$. Since f is nondecreasing in $[0, \infty)$, it vanishes in $[0, 2u]$. Repeat this argument to see that $f(t) = 0$ for $t \in [0, 2^n u], n \in \mathbb{N}$, whence $f(t) = 0, t \in [0, \infty)$. Thus either $f = 0$ or f does not vanish in $\mathbb{R}$.

3. We divide both sides of (5) by $f(t) > 0$ and integrate the resulting inequality over the interval $[a, t], (t > a)$:

$$\int_a^t \frac{f'(s)}{f(s)} \, ds \geq 2 \int_a^t \frac{ds}{s}, \tag{7}$$

i.e, $\ln f(t) - \ln f(a) \geq 2 \ln t - 2 \ln a$ and

$$f(t) \geq Kt^2, \qquad K := f(a)a^{-2} > 0 \qquad (t > a).$$

Doing the same things but in the interval $[t, a), 0 < t < a$, we obtain (cf. (7)) the inequality $\ln f(a) - \ln f(t) \geq 2 \ln a - 2 \ln t$ and

$$f(t) \leq Kt^2 \qquad (t \in (0, a)).$$

This inequality holds also in $[0, a]$, as $f(0) = 0$ and $a^{-2}f(a) = K$, cf. (1) and (*). Summarizing, we have proved that

$$f(t) \leq Kt^2 \qquad (0 \leq t \leq a), \tag{A}$$

$$f(t) \geq Kt^2 \qquad (t > a). \tag{B}$$

4. The derivative f'' exists, in fact, in $(-a, a)$, as f is even. To estimate f in $[0, a)$ from the other side than in (A) we apply inequality (4) which yields

$$\frac{f'(t) - f'(s)}{t - s} \geq 2 \frac{f(t - s)}{(t - s)^2} \qquad (t \neq s). \tag{8}$$

Let $s \in [0, a)$ and $t \to s$. Since (1) and (2) imply

$$\lim_{h \to 0} \frac{f(h)}{h^2} = \frac{f''(0)}{2}$$

$(= K$, cf. (*)), we have the inequality for the limits in (8):

$$f''(s) \geq f''(0) = 2K.$$

After integration over $[0, s] \subset [0, a)$, this inequality produces $f'(s) \geq 2Ks$ (cf. (1) and (2)), in turn $f(s) \geq Ks^2$; both inequalities holding in

$(0, a)$. The validity of the latter can be extended, by (1) to 0 and by (*) to a. Recalling (A) we get equality (Q) for $s \in [0, a]$.

5. It remains to prove that $f(t) \leq Kt^2$ for $t > a$. We apply (6), first for $2s \in (a, 2a]$. Then $s \in (a/2, a]$ and $f'(s) = 2Ks$, according to (Q) (already valid in $[0, a)$). Hence $2s2Ks \geq f(2s)$, i.e., $f(2s) \leq K(2s)^2$ and

$$f(t) \leq Kt^2 \tag{C}$$

for $t \in (a, 2a]$. This argument, when applied n times yields inequality (C) for $t \in (a, 2^n a], n \in N$, therefore for $t > a$. Having (B) and (C) in mind we see that (Q) holds true for $t > a$, too, whence for $t \geq 0$, and the proof is completed. $\square$

Remark (added in proof). Z. Kominek proved (private communication) S. Rolewicz's conjecture assuming additionally that $f'(1) = 2f(1)$. In a paper under preparation by Z. Kominek and the author the conjecture is confirmed with no further assumptions on f and related problems are also dealt with.

References

[1] S. Rolewicz, *On cyclic $\alpha(\cdot)$-monotone multifunctions*, Studia Math. **141** (2000), 263–272.

[2] S. Rolewicz, *φ-convex functions in metric spaces*, Int. Journal of Math Sci., Plenum (invited paper, in preparation).

A CHARACTERIZATION OF STATIONARY SETS FOR THE CLASS OF JENSEN CONVEX FUNCTIONS

Roman Ger

Institute of Mathematics, Silesian University,
ul. Bankowa 14, PL-40-007, Katowice, Poland
romanger@us.edu.pl

Kazimierz Nikodem

Department of Mathematics, Technical University,
ul. Willowa 2, PL-43-309 Bielsko-Biala, Poland
knik@pb.bielsko.pl

Abstract   Let D be a nonempty open and convex subset of a locally convex space X and $J(D)$ denote the family of all Jensen convex functions on D. A set $T \subset D$ is called *stationary* for the class $J(D)$ iff for every $f \in J(D)$ the condition $f|_T = 0$ implies that $f = 0$. We prove that a set T is stationary for $J(D)$ iff $D \subset \mathrm{cl\,conv}_{\mathbb{Q}} T$ and $T \in \mathcal{A}(X)$. This result extends a characterization of stationary sets for the class $J(\mathbb{R}^n)$ obtained recently by M. Babilonová [1] to the case open and convex subdomains of infinite dimensional spaces.

Moreover, we show that a set $T \subset \mathbb{R}^n$ admitting a point of symmetry is stationary for $J(D)$ iff $\mathrm{conv}_{\mathbb{Q}} T = D$.

Keywords: Jensen convex function, stationary set

Mathematics Subject Classification (2000): 39B72, 26B25

1.   INTRODUCTION

Let X be a nonempty set and $\mathcal{K}$ be a given class of functions $f : X \to \mathbb{R}$. We say that a set $T \subset X$ is *stationary* for the class $\mathcal{K}$ iff for every $f \in \mathcal{K}$ the condition $f|_T = 0$ implies that $f = 0$. To illustrate this notion let us consider the following examples.

Z. Daróczy and Z. Páles (eds.),
Functional Equations - Results and Advances, 25–28.
© 2002 *Kluwer Academic Publishers.*

Example 1. Let $\mathcal{K}_1$ be the family of all continuous functions $f : \mathbb{R} \to \mathbb{R}$. Then a set $T \subset \mathbb{R}$ is stationary for $\mathcal{K}_1$ iff it is dense in $\mathbb{R}$.

Example 2. Let $\mathcal{K}_2$ be the family of all polynomials $f : \mathbb{R} \to \mathbb{R}$ of degree at most n. Then a set $T \subset \mathbb{R}$ is stationary for $\mathcal{K}_2$ iff it has at least $n + 1$ elements.

Example 3. Let $\mathcal{K}_3$ be the family of all additive functions $f : \mathbb{R} \to \mathbb{R}$. Then a set $T \subset \mathbb{R}$ is stationary for $\mathcal{K}_3$ iff it contains a Hamel basis of $\mathbb{R}$ over $\mathbb{Q}$ (cf. e.g. [3, Theorem 2, p. 121]).

During the 23-rd Summer Symposium in Real Analysis in Łódź (Poland) in 1999 S. Marcus posed the following problem: Characterize the stationary sets for the class of Jensen convex functions.

Let $D \neq \emptyset$ be a convex and open subset of a real linear topological space X. Recall that a function $f : D \to \mathbb{R}$ is *Jensen convex* if

$$f\left(\frac{x+y}{2}\right) \leq \frac{f(x)+f(y)}{2} \qquad \text{for all} \quad x, y \in D.$$

We denote by $J(D)$ the family of all Jensen convex functions on D whereas the symbol $\text{conv}_{\mathbb{Q}} T$ will stand for $\mathbb{Q}$-convex hull of a set $T \subset X$. The following set-classes connected with Jensen convex functions were introduced (in the case where $X = \mathbb{R}^n$) by R. Ger and M. Kuczma [2] ; see also [3, Chapter IX]:

$\mathcal{A}(X) := \{T \subset X :$ every Jensen convex function $f : D \to \mathbb{R}$,

with $D \supset T$, upper bounded on T is continuous$\}$

and

$\mathcal{C}(X) := \{T \subset X :$ every additive function

$f : X \to \mathbb{R}$ bounded on T is continuous$\}$.

The following result concerning S. Marcus' problem stated above has recently been obtained by M. Babilonová [1].

Theorem. *A set $T \subset \mathbb{R}^n$ is stationary for the class $J(\mathbb{R}^n)$ if and only if* cl $\text{conv}_{\mathbb{Q}} T = \mathbb{R}^n$ *and* $T \in \mathcal{A}(\mathbb{R}^n)$.

The assumptions that the functions in question are defined on the *whole* space and that the space itself is just $\mathbb{R}^n$, are essential in the proof presented by M. Babilonová. In particular she is using the metric structure of $\mathbb{R}^n$, a characterization of the class $\mathcal{A}(\mathbb{R}^n)$ as well as the fact that nonpositive convex functions defined on the whole space are necessarily constant. However, with the aid of a different method, we can extend this result to much more general situation. Namely, we are able to characterize the stationary sets for the class of Jensen convex functions defined on an open convex subset of a locally convex space.

2.   RESULTS

Preserving the denotations introduced in the previous section we shall prove the following

Theorem 1. *Let $D \neq \emptyset$ be an open convex subset of a Hausdorff locally convex space X. A set $T \subset X$ is stationary for the class $J(D)$ if and only if $D \subset \mathrm{cl\,conv}_\mathbb{Q}\, T$ and $T \in \mathcal{A}(X)$.*

Proof. ($\Leftarrow$) By the Jensen convexity of a function $f : D \to \mathbb{R}$ such that $f|_T = 0$ we infer that $f(x) \le 0$ for all $x \in \mathrm{conv}_\mathbb{Q}\, T$. Since $T \in \mathcal{A}(X)$, the function f has to be continuous on D and, hence, $f \le 0$ on $(\mathrm{cl\,conv}_\mathbb{Q}\, T) \cap D = D$. Fix a point $x_0 \in T$ and take an arbitrary $x \in D$. Since $x_0 \in \mathrm{int}\, D$, there exist $y \in D$ and $q \in \mathbb{Q} \cap (0,1)$ such that $x_0 = qx + (1-q)y$. Then

$$0 = f(x_0) \le qf(x) + (1-q)f(y) \le qf(x) \le 0$$

and, consequently, $f(x) = 0$, i.e. $f = 0$ on D, as claimed.

($\Rightarrow$) Assume that T is a stationary set for $J(D)$ and suppose that D is not contained in $\mathrm{cl\,conv}_\mathbb{Q}\, T$, i.e. there exists a point $x_0 \in D \setminus \mathrm{cl\,conv}_\mathbb{Q}\, T$. Since the singleton $\{x_0\}$ is convex compact and the set $\mathrm{cl\,conv}_\mathbb{Q}\, T$ is convex and closed, there exist a continuous linear functional $x^* : X \to \mathbb{R}$ and a constant $c \in \mathbb{R}$ such that

$$x^*(x_0) > c > x^*(x) \qquad \text{for all} \quad x \in \mathrm{cl\,conv}_\mathbb{Q}\, T.$$

Consider the function $f : D \to \mathbb{R}$ defined by the formula $f(x) = \max\{x^*(x), c\} - c$, $x \in D$. Clearly, f is convex, $f|_T = 0$ and $f(x_0) > 0$, which contradicts our assumption.

Now, suppose that $T \notin \mathcal{A}(X)$. Then there exists a discontinuous Jensen convex function $g : D \to \mathbb{R}$ upper bounded on T, say $g(x) \le M$ for $x \in T$. Define a function $f : D \to \mathbb{R}$ by the formula $f(x) = \max\{g(x), M\} - M$. Obviously, f is Jensen convex and $f|_T = 0$. Moreover, f is discontinuous since, otherwise, f were bounded on a neighbourhood of some point whence so would be g and this would force g to be continuous, a contradiction. Thus, in particular, $f \neq 0$ which contradicts the stationarity of T and finishes the proof.   □

In the case where $X = \mathbb{R}^n$ and T is symmetric with respect to a point, we have much simpler criterion verifying the stationarity of T for the class $J(D)$. Namely, as a consequence of Theorem 1, we get the following characterization.

Theorem 2. *Let $\emptyset \neq D \subset \mathbb{R}^n$ be an open convex set and let T be a subset of D admitting a point of symmetry. Then T is stationary for the class $J(D)$ if and only if $\mathrm{conv}_\mathbb{Q}\, T = D$.*

Proof. ($\Leftarrow$) Since $\operatorname{conv}_{\mathbb{Q}} T = D$, we have $\operatorname{cl}\operatorname{conv}_{\mathbb{Q}} T \supset D$ and $T \in \mathcal{A}(\mathbb{R}^n)$ (see [3, Theorem 3, p. 210]). Thus, by Theorem 1, the set T is stationary for $J(D)$.

($\Rightarrow$) Assume that T is stationary for $J(D)$ and put $T_0 := \operatorname{conv}_{\mathbb{Q}} T$. Without loss of generality we may assume that T is symmetric with respect to 0, i.e. $T = -T$. By Theorem 1 we have $T \in \mathcal{A}(\mathbb{R}^n) \subset \mathcal{C}(\mathbb{R}^n)$ and $D \subset \operatorname{cl} T_0$. By means of J. Mościcki's characterization of the set-class $\mathcal{C}(\mathbb{R}^n)$ (cf. [4, Theorem 5] or [3, Theorem 2, p. 240]), we get $\operatorname{int} T_0 = \operatorname{int}\operatorname{conv}_{\mathbb{Q}}(T) = \operatorname{int}\operatorname{conv}_{\mathbb{Q}}(T \cup (-T)) \neq \emptyset$.

Fix arbitrarily a point $x \in D$. Since T_0 is dense in D, we can find points $y \in T_0$ and $z \in \operatorname{int} T_0$ and a number $q \in \mathbb{Q} \cap (0,1)$ such that $x = qy + (1-q)z$. Hence $x \in T_0$, and, consequently, $D \subset T_0$. Since the converse inclusion is obvious, this finishes the proof. $\square$

References

[1] M. Babilonová, *Solution of a problem of S. Marcus concerning J-convex functions*, Preprint Series in Mathematical Analysis, Preprint MA 15/1999, Silesian University Opava, Czech Republic.

[2] R. Ger and M. Kuczma, *On the boundedness and continuity of convex functions and additive functions*, Aequationes Math. 4 (1970), 157–162.

[3] M. Kuczma, *An Introduction to the Theory of Functional Equations and Inequalities*, Państwowe Wydawnictwo Naukowe & Uniwersytet Śląski, Warszawa–Kraków–Katowice, 1985.

[4] J. Mościcki, *Note on characterizations of some set-classes connected with the continuity of additive functions*, Prace Naukowe Uniwersytetu Śląskiego 425, Prace Matematyczne 12 (1982), 47–52.

ON THE CHARACTERIZATION
OF WEIERSTRASS'S SIGMA FUNCTION

Antal Járai

Department of Numerical Mathematics, Eötvös Loránd University,
H-1117 Budapest, Pázmány Péter sétány 1/D, Hungary
ajarai@moon.inf.elte.hu

Wolfgang Sander

TU-Braunschweig
D-38106 Braunschweig, Pockelsstr. 14, Germany
w.sander@tu-bs.de

Dedicated to Professor János Aczél on the occasion of his seventy-fifth
birthday

Abstract   It will be proved that measurable and not almost everywhere zero functions $f_1, f_2 : \mathbb{R}^n \to \mathbb{C}$ satisfying the functional equation

$$f_1(u+v)f_2(u-v) = \sum_{i=1}^{k} g_i(u)h_i(v) \qquad (u, v \in \mathbb{R}^n, \ k \in \mathbb{N})$$

are infinitely often differentiable. In the case $k \leq 2$ the equation is solved.

Keywords: functional equation, Weierstrass's sigma function

Mathematics Subject Classification (2000): 39B32, 33E05

1.   INTRODUCTION

Problem 1. On the Thirtieth International Symposium on Functional Equations (1992, Oberwolfach) W. Sander raised the following question [17]:

Z. Daróczy and Z. Páles (eds.),
Functional Equations - Results and Advances, 29–79.
© *2002 Kluwer Academic Publishers.*

Let $f, g_1, g_2, h_1, h_2 : \mathbb{R}^n \to \mathbb{C}$. Suppose that f is not everywhere zero and that g_1 and g_2 are linearly independent. If f, g_1, g_2, h_1, h_2 satisfy

$$f(u+v)f(u-v) = g_1(u)h_1(v) + g_2(u)h_2(v) \qquad (u, v \in \mathbb{R}^n), \qquad (1)$$

prove or disprove that the measurability of f implies the continuity of f.

This question is connected to the characterization of Weierstrass's sigma function by addition theorems. Assuming continuity a characterization of (1) was done by M. Bonk in his "Habilitationsschrift" and published in [3]. He was motivated by the survey paper "The state of the second part of Hilbert's fifth problem" [1] by J. Aczél.

In connection with his fifth problem Hilbert [9] pointed out that although the method of reduction to differential equations makes it possible to solve functional equations — in particular those solved in this way by Abel in an elegant way — the inherent differentiability assumptions are typically unnatural.

In his paper [1] J. Aczél gave a survey on results concerning functional equations treated by Abel in his work. The papers of Abel on functional equation are cited by Hilbert with the remark that it would be interesting to get Abel's results without differentiability assumptions. J. Aczél gives details on results obtained about different equations treated by Abel. Concerning the equations

$$f(u+v)f(u-v) = f(u)^2 g(v)^2 - f(v)^2 g(u)^2, \qquad (2)$$

$$g(u+v)g(u-v) = g(u)^2 g(v)^2 - c^2 f(u)^2 f(v)^2 \qquad (3)$$

he wrote that "Abel reduced this system to differential equations of up to fourth order, of course under supposition of differentiability (of fourth order), that is, since the functions are complex, of analyticity. Haruki [8] has found, also by reduction to differential equations of at most fourth order, the general (in fourth order) differentiable solutions of (2) *alone*. *The general continuous solution* either of (2) alone or of the system (2), (3) for all $u, v \in \mathbb{C}$ is not known (to me); even less are those under regularity conditions weaker than continuity or on subsets of $\mathbb{C}^2$."

M. Bonk [4] investigated the functional equation

$$f_1(u+v)f_2(u-v) = \sum_{i=1}^{k} g_i(u)h_i(v) \qquad (u, v \in \mathbb{R}^n), \qquad (4)$$

and the special case of it, the equation

$$f(u+v)f(u-v) = \sum_{i=1}^{k} g_i(u)h_i(v) \qquad (u, v \in \mathbb{R}^n). \qquad (5)$$

The functions are supposed to be complex valued. He proves a regularity result stating that all continuous solutions f_1, f_2, g_i, h_i of (4) with linearly independent g_i's and h_i's can be (uniquely) extended to analytic functions on $\mathbb{C}^n$ so that those extensions also satisfy (4) on $\mathbb{C}^n$. Using this result he presents all continuous solutions of (5) in the case $k \leq 2$. Of course, as a special case, it is possible to get the continuous solutions of (2) or of (3) and of the system (2), (3).

In his investigations M. Bonk uses Fourier transform and special estimates. Roughly speaking, his regularity theorem depends on the fact that a solutions f of (5) can be "renormalized" so that it becomes a function vanishing exponentially at infinity, and that the Fourier transform of such an f satisfies the same type of equation. Then the Laplace transform gives the extension to $\mathbb{C}^n$. The case $k \leq 1$ is elementary. In the case $k \leq 2$ solutions can be found using complex function theory, but this case is much more complicated as the case $k \leq 1$.

In 1992 R. Rochberg and L. A. Rubel [16] used different methods to find all analytic solutions of equation (4) in the case $n = 1$ and $k \leq 2$. They presented two ways to solve (4). The first method reduces the functional equation to a differential equation (by using a computer). The second method uses special symmetries of the equation and is a more geometrical proof. (They missed some solutions in both cases.)

Finally Bonk obtained in 1997 [5] all continuous solutions of equation (4) in the case $k \leq 2$.

The aim of this paper is to prove that every measurable solution (f_1, f_2) of (4) which is not almost everywhere zero is in $\mathcal{C}^\infty = \mathcal{C}^\infty(\mathbb{R}^n, \mathbb{C})$. In particular, the problem of W. Sander will be solved. On the other hand, we show that the results of M. Bonk can also be obtained using reduction to differential equations. Hence we obtain generalizations of the results of M. Bonk and the results of R. Rochberg and L. A. Rubel with a proof using *only generally applicable methods*.

The proof of the regularity part depends on some new general regularity theorems proved by A. Járai in 1994 (see [11]), and on some earlier regularity results of A. Járai [10]. By our method it is possible to prove that the not almost everywhere zero solutions f_1, f_2 of the more general equation

$$f_1(F_1(u, v))f_2(F_2(u, v)) = \sum_{i=1}^{k} g_i(u)h_i(v) \qquad ((u, v) \in D \subset \mathbb{R}^n \times \mathbb{R}^n),$$

(6)

are also in $\mathcal{C}^\infty$. Here the given $\mathcal{C}^\infty$ functions F_1, F_2 are supposed to satisfy some weak conditions, for example the system $x = F_1(u, v), \quad y = F_2(u, v)$ has to be solvable with respect to u, v. (The method of Bonk does not work in this generality.) Although by our method even more general equations can be treated, for simplicity we restrict ourselves only to equation (4). We obtain and solve differential equations like in the paper of R. Rochberg and L. A. Rubel, but the difference is instead of finding analytic solutions of ordinary differential equations we have to find complex valued C^∞ solutions of a system of partial differential equations.

We remark that in 1974 J. A. Baker [2] found the measurable and not almost everywhere zero complex valued solutions f_1, f_2 of the equation

$$f_1(x)f_2(y) = \prod_{i=1}^{m} g_i(a_i x + b_i y) \qquad (x, y \in \mathbb{R}, m \in \mathbb{N}). \tag{7}$$

With $m = 2$ his result gives the solutions of (4) in the special case $n = 1$, $k = 1$. In this last case K. Lajkó [14] also presented the general solution.

Remark 1. As already known, the answer to the question of W. Sander, as it stands, is negative, that is the measurability of f does not imply the continuity of f. This is shown by the following example:

Let $f = h_1 = h_2 = \xi_{\mathbb{Q}}$, the characteristic function of the set of the rationals, let g_1 be arbitrary, and let $g_2 = \xi_{\mathbb{Q}} - g_1$. Then (1) is satisfied since

$$\xi_{\mathbb{Q}}(u + v)\xi_{\mathbb{Q}}(u - v) = \xi_{\mathbb{Q}}(u)\xi_{\mathbb{Q}}(v).$$

Moreover, in the following example, not only g_1 and g_2, but h_1 and h_2 are also linearly independent: $f(x) = x\xi_{\mathbb{Q}}(x)$, $g_1(u) = u^2\xi_{\mathbb{Q}}(u)$, $g_2(u) = -\xi_{\mathbb{Q}}(u)$, $h_1(v) = \xi_{\mathbb{Q}}(v)$, $h_2(v) = v^2\xi_{\mathbb{Q}}(v)$.

In these examples it is important to have that $\{x : f(x) \neq 0\}$ is a subgroup of $\mathbb{R}$ and of Lebesgue measure zero. Similar examples can be constructed using other subgroups of the reals having measure zero. Counterexamples defined on $\mathbb{R}^n$ can be constructed for example by using projections and the counterexample in the one-dimensional case.

2.   REGULARITY RESULTS

The examples in Remark 1 show that we can expect a positive answer by a modification of the above problem assuming that f is not almost everywhere zero. This modified question is answered by our main theorem. For the proof, we want to apply a general "measurability implies continuity" type theorem, which is formulated for the convenience of the reader.

Theorem 1. *Let Z be a metric space, Z_0 be a σ-compact metric space and let Z_i $(i = 1, 2, \ldots, m)$ be separable metric spaces. Let X and Y be arbitrary metric spaces and let X_i $(i = 1, 2, \ldots, m)$ be locally compact metric spaces. Let ν be a Borel regular measure on Y and let μ_i $(i = 1, 2, \ldots, m)$ be Radon measures on X_i. Suppose that $\nu(Y) < \infty$, $\mu_i(X_i) < \infty$ $(i = 1, 2, \ldots, m)$, and let D be subset of $X \times Y$. Consider the functions $F : X \to Z$, $F_0 : Y \to Z_0$, $G_i : D \to X_i$, $F_i : X_i \to Z_i$ $(i = 1, 2, \ldots, m)$ and $H : D \times Z_0 \times Z_1 \times \ldots \times Z_m \to Z$. Let x_0 be a fixed element of X, and suppose that the following conditions hold:*

(1)   for each $(x, y) \in D$,

$$F(x) = H(x, y, F_0(y), F_1(G_1(x, y)), F_2(G_2(x, y)), \ldots, F_n(G_m(x, y)));$$

(2) F_i is μ_i measurable on X_i $(i = 1, 2, \ldots, m)$;

(3) H is uniformly continuous on $D \times C$ whenever C is a compact subset of $Z_0 \times Z_1 \times \cdots \times Z_m$;

(4) G_i is uniformly continuous on D $(i = 1, 2, \ldots, m)$;

(5)   there exists an $\eta > 0$ such that $D_x = \{y \in Y : (x, y) \in D\}$ is measurable and $\nu(D_x \cap D_{x_0}) \geq \eta$ whenever $x \in X$;

(6)   for each $\varepsilon > 0$ there exists a $\delta > 0$ for which $\mu_i(g_i(x, B)) \geq \delta$ whenever $1 \leq i \leq m$, $x \in X$, $B \subset D_x$ and $\nu(B) \geq \varepsilon$.

Then F is continuous at the point x_0.

Proof. This is Theorem 3.1 from Járai [10]. □

Theorem 2. *Suppose, that the functions $f_1, f_2, g_i, h_i : \mathbb{R}^n \to \mathbb{C}$ satisfy the functional equation*

$$f_1(u + v)f_2(u - v) = \sum_{i=1}^{k} g_i(u)h_i(v) \qquad (u, v \in \mathbb{R}^n), \qquad (1)$$

moreover, f_1, f_2 are measurable and not almost everywhere zero. Then f_1, f_2 are continuous.

Proof. We may suppose that g_i, h_i are linearly independent (if not, (1) is valid with k', $k' \leq k$). By the linear independency of the functions g_i and h_i there exist regular (k, k)-matrices $(a_{i,j})$ and $(b_{i,j})$ and fixed vectors $u_j, v_j \in \mathbb{R}^n$, $1 \leq j \leq k$ such that

$$h_i(v) = \sum_{j=1}^{k} a_{i,j} f_1(u_j + v) f_2(u_j - v) \qquad (v \in \mathbb{R}^n) \qquad (2)$$

and

$$g_i(u) = \sum_{j=1}^{k} b_{i,j} f_1(u + v_j) f_2(u - v_j) \qquad (u \in \mathbb{R}^n). \qquad (3)$$

Hence the measurability of f_1 and f_2 implies the measurability of h_i and g_i. Introducing the new variables $x = u + v$ and $y = u - v$, we obtain from (1)

$$f_1(x) f_2(y) = \sum_{i=1}^{k} g_i \left(\frac{x+y}{2} \right) h_i \left(\frac{x-y}{2} \right) \qquad (x, y \in \mathbb{R}^n).$$

Let us choose a compact set Y with positive measure such that $f_2(y) \neq 0$ whenever $y \in Y$. Then we have

$$f_1(x) = \frac{1}{f_2(y)} \sum_{i=1}^{k} g_i \left(\frac{x+y}{2} \right) h_i \left(\frac{x-y}{2} \right) \qquad (x \in \mathbb{R}^n, y \in Y).$$

Now we can apply Theorem 1 with $n = 2k$,

$$H(x, y, z_0, \dots, z_{2k}) = \frac{1}{z_0} \sum_{i=1}^{k} z_i z_{k+i},$$

$G_i(x, y) = (x + y)/2$, $G_{k+i}(x, y) = (x - y)/2$, $F_i = g_i$, $F_{k+i} = h_i$, $1 \leq i \leq k$. For arbitrary $x_0 \in \mathbb{R}^n$ choose a compact neighbourhood X of x_0. Letting $D = X \times Y$ and choosing suitable compact sets as X_i the previous theorem shows that f_1 is continuous at x_0. Since x_0 was arbitrary, this means that f_1 is continuous everywhere. Similarly we obtain that f_2 is continuous everywhere. □

Notice that we used only that f_2 is not almost everywhere zero and there is a point in which f_1 is not zero.

Remark 2. Similar regularity theorems do not remain true for the more general functional equation

$$\prod_{j=1}^{m} f(a_j u + b_j v) = \sum_{i=1}^{k} g_i(u) h_i(v) \qquad (u, v \in \mathbb{R}^n).$$

Consider the functional equation

$$f(a_1 u + b_1 v) f(a_2 u + b_2 v) f(a_3 u + b_3 v) = 0$$

where f is an arbitrary measurable function which is zero outside $]1, 2[$. Now let $a_1 = b_1 = a_2 = 1$, $b_2 = b_3 = -1$, $a_3 = 3$. If $x = u + v \in \,]1, 2[$

and $y = u - v \in\ ]1, 2[$ then $a_3 u + b_3 v = x + 2y \in\ ]3, 6[$, hence the product is always zero. Thus f satisfies the more general functional equation, f is measurable, but f could be not continuous.

To prove our main theorem we shall use a recent result from Járai [11]. For the readers' convenience we shall formulate here what we need.

Theorem 3. *Let Z be an open subset of a Euclidean space. Let X and Y be open subsets of $\mathbb{R}^s$ and $\mathbb{R}^t$, respectively. Let D be an open subset of $X \times Y$, and C a compact subset of X. Consider the functions $F : X \to Z$, $G_i : D \to X$ $(i = 1, \dots, m)$, $H : D \times Z^m \to Z$. Suppose, that*

(1)   for each $(x, y) \in D$,

$$F(x) = H(x, y, F(G_1(x, y)), \dots, F(G_m(x, y)));$$

(2) H is C^∞;

(3) G_i is C^∞ and for each $x \in X$ there exists a y such that $(x, y) \in D$, $G_i(x, y) \in C$ and $\dfrac{\partial G_i}{\partial y}(x, y)$ has rank s for $i = 1, \dots, m$;

(4)   the function F is continuous.

Then F is C^∞ on X.

Proof. This theorem follows from Theorem 2 in Járai [11] and Theorem 1.5 in Járai [10]. □

Theorem 4. *Suppose, that the functions $f_1, f_2, g_i, h_i : \mathbb{R}^n \to \mathbb{C}$ satisfy the functional equation*

$$f_1(u + v) f_2(u - v) = \sum_{i=1}^{k} g_i(u) h_i(v) \qquad (u, v \in \mathbb{R}^n),$$

moreover, f_1, f_2 are measurable and not almost everywhere zero. Then f_1, f_2 are C^∞.

Proof. Theorem 2 above shows that f_1 and f_2 are continuous. Like in the proof of Theorem 2, we may suppose that the functions g_i and the functions h_i are linearly independent. Using (2) and (3) of Theorem 2 we obtain that there exist complex constants $c_{i,j}$ and $2k$ fixed vectors $u_i, v_j \in \mathbb{R}^n$ such that

$$f_1(u + v) f_2(u - v) = \sum_{i,j=1}^{k} c_{i,j} f_1(u_i + v) f_2(u_i - v) f_1(u + v_j) f_2(u - v_j)$$

$$(u, v \in \mathbb{R}^n).$$

Substituting $x = u + v$ and $y = u - v$ we obtain that

$$f_1(x)f_2(y) = \sum_{i,j=1}^{k} c_{i,j} f_1\left(u_i + \frac{x-y}{2}\right) f_2\left(u_i - \frac{x-y}{2}\right)$$
$$\cdot f_1\left(\frac{x+y}{2} + v_j\right) f_2\left(\frac{x+y}{2} - v_j\right) \tag{1}$$

whenever $x, y \in \mathbb{R}^n$. Let us first treat the special case $f_1 = f_2 = F$. First choose a bounded open set Y on which F is nonzero. Then (1) can be rewritten as

$$F(x) = \frac{1}{F(y)} \sum_{i,j=1}^{k} c_{i,j} F\left(u_i + \frac{x-y}{2}\right) F\left(u_i - \frac{x-y}{2}\right)$$
$$\cdot F\left(\frac{x+y}{2} + v_j\right) F\left(\frac{x+y}{2} - v_j\right).$$

Now we apply Theorem 3 with $s = t = n$, $m = 4k$,

$$H(x, y, z_0, z_1, \dots, z_{4k}) = \frac{1}{z_0} \sum_{i,j=1}^{k} z_i z_{k+i} z_{2k+j} z_{3k+j} \tag{2}$$

and $G_i(x,y) = u_i + (x-y)/2$, $G_{k+i}(x,y) = u_i - (x-y)/2$, $G_{2k+i}(x,y) = (x+y)/2 + v_i$, $G_{3k+i}(x,y) = (x+y)/2 - v_i$, $1 \le i \le k$. Then for any upper bound K of the norms of all fixed vectors u_i, v_i and $y \in Y$, the conditions of the previous theorem are satisfied by the functional equation for $x \in X$ and $y \in Y$ where X is the open ball around the origin having radius $5K$ and C is the closed ball around the origin having radius $4K$. Hence, F is C^∞ on X, and because K may be arbitrary large, everywhere on $\mathbb{R}^n$.

It would be possible to reduce the general case to this special one. Instead of this we shall apply a trick which makes it possible to reduce the regularity problem of a system of functional equations with several unknown function to the regularity problem of one functional equation with one unknown function.

Let us choose bounded open sets Y_1 and Y_2 such that f_1 is nonzero on Y_1 and f_2 is nonzero on Y_2. Introducing new variables and letting $X_1 = X_2$ be the open ball around the origin having radius $5K$ where K is an upper bound of the norm of all vectors $u_i, v_i, y \in Y_1$ and $y \in Y_2$, we obtain from equation (1) the following system of functional equations:

$$f_1(x_1) = \frac{1}{f_2(y_1)} \sum_{i,j=1}^{k} c_{i,j} f_1\left(u_i + \frac{x_1 - y_1}{2}\right) f_2\left(u_i - \frac{x_1 - y_1}{2}\right)$$
$$\cdot f_1\left(\frac{x_1 + y_1}{2} + v_j\right) f_2\left(\frac{x_1 + y_1}{2} - v_j\right),$$

$$f_2(x_2) = \frac{1}{f_1(y_2)} \sum_{i,j=1}^{k} c_{i,j} f_1\left(u_i + \frac{x_2 - y_2}{2}\right) f_2\left(u_i - \frac{x_2 - y_2}{2}\right)$$
$$\cdot f_1\left(\frac{x_2 + y_2}{2} + v_j\right) f_2\left(\frac{x_2 + y_2}{2} - v_j\right),$$

where $x_1 \in X_1$, $y_1 \in Y_2$, $x_2 \in X_2$ and $y_2 \in Y_1$. Now let $X = X_1 \times X_2$ and let f be defined by $f(x) = (f_1(x_1), f_2(x_2))$ where $x = (x_1, x_2)$. The simple idea is that f satisfies a functional equation

$$f(x) = H'(f(G'_0(y)), f(G'_1(x,y)), \dots, f(G'_{4k}(x,y))) \qquad ((x,y) \in D)$$

where $y = (y_1, y_2)$, with appropriate D, G'_i and H'. Indeed, let $Y = Y_2 \times Y_1$, $D = X \times Y$, $G'_0(y_1, y_2) = (y_2, y_1)$ and for $1 \le i \le k$ let

$$G'_i(x_1, x_2, y_1, y_2) = \left(u_i + \frac{x_1 - y_1}{2}, u_i - \frac{x_2 - y_2}{2}\right),$$

$$G'_{k+i}(x_1, x_2, y_1, y_2) = \left(u_i + \frac{x_2 - y_2}{2}, u_i - \frac{x_1 - y_1}{2}\right),$$

$$G'_{2k+i}(x_1, x_2, y_1, y_2) = \left(\frac{x_1 + y_1}{2} + v_i, \frac{x_2 + y_2}{2} - v_i\right),$$

$$G'_{3k+i}(x_1, x_2, y_1, y_2) = \left(\frac{x_2 + y_2}{2} + v_i, \frac{x_1 + y_1}{2} - v_i\right).$$

Moreover, let $H' = (H_1, H_2)$ where H_1 and H_2 are appropriate complex valued functions similar to (2).

Now letting $C_1 = C_2$ be the closed ball around the origin and having radius $4K$, and letting $C = C_1 \times C_2$, we may apply the previous theorem to get that f is in C^∞ on X. This implies that f_1 and f_2 are C^∞ on the open ball having radius $5K$ and centered at the origin. Because K can be arbitrary large, f_1 and f_2 are C^∞. $\qquad \square$

3. CHARACTERIZATION OF THE SIGMA FUNCTION

We shall prove that the measurable and not almost everywhere zero solutions of the functional equation (4) in Problem 1 can be given in the case $k = 2$ with Weierstrass's sigma function. Here we only sum up the most important properties of Weierstrass's functions in the form we shall use them. The proofs can be found in the book of Saks and Zygmund [18].

Remark 3. Weierstrass's functions. Let ω_1, ω_2 complex numbers, different from zero, and such that $\omega_1/\omega_2 \notin \mathbb{R}$. Let $\Omega = \omega_1\mathbb{Z} + \omega_2\mathbb{Z}$. *Weierstrass's sigma function* is defined by the absolutely convergent product

$$\sigma(z, \omega_1, \omega_2) = z \prod_{0 \neq \omega \in \Omega} \left(1 - \frac{z}{\omega}\right) \exp\left(\frac{z}{\omega} + \frac{1}{2}\left(\frac{z}{\omega}\right)^2\right).$$

σ is an entire function of the variable z and its (simple) roots are the points of Ω. The function

$$\sigma(z, \omega_1, \infty) = z \prod_{0 \neq \omega \in \Omega} \left(1 - \frac{z}{\omega}\right) e^{z/\omega} = \frac{\omega_1}{\pi} \sin(\pi z/\omega_1),$$

where $\Omega = \omega_1\mathbb{Z}$, and the function

$$\sigma(z, \infty, \infty) = z \prod_{0 \neq \omega \in \Omega} \left(1 - \frac{z}{\omega}\right) = z,$$

where $\Omega = \{0\}$ can be considered as limit cases. The pairs ω_1, ω_2 will be called *lattice parameters*; i. e. $\omega_1 = \infty$ and $\omega_2 = \infty$, or $0 \neq \omega_1 \in \mathbb{C}$ and $\omega_2 = \infty$, or $0 \neq \omega_1 \in \mathbb{C}$, $0 \neq \omega_2 \in \mathbb{C}$ and $\omega_1/\omega_2 \notin \mathbb{R}$. The derivative of the Weierstrass's functions with respect to the first variable will be simply denoted by a prime. It is clear from the definition that for any complex number $\alpha \neq 0$ we have

$$\sigma(\alpha z, \alpha\omega_1, \alpha\omega_2) = \alpha\sigma(z, \omega_1, \omega_2).$$

Weierstrass's ζ function is defined as the logarithmic derivative of the sigma function:

$$\zeta(z, \omega_1, \omega_2) = \frac{\sigma'(z, \omega_1, \omega_2)}{\sigma(z, \omega_1, \omega_2)}.$$

ζ is a meromorphic function in the first variable, its poles are the points of Ω. In the above limit cases it is known that

$$\zeta(z, \infty, \infty) = \frac{1}{z}$$

and

$$\zeta(z, \omega_1, \infty) = \frac{\pi}{\omega_1} \cot\left(\frac{\pi}{\omega_1} z\right) = \frac{1}{z} + \sum_{0 \neq \omega \in \Omega} \left(\frac{1}{z - \omega} + \frac{1}{\omega}\right),$$

otherwise

$$\zeta(z,\omega_1,\omega_2) = \frac{1}{z} + \sum_{0 \neq \omega \in \Omega} \left(\frac{1}{z-\omega} + \frac{1}{\omega} + \frac{z}{\omega^2} \right).$$

Here it is also clear from the definition that for any complex number $\alpha \neq 0$ we have

$$\zeta(\alpha z, \alpha \omega_1, \alpha \omega_2) = \frac{\zeta(z,\omega_1,\omega_2)}{\alpha}.$$

Weierstrass's $\mathcal{P}$ function is defined by

$$\mathcal{P}(z,\omega_1,\omega_2) = -\zeta'(z,\omega_1,\omega_2).$$

In the special cases we get

$$\mathcal{P}(z,\infty,\infty) = \frac{1}{z^2}$$

and

$$\mathcal{P}(z,\omega_1,\infty) = \frac{\pi^2}{\omega_1^2 \sin^2(\pi z/\omega_1)} = \frac{1}{z^2} + \sum_{0 \neq \omega \in \Omega} \frac{1}{(z-\omega)^2},$$

otherwise

$$\mathcal{P}(z,\omega_1,\omega_2) = \frac{1}{z^2} + \sum_{0 \neq \omega \in \Omega} \left(\frac{1}{(z-\omega)^2} - \frac{1}{\omega^2} \right).$$

The $\mathcal{P}$ function is periodic in its first variable, and Ω gives the set of all periods. By the definition for any complex number $\alpha \neq 0$ we have

$$\mathcal{P}(\alpha z, \alpha \omega_1, \alpha \omega_2) = \frac{\mathcal{P}(z,\omega_1,\omega_2)}{\alpha^2}. \tag{1}$$

We shall use the differential equations of the $\mathcal{P}$ function. In the special cases we have that

$$\mathcal{P}'(z,\infty,\infty)^2 = 4\mathcal{P}(z,\infty,\infty)^3,$$

and

$$\mathcal{P}'(z,\omega_1,\infty)^2 = 4\mathcal{P}(z,\omega_1,\infty)^3 - 4\frac{\pi^2}{\omega_1^2}\mathcal{P}(z,\omega_1,\infty)^2,$$

otherwise

$$\mathcal{P}'(z,\omega_1,\omega_2)^2 = 4\mathcal{P}(z,\omega_1,\omega_2)^3 - g_2\mathcal{P}(z,\omega_1,\omega_2) - g_3,$$

where

$$g_2 = g_2(\omega_1,\omega_2) = 60 \sum_{0 \neq \omega \in \Omega} \frac{1}{\omega^4} \quad \text{and} \quad g_3 = g_3(\omega_1,\omega_2) = 140 \sum_{0 \neq \omega \in \Omega} \frac{1}{\omega^6}.$$

It is not hard to prove that $g_2^3 - 27g_3^2 \neq 0$. It can be proven (see Saks and Zygmund [18], Ch. VIII, § 13), that if $g_2^3 - 27g_3^2 \neq 0$, then there exist lattice constants ω_1, ω_2 different from infinity such that the corresponding g_2, g_3 are the given ones.

We introduce unified notations: if $\omega_1 = \omega_2 = \infty$, then let $g_1 = g_2 = g_3 = 0$; if $\omega_1 \neq \infty$, but $\omega_2 = \infty$, then let $g_1 = 4\pi^2/\omega_1^2$ and $g_2 = g_3 = 0$; otherwise let $g_1 = 0$. With these notations we always have $g_1 = 0$ or $g_2 = g_3 = 0$, or both, but the differential equation of the $\mathcal{P}$ functions can be written in a unified way as

$$\mathcal{P}'(z, \omega_1, \omega_2)^2 = 4\mathcal{P}(z, \omega_1, \omega_2)^3 - g_1\mathcal{P}(z, \omega_1, \omega_2)^2 - g_2\mathcal{P}(z, \omega_1, \omega_2) - g_3.$$

Let us note that the $\mathcal{P}$ function satisfies the second order differential equation

$$\mathcal{P}''(z, \omega_1, \omega_2) = 6\mathcal{P}(z, \omega_1, \omega_2)^2 - g_1\mathcal{P}(z, \omega_1, \omega_2) - \frac{g_2}{2}$$

too. The numbers g_1, g_2, g_3 are usually called *invariants*, because they only depend on Ω, but not on ω_1, ω_2.

Because of the above first and second order differential equations are autonomous, with any shift value $e \in \mathbb{C}$ the function $\mathcal{P}(z + e, \omega_1, \omega_2)$ satisfies the same differential equation as the function $\mathcal{P}(z, \omega_1, \omega_2)$.

Using $\mathcal{P}$ function we can get solutions of any (complex) differential equation $y'^2 = 4y_3 + c_1 y^2 + c_2 y + c_3$, namely in the form $\mathcal{P}(z+e, \omega_1, \omega_2) + c$ with appropriate lattice constants ω_1, ω_2 and complex constants e, c. Indeed, if all the zeros e_1, e_2, e_3 of the polynomial $4y_3 + c_1 y^2 + c_2 y + c_3$ are different, then let $c = (e_1 + e_2 + e_3)/3 = c_1/12$, and we introduce the new variable $\tilde{y} = y + c$. Then with appropriate complex constants g_2, g_3, for which $g_2^3 - 27g_3^2 \neq 0$, we obtain

$$4\left(\tilde{y} - \frac{c_1}{12}\right)^3 + c_1\left(\tilde{y} - \frac{c_1}{12}\right)^2 + c_2\left(\tilde{y} - \frac{c_1}{12}\right) + c_3 = 4\tilde{y}^3 - g_2\tilde{y} - g_3.$$

With the corresponding lattice constants ω_1, ω_2 different from infinity and with any shift $e \in \mathbb{C}$ the function $z \mapsto \mathcal{P}(z + e, \omega_1, \omega_2) + c$ is a solution of the differential equation $y'^2 = 4y_3 + c_1 y^2 + c_2 y + c_3$.

If not all of the zeros e_1, e_2, e_3 of the polynomial $4y_3 + c_1 y^2 + c_2 y + c_3$ are different, then the special cases will give the solutions. If all the three zeros coincide, then again let $c = (e_1 + e_2 + e_3)/3 = e_1 = e_2 = e_3 = c_1/12$. Introducing the new variable $\tilde{y} = y + c$ we have

$$4\left(\tilde{y} - \frac{c_1}{12}\right)^3 + c_1\left(\tilde{y} - \frac{c_1}{12}\right)^2 + c_2\left(\tilde{y} - \frac{c_1}{12}\right) + c_3 = 4\tilde{y}^3.$$

If only two zeros, say e_1 and e_2 coincide, then let $c = (e_1 + e_2)/2 = e_1 = e_2$, and introducing the new variable $\tilde{y} = y + c$ we have

$$4\left(\tilde{y} - \frac{c_1}{12}\right)^3 + c_1\left(\tilde{y} - \frac{c_1}{12}\right)^2 + c_2\left(\tilde{y} - \frac{c_1}{12}\right) + c_3 = 4\tilde{y}^3 - 4e_3\tilde{y}^2.$$

Lemma 1. *With the above notations, let ω_1, ω_2 be lattice constants and let g_1, g_2, g_3 be the corresponding invariants. Let us consider the equation*

$$y^2 = 4x^3 - g_1x^2 - g_2x - g_3.$$

For all solution $(x, y) \in \mathbb{C} \times \mathbb{C}$, $y \neq 0$ of this equation there exists a $z \in \mathbb{C}$ such that

$$\mathcal{P}(z, \omega_1, \omega_2) = x \qquad \text{and} \qquad \mathcal{P}'(z, \omega_1, \omega_2) = y.$$

For any other solution $z' \in \mathbb{C}$ we have $z' - z \in \Omega$.

Proof. For the case $\omega_2 \neq \infty$ the proof can be found in the book of Saks and Zygmund [18], Ch. VIII. 14.5. The case $\omega_2 = \infty$ can be proved by direct computation. $\square$

Lemma 2. *Using the notations of Remark 3, let us suppose that with some complex constants c, $\tilde{c}$, $d \neq 0$, e, $\tilde{e}$ and with some lattice constants ω_1, ω_2 and $\tilde{\omega}_1, \tilde{\omega}_2$, respectively, we have*

$$\mathcal{P}(dx + e, \omega_1, \omega_2) + c = \mathcal{P}(dx + \tilde{e}, \tilde{\omega}_1, \tilde{\omega}_2) + \tilde{c} \tag{1}$$

for all x from a non-empty open interval of the real line. Then ω_1, ω_2 and $\tilde{\omega}_1, \tilde{\omega}_2$ generate the same lattice Ω, the functions

$$z \mapsto \mathcal{P}(z, \omega_1, \omega_2) \qquad \text{and} \qquad z \mapsto \mathcal{P}(z, \tilde{\omega}_1, \tilde{\omega}_2)$$

coincide and we have $c = \tilde{c}$, and $e - \tilde{e} \in \Omega$.

Proof. Let Ω and $\tilde{\Omega}$ denote the lattice generated by ω_1, ω_2 and $\tilde{\omega}_1, \tilde{\omega}_2$, respectively. The left hand side of the equation (1) is analytic on the open set $\mathbb{C} \setminus ((\Omega - e)/d)$, and the right hand side is analytic on the open set $\mathbb{C} \setminus ((\tilde{\Omega} - \tilde{e})/d)$, and these sets are equal on an interval of the real line. Hence by the principle of analytic continuation (see Dieudonné [6], 9.4.2) they are equal on the intersection of these open sets. But the set $((\Omega - e) \cup (\tilde{\Omega} - \tilde{e}))/d$ contains only isolated points, hence all points of this set are poles of both functions, i. e. $(\Omega - e)/d = (\tilde{\Omega} - \tilde{e})/d$. Hence $\Omega = \tilde{\Omega}$ and $e - \tilde{e} \in \Omega$. The sum representation of the $\mathcal{P}$ function implies that $z \mapsto \mathcal{P}(z, \omega_1, \omega_2)$ and $z \mapsto \mathcal{P}(z, \tilde{\omega}_1, \tilde{\omega}_2)$ are equal. Because $e - \tilde{e}$ is a period of this function, for any x for which $dx + e$ is not a pole we have that

$$\mathcal{P}(dx + e, \omega_1, \omega_2) = \mathcal{P}(dx + \tilde{e}, \tilde{\omega}_1, \tilde{\omega}_2),$$

hence $c = \tilde{c}$.   □

Lemma 3. *Using the notations of Remark 3, suppose that ω_1, ω_2 and $\tilde{\omega}_1, \tilde{\omega}_2$ are lattice constants, and the corresponding invariants g_i and $\tilde{g}_i$ are the same, i. e. $g_1 = \tilde{g}_1$, $g_2 = \tilde{g}_2$ and $g_3 = \tilde{g}_3$. Then*

$$\mathcal{P}(z, \omega_1, \omega_2) = \mathcal{P}(z, \tilde{\omega}_1, \tilde{\omega}_2)$$

for all $z \in \mathbb{C}$.

Proof. If $g_2 = \tilde{g}_2 = 0$ and $g_3 = \tilde{g}_3 = 0$, then the statement directly follows from the definition of $\mathcal{P}(z, \omega_1, \infty)$. Otherwise let e be an arbitrary real number for which $y = \mathcal{P}'(e, \omega_1, \omega_2) \neq 0$ is defined and let $x = \mathcal{P}(e, \omega_1, \omega_2)$. Then

$$y^2 = 4x^3 - g_1 x^2 - g_2 x - g_3,$$

hence by Lemma 1 there exists $\tilde{e} \in \mathbb{C}$, for which

$$\mathcal{P}(\tilde{e}, \tilde{\omega}_1, \tilde{\omega}_2) = x \quad \text{and} \quad \mathcal{P}'(\tilde{e}, \tilde{\omega}_1, \tilde{\omega}_2) = y.$$

If s is an appropriate "square root", i. e. the inverse of the restriction of $z \mapsto z^2$ to an appropriate neighbourhood of y, then

$$t \mapsto \mathcal{P}(t + e, \omega_1, \omega_2) \quad \text{and} \quad t \mapsto \mathcal{P}(t + \tilde{e}, \tilde{\omega}_1, \tilde{\omega}_2)$$

also satisfy the differential equation

$$u' = s(4u^3 - g_1 u^2 - g_2 u - g_3)$$

and the initial condition $u(0) = x$, hence these function are equal for real numbers t from an appropriate neighbourhood of the origin. Hence the previous Lemma can be applied, and we obtain the statement.   □

Remark 4. Functional equations and sigma functions. We start with some simple remarks about the solutions $f_1, f_2, g_1, \ldots, g_k, h_1, \ldots h_k$ of the functional equation

$$f_1(u + v) f_2(u - v) = \sum_{i=1}^{k} g_i(u) h_i(v). \tag{1}$$

We shall consider only complex valued solutions defined on $\mathbb{K}^n$. ($\mathbb{K} = \mathbb{R}$ or $\mathbb{K} = \mathbb{C}$).

(2) *Solutions defined on $\mathbb{C}^n$ can be considered as solutions defined on $\mathbb{R}^{2n}$. The restriction of any solution to any linear subspace over $\mathbb{R}$ is a solution over the given subspace. Especially, the restriction of solutions over $\mathbb{C}^n$ are solutions over $\mathbb{R}^n$.*

(3) For any linear mapping $L : \mathbb{K}^m \to \mathbb{K}^n$ the functions $f_1 \circ L$, $f_2 \circ L$, $g_i \circ L$, $h_i \circ L$ are solutions over $\mathbb{K}^m$.

(4) For any constants $c_1, c_2 \in \mathbb{K}^n$ the shifted mappings $x \mapsto f_1(x + c_1)$, $y \mapsto f_2(y + c_2)$, $u \mapsto g_i(u + (c_1 + c_2)/2)$, $v \mapsto h_i(v + (c_1 - c_2)/2)$ are also solutions.

(5) If $\tilde{f}_1$, $\tilde{f}_2$, $\tilde{g}_1$, $\tilde{h}_1$ are arbitrary solutions in the case $k = 1$, then $f_1\tilde{f}_1$, $f_2\tilde{f}_2$, $g_i\tilde{g}_1$ and $h_i\tilde{h}_1$ $(i = 1, \ldots, k)$ are also solutions.

(6) If T is an arbitrary (complex) $k \times k$ matrix having an inverse and with transpose T', then f_1, f_2, $\tilde{g}_i$ and $\tilde{h}_i$ are solutions too, where $g = (g_1, \ldots, g_k)$, $h = (h_1, \ldots, h_k)$, $\tilde{g} = (\tilde{g}_1, \ldots, \tilde{g}_k)$, $\tilde{h} = (\tilde{h}_1, \ldots, \tilde{h}_k)$ and $\tilde{g}(u) = T^{-1}g(u)$, $\tilde{h}(v) = T'h(v)$.

Using the above transformations we may obtain several solutions from a given solution of equation (1). In the case $k = 1$ the complex analytic functions

$$f_1(z) = f_2(z) = \exp(z^2), \qquad g_1(z) = h_1(z) = \exp(2z^2) \qquad (7)$$

are solutions. Using the above transformations we obtain from these the $\mathbb{R}^n \to \mathbb{C}$ type solutions

$$\begin{aligned} f_1(x) &= \exp(a_1 + L_1(x) + Q(x)), \\ f_2(y) &= \exp(a_2 + L_2(y) + Q(y)), \end{aligned} \qquad (8)$$

where $a_1, a_2 \in \mathbb{C}$ are constants, $L_1, L_2 : \mathbb{R}^n \to \mathbb{C}$ are linear forms, and $Q : \mathbb{R}^n \to \mathbb{C}$ is a quadratic form.

Additional solutions in the case $k = 2$ in the case of lattice constant $\omega_1 \neq \infty$, $\omega_2 \neq \infty$ (see the previous point) are given by

$$\begin{aligned} f_1(z) &= f_2(z) = \sigma(z, \omega_1, \omega_2), \\ g_1(z) &= h_2(z) = \sigma(z, \omega_1, \omega_2)^2, \\ -g_2(z) &= h_1(z) = -\sigma(z, \omega_1, \omega_2)^2 \mathcal{P}(z, \omega_1, \omega_2) \\ &= \sigma(z, \omega_1, \omega_2)\sigma''(z, \omega_1, \omega_2) - \sigma'(z, \omega_1, \omega_2)^2 \end{aligned} \qquad (9)$$

(see Saks-Zygmund [18], Ch. VIII. § 7), in the case $\omega_2 = \infty$, i. e. if ω, ∞ are the lattice constants, the solution

$$\begin{aligned} f_1(z) &= f_2(z) = \sigma(z, \omega, \infty), \\ g_1(z) &= h_2(z) = \sigma(z, \omega, \infty)^2, \\ -g_2(z) &= h_1(z) = -\sigma'(z, \omega, \infty)^2, \end{aligned} \qquad (10)$$

moreover the solutions

$$f_1(z) = \sigma(z, \omega, \infty),$$
$$f_2(z) \equiv 1,$$
$$g_1(z) = h_2(z) = \sigma(z, \omega, \infty),$$
$$g_2(z) = h_1(z) = \sigma'(z, \omega, \infty)$$

(11)

and

$$f_1(z) \equiv 1,$$
$$f_2(z) = \sigma(z, \omega, \infty),$$
$$g_1(z) = h_2(z) = \sigma(z, \omega, \infty),$$
$$-g_2(z) = h_1(z) = \sigma'(z, \omega, \infty);$$

(12)

these are $\mathbb{C} \to \mathbb{C}$ type solutions. Note that in all these cases the functions g_1, g_2 and the functions h_1, h_2 too, are linearly independent on any neighbourhood of any point of the complex plane. From these solutions we obtain by transformations the solutions

$$f_1(x) = \exp(a_1 + L_1(x) + Q(x))\sigma(L(x) + e_1, \omega_1, \omega_2),$$
$$f_2(y) = \exp(a_2 + L_2(y) + Q(y))\sigma(L(x) + e_2, \omega_1, \omega_2),$$

(13)

the solutions

$$f_1(x) = \exp(a_1 + L_1(x) + Q(x))\sigma(L(x) + e_1, \omega, \infty),$$
$$f_2(y) = \exp(a_2 + L_2(y) + Q(y)),$$

(14)

and the solutions

$$f_1(x) = \exp(a_1 + L_1(x) + Q(x)),$$
$$f_2(y) = \exp(a_2 + L_2(y) + Q(y))\sigma(L(x) + e_2, \omega, \infty),$$

(15)

having the type $\mathbb{R}^n \to \mathbb{C}$, where ω_1, ω_2 and ω, ∞ are lattice constants, $a_1, a_2 \in \mathbb{C}$, $e_1, e_2 \in \mathbb{R}^n$ are constants, $L, L_1, L_2 : \mathbb{R}^n \to \mathbb{C}$ are linear forms, and $Q : \mathbb{R}^n \to \mathbb{C}$ is a quadratic form. Note that if $L \not\equiv 0$, then the corresponding functions g_1, g_2 and h_1, h_2 are linearly independent on any neighbourhood of any point of $\mathbb{R}^n$.

Or main result will state that from the complex-to-complex "basic solutions" given by (7) and (9)–(12) using the above transformations we can obtain any solutions. To be more exact, in the case $k = 2$ the pairs $f_1, f_2 : \mathbb{R}^n \to \mathbb{C}$ given by (8) and (13)–(15) are all solutions of the functional equation (1) for which f_1 and f_2 are measurable and none of them is almost everywhere zero. In (8) and in (13)–(15) we have not listed the functions g_i, h_i corresponding to a given pair f_1, f_2, but they can be obtained easily (cf. [5]).

Moreover, as shown by the following lemma (for a proof see Theorem 2.3.1 in T. M. Rassias, J. Šimša [15] and H. Gauchman and L. A. Rubel [7]), it is enough to find for a given pair f_1, f_2 one system of solutions g_i, h_i $i = 1, 2, \ldots, k$. The importance of this lemma is that it shows that by solving equation (4) of Problem 1 we may concentrate in finding the form of the functions f_1, f_2.

Lemma 4. *Let U, V be sets, let $g_i, \tilde{g}_i : U \to \mathbb{C}$, $h_i, \tilde{h}_i : V \to \mathbb{C}$ be functions, and let us suppose that the functions $g_1, \ldots, g_k$, the functions $h_1, \ldots, h_k$ are linearly independent, respectively, moreover that*

$$F(u, v) = \sum_{i=1}^{k} g_i(u) h_i(v)$$

and

$$F(u, v) = \sum_{i=1}^{\tilde{k}} \tilde{g}_i(u) \tilde{h}_i(v)$$

are two representations of the function $F : U \times V \to \mathbb{C}$. Then $\tilde{k} \geq k$. Moreover, $\tilde{k} = k$ if and only if $\tilde{g}_1, \ldots \tilde{g}_k$ and $\tilde{h}_1, \ldots \tilde{h}_k$ are linearly independent. If $\tilde{k} = k$, then there exists a uniquely determined constant $k \times k$ matrix T having an inverse such that with the notations $g = (g_1, \ldots, g_k)$, $\tilde{g} = (\tilde{g}_1, \ldots, \tilde{g}_k)$, $h = (h_1, \ldots, h_k)$ and $\tilde{h} = (\tilde{h}_1, \ldots, \tilde{h}_k)$ we have $\tilde{g}(u) = T^{-1} g(u)$ and $h(v) = T' h(v)$ for all $u \in U$, $v \in V$. Here T' is the transpose of T.

The following lemma reduces still once more the problem to solve. Roughly speaking, it tells us that it is enough to find the solution locally if you want to obtain the global solution.

Lemma 5. *Let $f_1, f_2, g_i, h_i, \tilde{g}_i, \tilde{h}_i : \mathbb{R}^n \to \mathbb{C}$ $(i = 1, \ldots, k)$ be functions, and let us suppose that the functions g_i, h_i are analytic and there exists a $\delta > 0$ such that*

$$f_1(u + v) f_2(u - v) = \sum_{i=1}^{k} g_i(u) h_i(v), \tag{1}$$

whenever $|u|, |v| < \delta$. Suppose that for any $\varepsilon > 0$ the functions g_i, $i = 1, 2, \ldots, k$ and the functions h_i, $i = 1, 2, \ldots, k$ are linearly independent, respectively, on the open neighbourhood of the origin having radius ε. Suppose moreover that

$$f_1(u + v) f_2(u - v) = \sum_{i=1}^{k} \tilde{g}_i(u) \tilde{h}_i(v) \tag{2}$$

is satisfied for all $u, v \in \mathbb{R}^n$. Then (1) is satisfied everywhere, and f_1, f_2 are analytic.

Proof. It is clear, that there is a maximal δ among those extended real numbers $0 < \delta \leq \infty$ for which (1) is satisfied whenever $|u|, |v| < \delta$. We have to prove that this maximal value is ∞. Suppose to the contrary that the maximal δ is finite.

We shall prove that f_2 is not zero everywhere on any neighbourhood of the origin. If f_2 would be zero everywhere on a ball having center zero and radius $\varepsilon \leq 2\delta$, then equation (1) would imply $0 = \sum_{i=1}^{k} g_i(u)h_i(v)$ for $|u|, |v| < \varepsilon/2$ contradicting the linear independence of the functions g_i, h_i. Substituting new variables in (1) we obtain that

$$f_1(x)f_2(y) = \sum_{i=1}^{k} g_i\left(\frac{x+y}{2}\right)h_i\left(\frac{x-y}{2}\right) \tag{3}$$

for all pairs $x, y \in \mathbb{R}^n$ for which $|x + y| < 2\delta$ and $|x - y| < 2\delta$. Let us fix a y_0 for which $f_2(y_0) \neq 0$. Dividing both sides of (3) by $f_2(y_0)$ we obtain that

$$f_1(x) = \frac{1}{f_2(y_0)} \sum_{i=1}^{k} g_i\left(\frac{x+y_0}{2}\right)h_i\left(\frac{x-y_0}{2}\right)$$

whenever $|x| < 2\delta - |y_0|$. Hence f_1 is analytic on the ball around the origin and having radius $2\delta - |y_0|$. Because $|y_0|$ can be arbitrary small, we obtain that f_1 is analytic on a ball around the origin having radius 2δ. Similarly we obtain that f_2 also analytic on this ball.

Now we use equation (2). Lemma 4 implies that $\tilde{g}_i$ and $\tilde{h}_i$ are linearly independent on any open ball around the origin. Let $\varepsilon > 0$ and let us choose vectors v_j, $j = 1, \ldots, k$ such that $|v_j| < \varepsilon$ and $\det((h_i(v_j))_{i,j=1}^{k}) \neq 0$. The equations

$$f_1(u + v_j)f_2(u - v_j) = \sum_{i=1}^{k} \tilde{g}_i(u)\tilde{h}_i(v_j)$$

are satisfied for all $u \in \mathbb{R}^n$, $j = 1, \ldots, k$. From these equations all $\tilde{g}_j(u)$ can be expressed as linear combinations with constant coefficients of the terms $f_1(u + v_j)f_2(u - v_j)$ (cf. (3) of Theorem 2). Hence $\tilde{g}_i$ is analytic on the open ball around the origin having radius $2\delta - \varepsilon$. Because ε can be arbitrary small, each $\tilde{g}_i$ is analytic on the open ball around the origin having radius 2δ. Now similarly as in the previous step, we obtain that f_1 and f_2 are analytic on the open ball around the origin having radius

4δ. Hence the left and the right hand side of equation (1) are both analytic on the open set

$$\{(u,v) : u, v \in \mathbb{R}^n, |u|, |v| < 2\delta\}$$

and they are equal on the subset

$$\{(u,v) : u, v \in \mathbb{R}^n, |u|, |v| < \delta\}$$

of this set. By the principle of the analytic continuation (see Dieudonné [6], 9.4.2) equation (1) is satisfied whenever $|u|, |v| < 2\delta$. This contradicts the maximality of δ, i. e. we have proved that $\delta = \infty$. □

Remark 5. Determinants. Let

$$F(u,v) = \sum_{i=1}^{k} g_i(u) h_i(v) \qquad \text{if} \quad u, v \in \mathbb{R}^n,$$

where $g_1, \ldots, g_k$ and $h_1, \ldots, h_k$ are linearly independent sets in the linear function spaces U and V, respectively. Moreover let $\Lambda_u^0, \ldots, \Lambda_u^k$ be $k + 1$ arbitrary linear functionals acting on functions of U and let $\Lambda_v^0, \ldots, \Lambda_v^k$ be $k + 1$ arbitrary linear functional acting on functions of V. Since the functionals are arbitrary, they can — for example — associate to a function the value of the function in a given point, the value of some partial derivative or some directional derivative, some linear combinations of these, etc. Now the $k + 1$ functions $v \mapsto \Lambda_u^j F(u,v)$, $j = 0, 1, \ldots, k$ are for each j linear combinations of the k functions h_i, hence they cannot be linearly independent. Thus there exist constants c_j which are not all zero such that $\sum_{j=0}^{k} c_j \Lambda_u^j F(u,v) = 0$. Applying the linear functionals Λ_v^i to this equation we obtain that

$$\sum_{j=0}^{k} c_j \Lambda_v^i \Lambda_u^j F(u,v) = 0 \qquad \text{if} \quad i = 0, \ldots, k.$$

This means that

$$\det \left(\Lambda_v^i \Lambda_u^j F(u,v) \right)_{i,j=0}^{k} = 0. \tag{1}$$

We shall use this equation to obtain a differential equation for the functions f_1, f_2 in (4) of Problem 1.

Remark 6. Differential equations. We shall use the notations of the previous point. Let us consider first the more simple one dimensional case $n = 1$. To simplify the notations, let f_u, f_v, f_{uu}, f_{uv}, etc. denote the partial derivates of a function f with respect to the variables u, v. Let us fix u and v. Then choosing $\Lambda_u^0(g) = g(u)$, $\Lambda_u^1(g) = g'(u)$, $\Lambda_u^2(g) = g''(u)$, $\Lambda_v^0(h) = h(v)$, $\Lambda_v^1(h) = h'(v)$, $\Lambda_v^2(h) = h''(v)$, (1) of Remark 5 goes over into

$$\det \begin{pmatrix} F & F_u & F_{uu} \\ F_v & F_{uv} & F_{uuv} \\ F_{vv} & F_{uvv} & F_{uuvv} \end{pmatrix} = 0.$$

The only interesting case is the case when $F(u, v) \neq 0$. Let $\ln$ denote the inverse of the restriction of the function $\exp$ to the set $\{z : |\Im(z)| < \pi\}$. Suppose, that $F(u_0, v_0) \neq 0$. Then — in a neighbourhood of the origin — the function $G(u, v) = \ln(F(u + u_0, v + v_0)/F(u_0, v_0))$ is defined and

$$F(u + u_0, v + v_0) = F(u_0, v_0) \exp(G(u, v)).$$

The determinant above can be easily expressed by G. Dividing by the common factors, and subtracting appropriate multiples of the first and second row and column from other rows and columns, respectively, we obtain that

$$\det \begin{pmatrix} 1 & 0 & 0 \\ 0 & G_{uv} & 2G_u G_{uv} + G_{uuv} \\ 0 & 2G_{uv}G_v + G_{uvv} & * \end{pmatrix} = 0,$$

where $* = 4G_u G_v G_{uv} + 2G_u G_{uvv} + 2G_v G_{uuv} + 2G_{uv}^2 + G_{uuvv}$. Subtracting an appropriate multiple of the second row and column from the third row and column, we arrive at

$$\det \begin{pmatrix} 1 & 0 & 0 \\ 0 & G_{uv} & G_{uuv} \\ 0 & G_{uvv} & 2G_{uv}^2 + G_{uuvv} \end{pmatrix} = 0,$$

i. e.,

$$2G_{uv}^3 + G_{uv}G_{uuvv} - G_{uuv}G_{uvv} = 0. \tag{1}$$

Remark 7. Partial differential equations. We want to obtain a similar equation for all cases $n \geq 1$. We shall use the notations of the previous point. If the directions (i. e. arbitrary vectors not necessarily having length 1) $u_1, u_2, \ldots$ are given, we denote by

$$f_{u_1}, f_{u_2}, \ldots, f_{u_1 u_1}, f_{u_1 u_2}, \ldots,$$

the directional derivates of the function f. As above, fixing the points u and v and using directional derivates with respect to arbitrary directions $u_1, u_2, \ldots$ and $v_1, v_2, \ldots$, choosing $\Lambda_u^0(g) = g(u)$, $\Lambda_u^1(g) = g_{u_1}(u)$,

$\Lambda_u^2(g) = g_{u_2 u_3}(u)$, $\Lambda_v^0(h) = h(v)$, $\Lambda_v^1(h) = h_{v_1}(v)$, $\Lambda_v^2(h) = h_{v_2 v_3}(v)$, (1)

of Remark 5 goes over into that the determinant of the matrix

$$\begin{pmatrix} 1 & 0 & 0 \\ 0 & G_{u_1 v_1} & G_{u_2} G_{u_3 v_1} + G_{u_3} G_{u_2 v_1} + G_{u_2 u_3 v_1} \\ 0 & G_{v_2} G_{u_1 v_3} + G_{v_3} G_{u_1 v_2} + G_{u_1 v_2 v_3} & ** \end{pmatrix} \quad (1)$$

is zero, where

$$\begin{aligned} ** = &\; G_{u_2} G_{v_2} G_{u_3 v_3} + G_{u_2} G_{v_3} G_{u_3 v_2} + G_{u_3} G_{v_2} G_{u_2 v_3} + G_{u_3} G_{v_3} G_{u_2 v_2} \\ &+ G_{u_2} G_{u_3 v_2 v_3} + G_{u_3} G_{u_2 v_2 v_3} + G_{v_2} G_{u_2 u_3 v_3} + G_{v_3} G_{u_2 u_3 v_2} \\ &+ G_{u_2 v_2} G_{u_3 v_3} + G_{u_2 v_3} G_{u_3 v_2} + G_{u_2 u_3 v_2 v_3}. \end{aligned}$$

Expanding the determinant, we obtain a partial differential equation, but this is much more complicated as the equation obtained in the case $n = 1$. To simplify it, let us look for other connections. Choosing $\Lambda_u^0(g) = g(u)$, $\Lambda_u^1(g) = g_{u_1}(u)$, $\Lambda_u^2(g) = g_{u_2}(u)$, $\Lambda_v^0(h) = h(v)$, $\Lambda_v^1(h) = h_{v_1}(v)$, $\Lambda_v^2(h) = h_{v_2}(v)$, similarly as above we obtain that

$$G_{u_1 v_1} G_{u_2 v_2} = G_{u_1 v_2} G_{u_2 v_1}. \quad (2)$$

Using this, and choosing $\Lambda_u^0(g) = g(u)$, $\Lambda_u^1(g) = g_{u_1}(u)$, $\Lambda_u^2(g) = g_{u_2}(u)$, $\Lambda_v^0(h) = h(v)$, $\Lambda_v^1(h) = g_{v_1}(v)$, $\Lambda_v^2(h) = h_{v_2 v_3}(v)$ we obtain that

$$G_{u_1 v_1} G_{u_2 v_2 v_3} = G_{u_2 v_1} G_{u_1 v_2 v_3}, \quad (3)$$

and similarly, choosing $\Lambda_u^0(g) = g(u)$, $\Lambda_u^1(g) = g_{u_1}(u)$, $\Lambda_u^2(g) = g_{u_2 u_3}(u)$, $\Lambda_v^0(h) = h(v)$, $\Lambda_v^1(h) = h_{v_1}(v)$, $\Lambda_v^2(h) = h_{v_2}(v)$ we obtain that

$$G_{u_1 v_1} G_{u_2 u_3 v_2} = G_{u_1 v_2} G_{u_2 u_3 v_1}. \quad (4)$$

From (2)–(4) we also get the equations

$$G_{u_1 v_1} G_{u_2 v_3} = G_{u_1 v_3} G_{u_2 v_1},$$
$$G_{u_1 v_1} G_{u_3 v_2} = G_{u_1 v_2} G_{u_3 v_1},$$
$$G_{u_1 v_1} G_{u_3 v_3} = G_{u_1 v_3} G_{u_3 v_1},$$
$$G_{u_2 v_2} G_{u_3 v_3} = G_{u_2 v_3} G_{u_3 v_2},$$
$$G_{u_1 v_1} G_{u_2 u_3 v_3} = G_{u_1 v_3} G_{u_2 u_3 v_1}.$$

Using these equations the equation which can be obtained expanding the determinant of the matrix (1) can be considerable simplified and we obtain that

$$2 G_{u_1 v_1} G_{u_2 v_2} G_{u_3 v_3} + G_{u_1 v_1} G_{u_2 u_3 v_2 v_3} - G_{u_2 u_3 v_1} G_{u_1 v_2 v_3} = 0. \quad (5)$$

Now with substitutions $x = u + v$, $y = u - v$, and notations $H(x) = \ln(f_1(x + x_0)/f_1(x_0))$, $K(y) = \ln(f_2(y + y_0)/f_2(y_0))$, and choosing $u_1 = e_1$, $v_1 = e'_1$, $u_2 = e_2$, $v_2 = e'_2$, $u_3 = e_3$, $v_3 = e'_3$, where $e_1, e'_1, e_2, e'_2, e_3, e'_3$ are arbitrary vectors from $\mathbb{R}^n$, we obtain that for all x, y from some given neighbourhood of the origin the following partial differential equations are satisfied: (note that we get $G(u, v) = H(x) + K(y)$ if $F(u, v) = f_1(u + v)f_2(u - v)$)

$$(H_{e_1 e'_1}(x) - K_{e_1 e'_1}(y))(H_{e_2 e'_2}(x) - K_{e_2 e'_2}(y))$$
$$= (H_{e_1 e'_2}(x) - K_{e_1 e'_2}(y))(H_{e_2 e'_1}(x) - K_{e_2 e'_1}(y)), \tag{6}$$

$$(H_{e_1 e'_1}(x) - K_{e_1 e'_1}(y))(H_{e_2 e'_2 e'_3}(x) + K_{e_2 e'_2 e'_3}(y))$$
$$= (H_{e_2 e'_1}(x) - K_{e_2 e'_1}(y))(H_{e_1 e'_2 e'_3}(x)) + K_{e_1 e'_2 e'_3}(y)), \tag{7}$$

$$(H_{e_1 e'_1}(x) - K_{e_1 e'_1}(y))(H_{e_2 e_3 e'_2}(x) - K_{e_2 e_3 e'_2}(y))$$
$$= (H_{e_1 e'_2}(x) - K_{e_1 e'_2}(y))(H_{e_2 e_3 e'_1}(x)) - K_{e_2 e_3 e'_1}(y)), \tag{8}$$

$$2(H_{e_1 e'_1}(x) - K_{e_1 e'_1}(y))(H_{e_2 e'_2}(x) - K_{e_2 e'_2}(y))(H_{e_3 e'_3}(x) - K_{e_3 e'_3}(y))$$
$$+ (H_{e_1 e'_1}(x) - K_{e_1 e'_1}(y))(H_{e_2 e_3 e'_2 e'_3}(x) + K_{e_2 e_3 e'_2 e'_3}(y))$$
$$- (H_{e_1 e'_2 e'_3}(x) + K_{e_1 e'_2 e'_3}(y))(H_{e_2 e_3 e'_1}(x) - K_{e_2 e_3 e'_1}(y)) = 0. \tag{9}$$

We shall solve this system of equations, first locally, and then globally. Local solutions will be found in the following three lemmas, and global solutions in our main result (Theorem 5). In the lemmas we only treat some of the logically possible cases, all other cases will be covered in the proof of the Main Theorem. The lemmas have completely local nature as well as the above considerations about the system of partial differential equations. Hence these considerations can be used by treating our functional equation (4) of Problem 1 or similar functional equations on subsets of $\mathbb{R}^n$ (or more general structures).

As a first step we consider the solutions along a direction. We shall use the notations of Remark 7 for directional derivatives, and the notations of Remark 3 concerning lattice constants and $\mathcal{P}$ function.

Lemma 6. Solution of a system of differential equations along a direction. *Let $f \in \mathbb{R}^n$ be a unit vector in $\mathbb{R}^n$. Suppose, that for some $\varepsilon > 0$ the functions H and K are complex valued C^∞ functions defined on an ε neighbourhood of the origin of $\mathbb{R}^n$ and that they satisfy the equations (6)–(9) of Remark 7 for each $|x|, |y| < \varepsilon$.*

(1) *If $H_{ff}(x) - K_{ff}(y) \neq 0$, $H_{fff}(x) \neq 0$ and $K_{fff}(y) \neq 0$ whenever $x, y \in \mathbb{R}^n$, $|x|, |y| < \varepsilon$, then there exist $0 < \varepsilon' < \varepsilon$ and $0 < \delta \leq \varepsilon - \varepsilon'$,*

such that for some $c \in \mathbb{C}$, for some lattice constants ω_1, ω_2, and for some complex valued continuous functions e, $\tilde{e}$ defined on an ε'-neighbourhood of the origin we have

$$H_{ff}(x + tf) = -\mathcal{P}(t + e(x), \omega_1, \omega_2) + c$$

and

$$K_{ff}(y + sf) = -\mathcal{P}(s + \tilde{e}(y), \omega_1, \omega_2) + c$$

whenever $x, y \in \mathbb{R}^n$, $|x|, |y| < \varepsilon'$, $t, s \in \mathbb{R}$, and $|t|, |s| < \delta$;

(2) If $H_{ff}(x) - K_{ff}(y) \neq 0$, $H_{fff}(x) \neq 0$ and $K_{fff}(y) = 0$, whenever $x, y \in \mathbb{R}^n$ and $|x|, |y| < \varepsilon$, then there exist $0 < \varepsilon' < \varepsilon$ and $0 < \delta \le \varepsilon - \varepsilon'$, such that for some $c \in \mathbb{C}$, for some lattice constants $\omega(x), \infty$ and for some complex valued function $e(x)$ defined on an ε'-neighbourhood of the origin we have

$$H_{ff}(x+tf) = -\mathcal{P}(t+e(x), \omega(x), \infty) + c \qquad \text{and} \qquad K_{ff}(y+sf) = c$$

whenever $x, y \in \mathbb{R}^n$, $|x|, |y| < \varepsilon'$, $t, s \in \mathbb{R}$, $|t|, |s| < \delta$;

(3) If $H_{ff}(x) - K_{ff}(y) \neq 0$, $H_{fff}(x) = 0$ and $K_{fff}(y) \neq 0$, whenever $x, y \in \mathbb{R}^n$ and $|x|, |y| < \varepsilon$, then there exist $0 < \varepsilon' < \varepsilon$ and $0 < \delta \le \varepsilon - \varepsilon'$, such that for some $c \in \mathbb{C}$, some lattice constants $\omega(y), \infty$ and for some continuous complex valued function $e(y)$ defined on the ε'-neighbourhood of the origin we have

$$H_{ff}(x+tf) = c \qquad \text{and} \qquad K_{ff}(y+sf) = -\mathcal{P}(s+e(y), \omega(y), \infty) + c$$

whenever $x, y \in \mathbb{R}^n$, $|x|, |y| < \varepsilon'$, $t, s \in \mathbb{R}$, $|t|, |s| < \delta$;

(4) If $H_{fff}(x) = 0$ and $K_{fff}(y) = 0$, whenever $x, y \in \mathbb{R}^n$ and $|x|, |y| < \varepsilon$, then for some $c \in \mathbb{C}$ and for all $0 < \varepsilon' < \varepsilon$ we have

$$H_{ff}(x + tf) = c \qquad \text{and} \qquad K_{ff}(y + sf) = c$$

whenever $x, y \in \mathbb{R}^n$, $|x|, |y| < \varepsilon'$ and $|t|, |s| < \delta = \varepsilon - \varepsilon'$.

Proof. Let $0 < \varepsilon'' < \varepsilon$ be arbitrary. For fixed x and y let $\mathcal{H}(t) = H_{ff}(x + tf)$ and $\mathcal{K}(s) = K_{ff}(y + sf)$. With these notations and chosing $e_1 = e_2 = e_3 = e'_1 = e'_2 = e'_3 = f$ we obtain from equation (9) of Remark 7 that

$$2(\mathcal{H}(t) - \mathcal{K}(s))^3 - (\mathcal{H}'(t)^2 - \mathcal{K}'(s)^2)$$
$$+ (\mathcal{H}(t) - \mathcal{K}(s))(\mathcal{H}''(t) + \mathcal{K}''(s)) = 0, \tag{5}$$

whenever $t, s \in \mathbb{R}$, $|t|, |s| < \varepsilon - \varepsilon''$.

First we shall treat case (1). With the notation $\mathcal{L}(t) = \mathcal{H}(t) - \mathcal{K}(0)$ we have $\mathcal{L}(0) \neq 0$ and $\mathcal{L}'(t) \neq 0$. From equation (5) we get

$$2\mathcal{L}(t)^3 - (\mathcal{L}'(t)^2 - \mathcal{K}'(0)^2) + \mathcal{L}(t)(\mathcal{L}''(t) + \mathcal{K}''(0)) = 0.$$

This means that function $\mathcal{L}$ satisfies the differential equation

$$\mathcal{L}''\mathcal{L} - \mathcal{L}'^2 + 2\mathcal{L}^3 + \mathcal{K}''(0)\mathcal{L} + \mathcal{K}'(0)^2 = 0. \tag{6}$$

Using well-known methods this differential equation can be reduced to the first order differential equation

$$\mathcal{L}'^2 = -4\mathcal{L}^3 - 2C\mathcal{L}^2 + 2\mathcal{K}''(0)\mathcal{L} + \mathcal{K}'(0)^2, \tag{7}$$

where C is a constant. Instead of making this reduction step exact, we shall prove that under the initial conditions

$$\mathcal{L}(0) = \mathcal{L}_0 \neq 0, \qquad \mathcal{L}'(0) = \mathcal{L}_1 \neq 0 \tag{8}$$

equations (6) and (7) are locally equivalent. To be more exact, any solution $\mathcal{L}$ of (7), for which $\mathcal{L}'$ is nowhere zero, is a solution of (6). In the other direction, for any solution $\mathcal{L}$ of (6) satisfying (8), too, there is a unique constant C satisfying

$$\mathcal{L}_1^2 = -4\mathcal{L}_0^3 - 2C\mathcal{L}_0^2 + 2\mathcal{K}''(0) + \mathcal{K}'(0)^2, \tag{9}$$

and there exists an open ball around the origin in $\mathbb{R}$ such that on this ball the solution $\mathcal{L}$ satisfies (7).

Let us consider first a solution of (7) with non-vanishing derivative. Differentiating equation (7) we obtain that

$$2\mathcal{L}''\mathcal{L}' + 12\mathcal{L}'\mathcal{L}^2 - 2\mathcal{K}''(0)\mathcal{L}' = -4C\mathcal{L}'\mathcal{L}.$$

Dividing both sides by $2\mathcal{L}'$ and multiplying by $\mathcal{L}$ we obtain that

$$\mathcal{L}''\mathcal{L} + 6\mathcal{L}^3 - \mathcal{K}''(0)\mathcal{L} = -2C\mathcal{L}^2.$$

Expressing $-2C\mathcal{L}^2$ from (7) and substituting into this equation we obtain (6).

The other direction is somewhat harder. Clearly there exists exactly one constant C satisfying (9). Hence under the given initial conditions C is uniquely determined. First we prove that solutions of (6) under the initial conditions (8) are locally unique in the sense that for any two solutions of (6) there exists a neighbourhood of the origin such that these solutions are equal on this neighbourhood. Both solutions satisfy on some neighbourhood of zero the explicit equation

$$\mathcal{L}'' = \frac{\mathcal{L}'^2}{\mathcal{L}} - 2\mathcal{L}^2 - \mathcal{K}''(0) - \frac{\mathcal{K}'(0)^2}{\mathcal{L}}. \tag{10}$$

Under the initial conditions (8) there is one and only one complete solution of (10). Hence any two local solutions of (6) coincide with this solution of (10) on some neighbourhood of the origin.

Now let us consider an arbitrary non-vanishing solution $\mathcal{L}$ of (6). Choosing the unique constant C defined by (9), any solution of (7) with non-vanishing derivative and satisfying (8) (there exists such a solution, as we shall see below), is equal to $\mathcal{L}$ on a neighbourhood of zero. Hence $\mathcal{L}$ satisfies (7) on a neighbourhood of zero.

The next step is to solve (7) under the initial conditions (8). By the properties of the $\mathcal{P}$ function (see Remark 3) there exists a constant c_1 and lattice constants ω_1 and ω_2 such that for any constant $e \in \mathbb{C}$ the function

$$t \mapsto -\mathcal{P}(t + e, \omega_1, \omega_2) + c_1$$

satisfies equation (7), except in at most countably many isolated points $t \in \mathbb{R}$. By Lemma 1 the complex constant e can be chosen so that the initial conditions (8) are satisfied. Note, that e depends only on the initial conditions, and although not unique, the difference of any two possible e's is in Ω. Because (6) and (7) are locally equivalent, we have proved that there exists a $\delta > 0$, such that with some constant c

$$\mathcal{H}(t) = -\mathcal{P}(t + e, \omega_1, \omega_2) + c$$

whenever $|t| < \delta$. We obtain similarly, that — for some $\tilde{\delta} > 0$ — there exist constants $\tilde{c}$ and $\tilde{e}$ and lattice constants $\tilde{\omega}_1$ and $\tilde{\omega}_2$, such that

$$\mathcal{K}(s) = -\mathcal{P}(s + \tilde{e}, \tilde{\omega}_1, \tilde{\omega}_2) + \tilde{c}$$

whenever $|s| < \tilde{\delta}$. Here also $\tilde{e}$ only depends upon the initial conditions and is uniquely determined up to an additive constant from $\tilde{\Omega}$.

Our considerations above are valid for fixed x and y. Let us investigate how the result depends upon x and y. If, say, $y = 0$, then for all x for which $|x| < \delta$, the above considerations can be applied, and we obtain that there exists lattice constants $\omega_1(x), \omega_2(x)$ and complex constants $c(x)$ and $e(x)$ and $\delta(x) > 0$, such that

$$H_{ff}(x + tf) = -\mathcal{P}(t + e(x), \omega_1(x), \omega_2(x)) + c(x) \qquad \text{if} \quad |t| < \delta(x) \quad (11)$$

whenever $|t| < \delta(x)$. Similarly, there exist lattice constants $\tilde{\omega}_1(y), \tilde{\omega}_2(y)$ and complex constants $\tilde{c}(y)$ and $\tilde{e}(y)$ and $\tilde{\delta}(y) > 0$, such that

$$K_{ff}(y + sf) = -\mathcal{P}(s + \tilde{e}(y), \tilde{\omega}_1(y), \tilde{\omega}_2(y)) + \tilde{c}(y) \qquad \text{if} \quad |s| < \tilde{\delta}(y) \quad (12)$$

whenever $|s| < \tilde{\delta}(y)$. We shall investigate the dependence of the parameters upon x and y, respectively.

Let $g_i(x)$ be defined by $\omega_1(x), \omega_2(x)$ and let $\tilde{g}_i(y)$ be defined by $\tilde{\omega}_1(y), \tilde{\omega}_2(y)$, respectively, where $i = 1, 2, 3$ (see Remark 3). To simplify notations we shall not indicate the dependence on x and y, respectively. Moreover we shall use the handy notation $\mathcal{P}(t) = \mathcal{P}(t + e, \omega_1, \omega_2)$ and $\tilde{\mathcal{P}}(s) = \mathcal{P}(s + \tilde{e}, \tilde{\omega}_1, \tilde{\omega}_2)$. The functions $\mathcal{P}$ and $\tilde{\mathcal{P}}$ satisfy the equations

$$\mathcal{P'}^2 = 4\mathcal{P}^3 - g_1\mathcal{P}^2 - g_2\mathcal{P} - g_3 \quad \text{and} \quad \mathcal{P}'' = 6\mathcal{P}^2 - g_1\mathcal{P} - g_2/2$$

and

$$\tilde{\mathcal{P}}'^2 = 4\tilde{\mathcal{P}}^3 - \tilde{g}_1\tilde{\mathcal{P}}^2 - \tilde{g}_2\tilde{\mathcal{P}} - \tilde{g}_3 \quad \text{and} \quad \tilde{\mathcal{P}}'' = 6\tilde{\mathcal{P}}^2 - \tilde{g}_1\tilde{\mathcal{P}} - \tilde{g}_2/2,$$

respectively. We substitute (11) and (12) into (5), use the last two equations and obtain (after cancellations)

$$
\begin{aligned}
0 = {}& 12(\tilde{c} - c)\mathcal{P}(t)\tilde{\mathcal{P}}(s) + (g_1 - \tilde{g}_1)\mathcal{P}(t)\tilde{\mathcal{P}}(s) \\
& + 6(c - \tilde{c})^2(\tilde{\mathcal{P}}(s) - \mathcal{P}(t)) + (c - \tilde{c})\left(g_1\mathcal{P}(t) + \tilde{g}_1\tilde{\mathcal{P}}(s)\right) \\
& + \frac{g_2 - \tilde{g}_2}{2}(\mathcal{P}(t) + \tilde{\mathcal{P}}(s)) \\
& + 2(c - \tilde{c})^3 + (c - \tilde{c})\frac{g_2 + \tilde{g}_2}{2} + (g_3 - \tilde{g}_3).
\end{aligned}
\tag{13}
$$

Differentiating with respect to t we obtain that

$$
\begin{aligned}
0 = {}& 12(\tilde{c} - c)\mathcal{P}'(t)\tilde{\mathcal{P}}(s) + (g_1 - \tilde{g}_1)\mathcal{P}'(t)\tilde{\mathcal{P}}(s) \\
& - 6(c - \tilde{c})^2\mathcal{P}'(t) + (c - \tilde{c})g_1\mathcal{P}'(t) + \frac{g_2 - \tilde{g}_2}{2}\mathcal{P}'(t).
\end{aligned}
\tag{14}
$$

Differentiation of (14) with respect to s yields

$$0 = 12(\tilde{c} - c)\mathcal{P}'(t)\tilde{\mathcal{P}}'(s) + (g_1 - \tilde{g}_1)\mathcal{P}'(t)\tilde{\mathcal{P}}'(s)$$

whenever $|t|$ and $|s|$ are sufficiently small. This is possible only if

$$g_1 - \tilde{g}_1 = 12(c - \tilde{c}). \tag{15}$$

Substituting (15) into (14) we get

$$0 = (g_1^2 - \tilde{g}_1^2 + 12g_2 - 12\tilde{g}_2)\mathcal{P}'(t),$$

which is possible only if

$$12(g_2 - \tilde{g}_2) = \tilde{g}_1^2 - g_1^2. \tag{16}$$

Putting (15) into (13), a comparison of terms not containing $\mathcal{P}(t)$ and $\tilde{\mathcal{P}}(s)$ leads to

$$(g_1 - \tilde{g}_1)^3 + 36(g_1 - \tilde{g}_1)(g_2 + \tilde{g}_2) + 864(g_3 - \tilde{g}_3) = 0. \tag{17}$$

We shall prove that $\tilde{c} = c$, $\tilde{g}_1 = g_1$, $\tilde{g}_2 = g_2$ and $\tilde{g}_3 = g_3$ are constant, hence we may choose $\tilde{\omega}_1 = \omega_1$ and $\tilde{\omega}_2 = \omega_2$ also to be constant.

First we prove that there exist no pair x, y such that $\tilde{\omega}_2(y) = \infty$, but $\omega_2(x) \neq \infty$, or that $\omega_2(x) = \infty$, but $\tilde{\omega}_2(y) \neq \infty$. We shall treat only the first case, the other can be treated similarly. If there would be such a pair, then, using that $\tilde{g}_2 = \tilde{g}_3 = g_1 = 0$, (cf. Remark 3) equation (16) would imply $12g_2 = \tilde{g}_1^2$, and (17) would give $\tilde{g}_1^3 = 216g_3$. But then $-\tilde{g}_1/12$ would be a zero of $4z^3 - g_2 z - g_3$ and its derivative, and hence (at least) double root of the equation $4z^3 - g_2 z - g_3 = 0$, contradicting that $\omega_2 \neq \infty$.

So there remain only two cases: $\omega_2(x) = \tilde{\omega}_2(y) = \infty$ for all x, y, or both values are finite for any pair x, y. In the first case $g_2(x) = g_3(x) = 0 = \tilde{g}_2(y) = \tilde{g}_3(y)$ everywhere, and hence from (17) we obtain that $g_1(x) = \tilde{g}_1(y)$ everywhere, i. e. both are constant and they are equal. Then by Lemma 3 we may choose $\omega_1 = \tilde{\omega}_1$ to be constant. In the second case $g_1(x) = 0 = \tilde{g}_1(y)$ everywhere, and hence from (16) we get that $g_2(x) = \tilde{g}_2(y)$, and from (17) we see that $g_3(x) = \tilde{g}_3(y)$ everywhere. Hence g_2 and $\tilde{g}_2$, moreover g_3 and $\tilde{g}_3$ are constant and they are equal, respectively. Using again Lemma 3 we may choose $\omega_1 = \tilde{\omega}_1$ and $\omega_2 = \tilde{\omega}_2$ to be constant. In both cases we obtain from (15) that $c(x)$ and $\tilde{c}(y)$ are constant and that they are equal.

Now let us investigate $e(x)$ and $\tilde{e}(y)$. As we have mentioned, $e(x)$ only depends upon the initial values $H_{ff}(x)$ and $H_{fff}(x)$ and is unique modulo Ω. We know that

$$H_{ff}(x) = -\mathcal{P}(e(x), \omega_1, \omega_2) + c$$

and

$$H_{fff}(x) = -\mathcal{P}'(e(x), \omega_1, \omega_2).$$

Let

$$X(x) = \mathcal{P}(e(x), \omega_1, \omega_2) \quad \text{and} \quad Y(x) = \mathcal{P}'(e(x), \omega_1, \omega_2).$$

By the above equations $X = c - H_{ff}$ and $Y = -H_{fff}$, thus they are in C^∞. The functions X and Y satisfy the equation $Y^2 = 4X^3 - g_1 X^2 - g_2 X - g_3$. Because Y is non-zero in the origin, Y is uniquely given by X around the origin. By the same reason the mapping $z \mapsto \mathcal{P}(z, \omega_1, \omega_2)$ has an inverse on some neighbourhood of $e(0)$. Let I denote its inverse and let $e^*(x) = I(c - H_{ff}(x))$. On some neighbourhood of the origin we have that

$$X(x) = \mathcal{P}(e^*(x), \omega_1, \omega_2).$$

Because Y is uniquely given by X, we have on some neighbourhood of the origin

$$Y(x) = \mathcal{P}'(e^*(x), \omega_1, \omega_2).$$

But this means by Lemma 1 that

$$e(x) - e^*(x) \in \Omega.$$

Hence we may suppose that $e \equiv e^*$, i. e. that $e \in C^\infty$. Similarly we obtain that we may suppose that $\tilde{e} \in C^\infty$ and thus $\tilde{e}$ is continuous.

Finally, we have to prove that $\delta(x)$ and $\tilde{\delta}(y)$ can be chosen to be independent from x and y, respectively and that they coincide. Let us choose $\delta(x)$ to be maximal between those numbers δ' which are not greater than ε'' and for which

$$H_{ff}(x + tf) = -\mathcal{P}(t + e(x), \omega_1, \omega_2) + c$$

whenever $|t| < \delta'$. We shall prove that if $|x| < \varepsilon' = \varepsilon''/2$, then $\delta(x) \geq \varepsilon'$. Suppose that for some x this does not hold, for example that arbitrary close to $\delta(x) < \varepsilon'$ there exists a $t > \delta(x)$ such that

$$H_{ff}(x + tf) \neq -\mathcal{P}(t + e(x), \omega_1, \omega_2) + c.$$

(The case when this is true for $t < -\delta(x)$ can be treated similarly.) Introducing the new variables $\tilde{t} = t - \delta(x)$ and $\tilde{x} = x + \delta(x)f$ we obtain that $H_{ff}(\tilde{x} + \tilde{t}f)$ and $-\mathcal{P}(\tilde{t} + e(x) + \delta(x), \omega_1, \omega_2) + c$ coincide on some left hand side neighbourhood of the origin but they are not identically equal on any right hand side neighbourhood of the origin. Since on the other hand

$$H_{ff}(\tilde{x} + \tilde{t}f) = -\mathcal{P}(\tilde{t} + e(\tilde{x}), \omega_1, \omega_2) + c,$$

whenever $|\tilde{t}| < \delta(\tilde{x})$, by Lemma 2 we obtain $e(\tilde{x}) - e(x) - \delta(x) \in \Omega$. But then we have $H_{ff}(\tilde{x} + \tilde{t}f) = -\mathcal{P}(\tilde{t} + e(x) + \delta(x), \omega_1, \omega_2) + c$ on a right hand side neighbourhood of the origin too, which is a contradiction.

In the next step, we shall treat case (2). Let us fix some x and y. In this case $\mathcal{K}' \equiv 0$ and $\mathcal{K}'' \equiv 0$, and hence like in case (1) we obtain that $\mathcal{L} = \mathcal{H} - \mathcal{K}(0)$ satisfies equation

$$\mathcal{L}'^2 = -4\mathcal{L}^3 - 2C\mathcal{L}^2 \tag{18}$$

on some neighbourhood of the origin (cf. (7)), moreover $\mathcal{L}$ satisfies the initial conditions (8). By the properties of the $\mathcal{P}$ function (see Remark 3) there exists a constant c_1 and there exist lattice constants ω, ∞ such that for any constant e the function

$$t \mapsto -\mathcal{P}(t + e, \omega, \infty) + c_1$$

satisfies equation (18) except in at most countably many isolated points $t \in \mathbb{R}$. Here also by Lemma 1 the complex constant e can be chosen so that the initial condition (8) is satisfied, moreover e depends only upon the initial conditions and although e is not uniquely determined, the difference of two suitable e's is in Ω. Hence we have proved, that there exists a $\delta > 0$ such that with some constant c we have

$$\mathcal{H}(t) = -\mathcal{P}(t + e, \omega, \infty) + c$$

whenever $|t| < \delta$. Moreover clearly $\mathcal{K}(s) = \tilde{c}$ for some constant $\tilde{c}$ on some neighbourhood of the origin.

Again we investigate the dependence of the result upon x and y. Here we also obtain that there exist lattice constants $\omega(x), \infty$ and complex constants $c(x)$ and $e(x)$ such that $\delta(x) > 0$ and

$$H_{ff}(x + tf) = -\mathcal{P}(t + e(x), \omega(x), \infty) + c(x) \quad \text{if} \quad |t| < \delta(x).$$

Moreover there exist complex constant $\tilde{c}(y)$ and $\tilde{\delta}(y) > 0$, such that

$$K_{ff}(y + sf) = \tilde{c}(y) \qquad \text{if} \quad |s| < \tilde{\delta}(y).$$

Let us now investigate the dependence of the parameters upon x and y.

Let $g_1(x)$ be defined by $\omega(x), \infty$ (see Remark 3). To simplify the notation we shall not indicate the dependence upon x and y, respectively. Moreover we shall use the handy notation $\mathcal{P}(t) = \mathcal{P}(t + e, \omega, \infty)$. The function $\mathcal{P}$ satisfies the differential equations

$$\mathcal{P}'^2 = 4\mathcal{P}^3 - g_1\mathcal{P}^2 \qquad \text{and} \qquad \mathcal{P}'' = 6\mathcal{P}^2 - g_1\mathcal{P}.$$

Substituting $\mathcal{H}$ and $\mathcal{K}$ into (5) and using the above equation we obtain that

$$0 = \left(g_1(c - \tilde{c}) - 6(c - \tilde{c})^2\right)\mathcal{P}(t) + 2(c - \tilde{c})^3. \tag{19}$$

Differentiating with respect to t we obtain that

$$0 = \left(g_1(c - \tilde{c}) - 6(c - \tilde{c})^2\right)\mathcal{P}'(t).$$

This is possible only if

$$g_1(c - \tilde{c}) = 6(c - \tilde{c})^2.$$

Substituting this into (19) we obtain that $c(x) = \tilde{c}(y)$, which is possible only if both are the same constant c.

We have to prove that $\delta(x)$ and $\tilde{\delta}(y)$ may be chosen independent from x and y, respectively, and equal to each other. This can be done in the same way like in case (1) using Lemma 2.

The proof of (3) is analogous to the proof of (2). Finally, (4) trivially follows from (5). $\qquad\qquad\square$

Lemma 7. Connections between different directional derivatives. *Suppose that $\delta > 0$, and e and f are vectors of $\mathbb{R}^n$. Suppose that H, K are complex valued C^∞ functions defined on an open δ neighbourhood of the origin of $\mathbb{R}^n$ satisfying equations (6)–(9) of Remark 7.*

(1) *If $H_{eee}(x)$, $K_{eee}(y)$ and $H_{ee}(x) - K_{ee}(y)$ are not zero for any $x, y \in \mathbb{R}^n$, $|x|, |y| < \delta$, then there exist complex constants d, c' and c'' such that*

$$H_{ef}(x) = dH_{ee}(x) + c' \quad \text{and} \quad K_{ef}(y) = dK_{ee}(y) + c',$$
$$H_{ff}(x) = d^2 H_{ee}(x) + c'' \quad \text{and} \quad K_{ff}(y) = d^2 K_{ee}(y) + c'',$$
$$H_{fff}(x) = d^3 H_{eee}(x) \quad \text{and} \quad K_{fff}(y) = d^3 K_{eee}(y)$$

whenever $x, y \in \mathbb{R}^n$ and $|x|, |y| < \delta$;

(2) *If $H_{eee}(x)$ and $H_{ee}(x) - K_{ee}(y)$ are not zero for any $x, y \in \mathbb{R}^n$, $|x|, |y| < \delta$, but $K_{eee}(y) = 0$, whenever $|y| < \delta$, then there exist complex constants d, c_{ee}, c_{ef} and c_{ff} such that*

$$K_{ee}(y) = c_{ee}, \quad K_{ef}(y) = c_{ef}, \quad K_{ff}(y) = c_{ff},$$
$$H_{ef}(x) - c_{ef} = d(H_{ee}(x) - c_{ee}),$$
$$H_{ff}(x) - c_{ff} = d^2(H_{ee}(x) - c_{ee}),$$
$$H_{fff}(x) = d^3 H_{eee}(x)$$

whenever $x, y \in \mathbb{R}^n$ and $|x|, |y| < \delta$;

(3) *If $K_{eee}(y)$ and $H_{ee}(x) - K_{ee}(y)$ are not zero for any $x, y \in \mathbb{R}^n$, $|x|, |y| < \delta$, but $H_{eee}(x) = 0$, whenever $|x| < \delta$, then there exist complex constants d, c_{ee}, c_{ef} and c_{ff} such that*

$$H_{ee}(x) = c_{ee}, \quad H_{ef}(x) = c_{ef}, \quad H_{ff}(x) = c_{ff},$$
$$K_{ef}(y) - c_{ef} = d(K_{ee}(y) - c_{ee}),$$
$$K_{ff}(y) - c_{ff} = d^2(K_{ee}(y) - c_{ee}),$$
$$K_{fff}(y) = d^3 K_{eee}(y)$$

whenever $x, y \in \mathbb{R}^n$ and $|x|, |y| < \delta$;

(4) *If H_{eee}, H_{fff}, K_{eee} and K_{fff} are zero everywhere on an open δ-neighbourhood of the origin, then there exist complex constants c_{ee}, c_{ef} and c_{ff} such that*

$$H_{ee}(x) = c_{ee} = K_{ee}(y),$$
$$H_{ef}(x) = c_{ef} = K_{ef}(y),$$
$$H_{ff}(x) = c_{ff} = K_{ff}(y)$$

whenever $x, y \in \mathbb{R}^n$ and $|x|, |y| < \delta$.

Proof. We first derive some additional equations from (7)–(9) in Remark 7. Let us write down again equation (9) of Remark 7, but changing the roll of e_1 and e_1', of e_2 and e_2' moreover of e_3 and e_3', respectively. Let us subtract from the new equation the original one. Then we arrive at

$$\left(H_{e_1 e_2' e_3'}(x) + K_{e_1 e_2' e_3'}(y) \right) \left(H_{e_1' e_2 e_3}(x) - K_{e_1' e_2 e_3}(y) \right)$$
$$= \left(H_{e_1' e_2 e_3}(x) + K_{e_1' e_2 e_3}(y) \right) \left(H_{e_1 e_2' e_3'}(x) - K_{e_1 e_2' e_3'}(y) \right).$$

Simplifying and changing again the role of e_1 and e_1' we obtain that

$$H_{e_1 e_2 e_3}(x) K_{e_1' e_2' e_3'}(y) = H_{e_1' e_2' e_3'}(x) K_{e_1 e_2 e_3}(y). \tag{5}$$

Further useful equations can be obtained from equation (7) and (8) of Remark 7. In equation (8) of Remark 7 let us substitute e_3 with e_3' and let us change e_1 with e_1' and e_2 with e_2'. If we add the resulting equation to (7) of Remark 7 or if we substract it from (7) of Remark 7 then we obtain two simpler equations. After changing e_1' with e_2 these simpler equations are

$$\left(H_{e_1 e_2}(x) - K_{e_1 e_2}(y) \right) H_{e_1' e_2' e_3'}(x)$$
$$= \left(H_{e_1' e_2}(x) - K_{e_1' e_2}(y) \right) H_{e_1 e_2' e_3'}(x), \tag{6}$$

and

$$\left(H_{e_1 e_2}(x) - K_{e_1 e_2}(y) \right) K_{e_1' e_2' e_3'}(y)$$
$$= \left(H_{e_1' e_2}(x) - K_{e_1' e_2}(y) \right) K_{e_1 e_2' e_3'}(y). \tag{7}$$

Let us start with the proof of (1). Choosing $e_1 = e_2 = e_3 = e_1' = e_2' = e$, $e_3' = f$ in (5) we obtain

$$H_{eee}(x) K_{eef}(y) = H_{eef}(x) K_{eee}(y).$$

Because H_{eee} and K_{eee} are nowhere zero we conclude that

$$\frac{K_{eef}(y)}{K_{eee}(y)} = \frac{H_{eef}(x)}{H_{eee}(x)} = d \quad \text{(say)}. \tag{8}$$

Now (6) implies

$$\left(H_{ef}(x) - K_{ef}(y) \right) H_{eee}(x) = \left(H_{ee}(x) - K_{ee}(y) \right) H_{eef}(x)$$
$$= \left(H_{ee}(x) - K_{ee}(y) \right) d H_{eee}(x).$$

Dividing both sides by H_{eee} we see that

$$H_{ef}(x) - d H_{ee}(x) = K_{ef}(y) - d K_{ee}(y),$$

which can be valid only if both sides are the same constant. Finally, from equation (6) of Remark 7 we have that

$$(H_{ff}(x) - K_{ff}(y))(H_{ee}(x) - K_{ee}(y))$$
$$= (H_{ef}(x) - K_{ef}(y))^2 = d^2 (H_{ee}(x) - K_{ee}(y))^2. \tag{9}$$

By hypothesis we may divide both sides by $H_{ee}(x) - K_{ee}(y)$, and obtain that $H_{ff}(x) - d^2 H_{ee}(x)$ and $K_{ff}(y) - d^2 K_{ee}(y)$ are the same constant. The last two equations in part (1) follow from the representation of H_{ff} and K_{ff}, using (8). Thus part (1) of the Lemma is proven.

To prove (2), we choose $e_1' = e_2' = e_3' = e$ in (5) and obtain that

$$H_{eee}(x)K_{e_1 e_2 e_3}(y) = H_{e_1 e_2 e_3}(x)K_{eee}(y).$$

Because the right hand side is zero everywhere, we have that $K_{e_1 e_2 e_3}$ is identically zero for arbitrary vectors e_1, e_2, e_3. Especially each partial derivative of K_{ee}, K_{ef} and K_{ff} is equal to zero, and hence K_{ee}, K_{ef} and K_{ff} are constants. Let us denote these constants by c_{ee}, c_{ef} and c_{ff}, respectively. Now we consider some directional derivative of the function

$$\frac{H_{ef}(x) - c_{ef}}{H_{ee}(x) - c_{ee}}$$

with respect to some direction g. This is

$$\frac{H_{efg}(x)(H_{ee}(x) - c_{ee}) - (H_{ef}(x) - c_{ef})H_{eeg}(x)}{(H_{ee}(x) - c_{ee})^2}.$$

By substituting $e_1 = e_2 = e_2' = e$, $e_1' = f$, $e_3' = g$ in (6) we see that the nominator is equal to zero. Hence each directional derivative is equal to zero, i. e.

$$\frac{H_{ef}(x) - c_{ef}}{H_{ee}(x) - c_{ee}} \equiv d$$

for some constant d. Using $K_{ff}(y) = c_{ff}$ and $K_{ee}(y) = c_{ee}$ we get from (9)

$$H_{ff}(x) - c_{ff} = d^2 (H_{ee}(x) - c_{ee}).$$

The proof of (3) is analogous to the proof of (2) (using (7) instead of (6)).

To prove part (4) of the Lemma we substitute $e_1 = e_2 = e_3 = e_1' = e_2' = e_3' = e$ into (9) of Remark 7 and get

$$2 (H_{ee}(x) - K_{ee}(y))^3 = 0.$$

Hence H_{ee} and K_{ee} are constant and they are equal. Similarly we obtain that H_{ff} and K_{ff} are also constant and they are equal. Now, using equation (6) of Remark 7, we arrive at

$$(H_{ef}(x) - K_{ef}(y))^2 = (H_{ee}(x) - K_{ee}(y))(H_{ff}(x) - K_{ff}(y)) = 0$$

which leads to $H_{ef}(x) = K_{ef}(y) = c_{ef}$ (say). $\qquad\square$

Lemma 8. Local solutions. *Suppose that H and K are complex valued C^∞ functions defined on an δ-neighbourhood of the origin of $\mathbb{R}^n$ satisfying the equations (6)–(9) of Remark 7. Moreover assume that*

(1) *for any index $1 \le j \le n$, the functions $\partial_j^3 H$ and $\partial_j^3 K$ are either zero everywhere in the given neighbourhood or are never zero in the given neighbourhood;*

(2) *for any such index j, for which $\partial_j^3 H$ or $\partial_j^3 K$ is never zero in the given neighbourhood, the difference $\partial_j^2 H(x) - \partial_j^2 K(y)$ is also never zero for any x and y from the given neighbourhood;*

(3) *there exists an index j, such that $\partial_j^3 H$ or $\partial_j^3 K$ or both are never zero on the given neighbourhood.*

Then, with the notations $x = (x_1, \ldots, x_n)$, $y = (y_1, \ldots, y_n)$, and with the notations of Remark 3 we have

(4) *if for each index j, $\partial_j^3 H$ and $\partial_j^3 K$ are never zero in the given neighbourhood, or if both of these functions is zero in the given neighbourhood, then we have for some lattice constants ω_1, ω_2 and for some complex numbers $a, \tilde{a}, b_i, \tilde{b}_i, c, \tilde{c}, c_{i,j} = c_{j,i}, d_i, e, \tilde{e}$ on some neighbourhood of the origin*

$$H(x) = a + \sum_{i=1}^{n} b_i x_i$$

$$+ \sum_{i,j=1}^{n} c_{i,j} x_i x_j + \ln\big(c\,\sigma(d_1 x_1 + \ldots + d_n x_n + e; \omega_1, \omega_2)\big)$$

and

$$K(y) = \tilde{a} + \sum_{i=1}^{n} \tilde{b}_i y_i$$

$$+ \sum_{i,j=1}^{n} c_{i,j} y_i y_j + \ln\big(\tilde{c}\,\sigma(d_1 y_1 + \ldots + d_n y_n + \tilde{e}; \omega_1, \omega_2)\big),$$

moreover d_i is zero if and only if $\partial_i^3 H$ and $\partial_i^3 K$ are zero everywhere;

(5) if for each index j either $\partial_j^3 H$ is never zero and $\partial_j^3 K$ is everywhere zero on the given neighbourhood, or if both are zero everywhere on the given neighbourhood, then we have for some lattice constants ω, ∞ and for some complex numbers $a, \tilde{a}, b_i, \tilde{b}_i, c, c_{i,j} = c_{j,i}, d_i, e, \tilde{e}$ on some neighbourhood of the origin

$$H(x) = a + \sum_{i=1}^{n} b_i x_i$$

$$+ \sum_{i,j=1}^{n} c_{i,j} x_i x_j + \ln\big(c\,\sigma(d_1 x_1 + \ldots + d_n x_n + e; \omega, \infty)\big)$$

and

$$K(y) = \tilde{a} + \sum_{i=1}^{n} \tilde{b}_i y_i + \sum_{i,j=1}^{n} c_{i,j} y_i y_j,$$

moreover d_i is zero if and only if $\partial_i^3 H$ is zero;

(6) if for every index j either $\partial_j^3 H$ is zero everywhere and $\partial_j^3 K$ is never zero or if both are zero everywhere on the given neighbourhood, then with some lattice constants ω, ∞ and with some complex constants $a, \tilde{a}, b_i, \tilde{b}_i, \tilde{c}, c_{i,j} = c_{j,i}, d_i, e, \tilde{e}$ on some neighbourhood of the origin

$$H(x) = a + \sum_{i=1}^{n} b_i x_i + \sum_{i,j=1}^{n} c_{i,j} x_i x_j$$

and

$$K(y) = \tilde{a} + \sum_{i=1}^{n} \tilde{b}_i y_i$$

$$+ \sum_{i,j=1}^{n} c_{i,j} y_i y_j + \ln\big(\tilde{c}\,\sigma(d_1 y_1 + \cdots + d_n y_n + \tilde{e}; \omega, \infty)\big),$$

moreover d_i is zero if and only if $\partial_i^3 K$ is zero;

(7) in all other cases possible under conditions (1), (2) (3) there is no solution on any neighbourhood of the origin.

Proof. By Lemma 7 it is clear that if there exists an index j such that $\partial_j^3 H$ is nowhere zero, but $\partial_j^3 K$ is identically zero on the given neighbourhood, then $\partial_i^3 K$ is identically zero for all $1 \le i \le n$ on the given neighbourhood. Similarly, if there exists an index j that $\partial_j^3 K$ is never zero but $\partial_j^3 H$ is identically zero on the given neighbourhood, then $\partial_i^3 H$

is identically zero for all $1 \leq i \leq n$ on the given neighbourhood. Hence there is a solution on the given neighbourhood only in the cases given by (4)–(6) and hence we have proved (7).

For the proof of (4)–(6) we use induction with respect to the dimension. Without loss of generality we may suppose that those indices i for which $\partial_i^3 H$ and $\partial_i^3 K$ are identically zero are greater than those for which one or both are non-zero.

Let us start with the proof of (4). In the case $n = 1$ the statement follows from Lemma 6. By induction we suppose that the statement is true for $n - 1$, i. e. there exist lattice constants ω_1, ω_2 and complex constants $a, \tilde{a}, b_i, \tilde{b}_i, c, \tilde{c}, c_{i,j} = c_{j,i}, d_i, e, \tilde{e}, (i, j < n)$ such that on a neighbourhood of the zero vector

$$
\begin{aligned}
H(x) = a + \sum_{i<n} b_i x_i + \sum_{i,j<n} c_{i,j} x_i x_j \\
+ \ln\bigl(c\,\sigma(d_1 x_1 + \ldots + d_{n-1} x_{n-1} + e; \omega_1, \omega_2)\bigr)
\end{aligned}
\tag{8}
$$

and

$$
\begin{aligned}
K(y) = \tilde{a} + \sum_{i<n} \tilde{b}_i y_i + \sum_{i,j<n} c_{i,j} y_i y_j \\
+ \ln\bigl(\tilde{c}\,\sigma(d_1 y_1 + \ldots + d_{n-1} y_{n-1} + \tilde{e}; \omega_1, \omega_2)\bigr),
\end{aligned}
\tag{9}
$$

whenever we are in the $n - 1$ dimensional subspace spanned by the first $n - 1$ standard base vectors $f_1, f_2, \ldots, f_{n-1}$, moreover for each $i < n$ it is true that $d_i = 0$ if and only if $\partial_i^3 H \equiv 0 \equiv \partial_i^3 K$.

To obtain handy notations let us introduce

$$
x' = (x_1, x_2, \ldots, x_{n-1}, 0) \qquad \text{and} \qquad y' = (y_1, y_2, \ldots, y_{n-1}, 0).
$$

First we shall investigate the case when $\partial_n^3 H$ and $\partial_n^3 K$ are never zero. From Lemma 6 we have that

$$
\partial_n^2 H(x' + x_n f_n) = -\mathcal{P}(x_n + e_1(x'), \tilde{\omega}_1, \tilde{\omega}_2) + c_1
\tag{10}
$$

and

$$
\partial_n^2 K(y' + y_n f_n) = -\mathcal{P}(y_n + \tilde{e}_1(y'), \tilde{\omega}_1, \tilde{\omega}_2) + c_1
\tag{11}
$$

for sufficiently small x', y', x_n and y_n, where x_n, y_n are real numbers and x', y' are vectors from $\mathbb{R}^n$ with last coordinate equal to zero.

The main idea of the proof is now to show that there is a constant d_n and there are continuous functions $e_i(x_i, \ldots, x_{n-1})$ and $\tilde{e}_i(y_i, \ldots, y_{n-1})$, $i = 2, \ldots, n$ satisfying

$$
d_n = D_i d_i, \quad i < n,
\tag{12}
$$

$$e_i(x_i, \ldots, x_{n-1}) = \frac{d_i x_i}{d_n} + e_{i+1}(x_{i+1}, \ldots, x_{n-1}), \tag{13}$$

and

$$\tilde{e}_i(y_i, \ldots, y_{n-1}) = \frac{d_i y_i}{d_n} + \tilde{e}_{i+1}(y_{i+1}, \ldots, y_{n-1}), \tag{14}$$

$i < n$. Let us observe, that e_n and $\tilde{e}_n$ are constants.

¿From (1) of Lemma 7 we obtain that for $i < n$ with some constants D_i and c_i', c_i'' we have

$$\partial_i \partial_n H(x) = D_i \partial_i^2 H(x) + c_i' \quad \text{and} \quad \partial_i \partial_n K(y) = D_i \partial_i^2 K(y) + c_i',$$
$$\partial_n^2 H(x) = D_i^2 \partial_i^2 H(x) + c_i'' \quad \text{and} \quad \partial_n^2 K(y) = D_i^2 \partial_i^2 K(y) + c_i'',$$
$$\partial_n^3 H(x) = D_i^3 \partial_i^3 H(x) \quad \text{and} \quad \partial_n^3 K(y) = D_i^3 \partial_i^3 K(y)$$
$$\tag{15}$$

for all sufficiently small x, y. From the last line follows that $D_i \neq 0$, $i < n$. Differentiating expression (10) with respect to x_n and then substituting $x_n = 0$ and using the last line of (15) for $i = 1$ and the induction hypothesis we get (using (8))

$$\mathcal{P}'(e_1(x_1, \ldots, x_{n-1}), \tilde{\omega}_1, \tilde{\omega}_2) = -\partial_n^3 H(x') = -D_1^3 \partial_1^3 H(x')$$
$$= D_1^3 d_1^3 \mathcal{P}'(d_1 x_1 + \ldots + d_{n-1} x_{n-1} + e, \omega_1, \omega_2). \tag{16}$$

On the other hand, using (10) and the second line of (15) and (8) we arrive at

$$\mathcal{P}(e_1(x_1, \ldots, x_{n-1}), \tilde{\omega}_1, \tilde{\omega}_2) - c_1$$
$$= -\partial_n^2 H(x') = -D_1^2 \partial_1^2 H(x') - c_1'' \tag{17}$$
$$= D_1^2 d_1^2 \mathcal{P}(d_1 x_1 + \ldots + d_{n-1} x_{n-1} + e, \omega_1, \omega_2) - 2D_1^2 c_{11} - c_1''$$

for all sufficiently small x'. The implicit function theorem implies that for any fixed $x_2, x_3, \ldots, x_{n-1}$ the function e_1 is continuously differentiable with respect to x_1. Differentiation of (17) with respect to x_1 yields

$$\mathcal{P}'(e_1(x_1, \ldots, x_{n-1}), \tilde{\omega}_1, \tilde{\omega}_2) \partial_1 e_1(x_1, \ldots, x_{n-1})$$
$$= D_1^2 d_1^3 \mathcal{P}'(d_1 x_1 + \ldots + d_{n-1} x_{n-1} + e, \omega_1, \omega_2). \tag{18}$$

Hence using (16) we see that

$$D_1^3 d_1^3 \mathcal{P}'(d_1 x_1 + \ldots + d_{n-1} x_{n-1} + e, \omega_1, \omega_2) \partial_1 e_1(x_1, \ldots, x_{n-1})$$
$$= D_1^2 d_1^3 \mathcal{P}'(d_1 x_1 + \ldots + d_{n-1} x_{n-1} + e, \omega_1, \omega_2),$$

i. e. $\partial_1 e_1(x_1, \ldots, x_{n-1}) = 1/D_1$. This means that for some continuous function $e_2(x_2, \ldots, x_{n-1})$ we have $e_1(x_1, x_2, \ldots, x_{n-1}) = x_1/D_1 +$

$e_2(x_2, \ldots, x_{n-1})$, that is (13) with $i = 1$. Defining $d_n = D_1 d_1$ (note that $d_n \neq 0$ because $D_1 \neq 0$ and $d_1 \neq 0$), using (17) and (1) of Remark 3 we obtain

$$
\begin{aligned}
\mathcal{P}&\left(\frac{d_1 x_1}{d_n} + e_2(x_2, \ldots, x_{n-1}), \tilde{\omega}_1, \tilde{\omega}_2\right) - c_1 \\
&= d_n^2 \mathcal{P}(d_1 x_1 + \ldots + d_{n-1} x_{n-1} + e, \omega_1, \omega_2) - 2D_1^2 c_{11} - c_1'' \\
&= \mathcal{P}\left(\frac{d_1 x_1}{d_n} + \ldots + \frac{d_{n-1} x_{n-1}}{d_n} + \frac{e}{d_n}, \frac{\omega_1}{d_n}, \frac{\omega_2}{d_n}\right) \\
&\quad - 2c_{11} D_1^2 - c_1''.
\end{aligned}
\tag{19}
$$

Putting $x_2 = \cdots = x_{n-1} = 0$ into this equation Lemma 2 implies that $c_1 = 2c_{11} D_1^2 + c_1''$ and that $\mathcal{P}(z, \tilde{\omega}_1, \tilde{\omega}_2) = \mathcal{P}(z, \omega_1/d_n, \omega_2/d_n)$. Defining $c_{nn} = c_{11} D_1^2 + c_1''/2$ we conlude from (10)

$$
\begin{aligned}
\partial_n^2 &H(x' + x_n f_n) \\
&= 2c_{nn} - \mathcal{P}\left(x_n + \frac{d_1 x_1}{d_n} + e_2(x_2, \ldots, x_{n-1}), \frac{\omega_1}{d_n}, \frac{\omega_2}{d_n}\right) \\
&= 2c_{nn} - d_n^2 \mathcal{P}(d_1 x_1 + d_n e_2(x_2, \ldots, x_{n-1}) + d_n x_n, \omega_1, \omega_2).
\end{aligned}
\tag{20}
$$

Now let $x_n = 0$ in (20) and use (17) to get

$$
\begin{aligned}
\mathcal{P}(d_1 x_1 &+ d_n e_2(x_2, \ldots, x_{n-1}), \omega_1, \omega_2) \\
&= \mathcal{P}(d_1 x_1 + \ldots + d_{n-1} x_{n-1} + e, \omega_1, \omega_2).
\end{aligned}
\tag{21}
$$

In exactly the same manner we obtain (because of the analogies between H and K in (15))

$$
\begin{aligned}
\mathcal{P}(d_1 y_1 &+ d_n \tilde{e}_2(y_2, \ldots, y_{n-1}), \omega_1, \omega_2) \\
&= \mathcal{P}(d_1 y_1 + \ldots + d_{n-1} y_{n-1} + \tilde{e}, \omega_1, \omega_2).
\end{aligned}
\tag{22}
$$

Let us remark that substituting $y_2 = \cdots = y_{n-1} = 0$ into this equation and substituting $x_2 = \cdots = x_{n-1} = 0$ into equation (21) moreover using Lemma 2 it follows that $d_n e_2(0, \ldots, 0) - e$ and $d_n \tilde{e}_2(0, \ldots, 0) - \tilde{e}$ are in the lattice Ω generated by ω_1, ω_2.

In the first step we arrived at (21) and (22) using (15) for $i = 1$. Now we continue similarly, but now we apply (15) with $i = 2$. Substituting $x_n = 0$ in (20) and using the second line of (15) for $i = 2$ and the induction hypothesis we obtain that

$$
\begin{aligned}
d_n^2 \mathcal{P}(d_1 x_1 &+ d_n e_2(x_2, \ldots, x_{n-1}), \omega_1, \omega_2) - 2c_{nn} \\
&= -\partial_n^2 H(x') = -D_2^2 \partial_2^2 H(x') - c_2'' \\
&= D_2^2 d_2^2 \mathcal{P}(d_1 x_1 + \ldots + d_{n-1} x_{n-1} + e, \omega_1, \omega_2) \\
&\quad - 2D_2^2 d_2^2 c_{22} - c_2''.
\end{aligned}
\tag{23}
$$

Using (21) this equation implies that $D_2 d_2 = d_n$. Differentiating (20) with respect to x_n and then substituting $x_n = 0$, and using the last line of (15) and the induction hypothesis we conclude that

$$
\begin{aligned}
d_n^3 \mathcal{P}'(d_1 x_1 + d_n e_2(x_2, \dots , x_{n-1}), \omega_1, \omega_2) \\
= -\partial_n^3 H(x') = -D_2^3 \partial_2^3 H(x') \\
= D_2^3 d_2^3 \mathcal{P}'(d_1 x_1 + \dots + d_{n-1} x_{n-1}, \omega_1, \omega_2),
\end{aligned}
\tag{24}
$$

i. e. because of $D_2 d_2 = d_n$,

$$
\begin{aligned}
\mathcal{P}'(d_1 x_1 + d_n e_2(x_2, \dots , x_{n-1}), \omega_1, \omega_2) \\
= \mathcal{P}'(d_1 x_1 + \dots + d_{n-1} x_{n-1}, \omega_1, \omega_2).
\end{aligned}
\tag{25}
$$

By (21) for any fixed $x_3, \dots , x_{n-1}$ we get from the implicit function theorem that the function e_2 is continuously differentiable with respect to x_2. Differentiating (21) with respect to x_2 we obtain that

$$
\begin{aligned}
d_n \mathcal{P}'(d_1 x_1 + d_n e_2(x_2, \dots , x_{n-1}), \omega_1, \omega_2) \partial_1 e_2(x_2, \dots , x_{n-1}) \\
= d_2 \mathcal{P}'(d_1 x_1 + \dots + d_{n-1} x_{n-1}, \omega_1, \omega_2).
\end{aligned}
\tag{26}
$$

Hence using (25) follows that

$$
\partial_1 e_2(x_2, x_3, \dots , x_{n-1}) = \frac{d_2}{d_n}
$$

everywhere. This means that for some continuous function

$$
e_3(x_3, \dots , x_{n-1})
$$

we have

$$
e_2(x_2, \dots , x_{n-1}) = \frac{d_2 x_2}{d_n} + e_3(x_3, \dots , x_{n-1}),
$$

which is (13) with $i = 2$. In the same manner if follows that for some continuous function $\tilde{e}_3(y_3, \dots , y_{n-1})$ we have

$$
\tilde{e}_2(y_2, \dots , y_{n-1}) = \frac{d_2 y_2}{d_n} + \tilde{e}_3(y_3, \dots , y_{n-1}).
$$

Continuing the induction in this way we finally obtain (12)–(14) and thus also

$$
\partial_n^2 H(x' + x_n f_n) = 2c_{nn} - d_n^2 \mathcal{P}(d_1 x_1 + d_2 x_2 + \dots + d_n x_n + d_n e_n, \omega_1, \omega_2)
$$

and

$$
\partial_n^2 K(y' + y_n f_n) = 2c_{nn} - d_n^2 \mathcal{P}(d_1 y_1 + d_2 y_2 + \dots + d_n y_n + d_n \tilde{e}_n, \omega_1, \omega_2).
$$

Here both $d_n e_n - e = d_n e_2(0, \dots, 0) - e$ and $d_n \tilde{e}_n - \tilde{e} = d_n \tilde{e}_2(0, \dots, 0) - \tilde{e}$ are in the lattice Ω and since the points of Ω are periods, we have

$$\partial_n^2 H(x' + x_n f_n) = 2c_{nn} - d_n^2 \mathcal{P}(d_1 x_1 + d_2 x_2 + \cdots + d_n x_n + e, \omega_1, \omega_2)$$

and

$$\partial_n^2 K(y' + y_n f_n) = 2c_{nn} - d_n^2 \mathcal{P}(d_1 x_1 + d_2 x_2 + \cdots + d_n x_n + \tilde{e}, \omega_1, \omega_2).$$

Now let us integrate these equations. We obtain that

$$\partial_n H(x' + x_n f_n) = 2c_{nn} x_n + d_n \zeta(d_1 x_1 + d_2 x_2 + \cdots + d_n x_n + e, \omega_1, \omega_2)$$
$$+ B_1(x_1, \dots, x_{n-1})$$

and

$$\partial_n K(y' + y_n f_n) = 2c_{nn} y_n + d_n \zeta(d_1 y_1 + d_2 y_2 + \cdots + d_n y_n + \tilde{e}, \omega_1, \omega_2)$$
$$+ \tilde{B}_1(y_1, \dots, y_{n-1}),$$

where the functions B_1 and $\tilde{B}_1$ are in C^∞. Differentiating these equations with respect to x_1 and y_1 and then substituting $x_n = 0$ and $y_n = 0$, respectively, using the first line of (15), the equations (8) and (9), moreover that $D_i d_i = d_n$ whenever $i < n$, we obtain that $\partial_1 B_1 \equiv \partial_1 \tilde{B}_1$ is the same constant. Let us denote this constant by $2c_{1,n}$. Thus we have

$$\partial_n H(x' + x_n f_n) = 2c_{nn} x_n + d_n \zeta(d_1 x_1 + d_2 x_2 + \cdots + d_n x_n + e, \omega_1, \omega_2)$$
$$+ 2c_{1,n} x_1 + B_2(x_2, \dots, x_{n-1})$$

and

$$\partial_n K(y' + y_n f_n) = 2c_{nn} y_n + d_n \zeta(d_1 y_1 + d_2 y_2 + \cdots + d_n y_n + \tilde{e}, \omega_1, \omega_2)$$
$$+ 2c_{1,n} y_1 + \tilde{B}_2(y_2, \dots, y_{n-1}).$$

Continuing by induction, we finally arrive at

$$\partial_n H(x' + x_n f_n) = 2c_{nn} x_n + d_n \zeta(d_1 x_1 + d_2 x_2 + \cdots + d_n x_n + e, \omega_1, \omega_2)$$
$$+ 2c_{1,n} x_1 + \cdots + 2c_{n-1,n} x_{n-1} + b_n$$

and

$$\partial_n K(y' + y_n f_n) = 2c_{nn} y_n + d_n \zeta(d_1 y_1 + d_2 y_2 + \cdots + d_n y_n + \tilde{e}, \omega_1, \omega_2)$$
$$+ 2c_{1,n} y_1 + \cdots + 2c_{n-1,n} y_{n-1} + \tilde{b}_n,$$

where we have defined $b_n = B_n$ and $\tilde{b}_n = \tilde{B}_n$, respectively.

Finally, we integrate these equations. By the induction hypothesis

$$\ln\bigl(c\,\sigma(d_1 x_1 + d_2 x_2 + \cdots + d_n x_n + e, \omega_1, \omega_2)\bigr) \quad \text{and}$$

$$\ln\bigl(\tilde{c}\,\sigma(d_1 y_1 + d_2 y_2 + \cdots + d_n y_n + \tilde{e}, \omega_1, \omega_2)\bigr)$$

are defined on a neighbourhood of the origin. Hence by integration of the previous equations we see that in a small open ball around the origin we have

$$H(x' + x_n f_n) = \ln\bigl(c\,\sigma(d_1 x_1 + d_2 x_2 + \cdots + d_n x_n + e, \omega_1, \omega_2)\bigr)$$

$$+ c_{nn} x_n^2 + \sum_{i=1}^{n-1} 2 c_{i,n} x_i x_n + b_n x_n + a_1(x_1, \dots, x_{n-1})$$

and

$$K(y' + y_n f_n) = \ln\bigl(\tilde{c}\,\sigma(d_1 y_1 + d_2 y_2 + \cdots + d_n y_n + \tilde{e}, \omega_1, \omega_2)\bigr)$$

$$+ c_{nn} y_n^2 + \sum_{i=1}^{n-1} 2 c_{i,n} y_i y_n + \tilde{b}_n y_n + \tilde{a}_1(y_1, \dots, y_{n-1}).$$

Substituting $x_n = 0$ and $y_n = 0$ and using initial conditions (8) and (9), respectively, we obtain that

$$a_1(x_1, \dots, x_{n-1}) = a + \sum_{i<n} b_i x_i + \sum_{i,j<n} c_{ij} x_i x_j \quad \text{and}$$

$$\tilde{a}_1(y_1, \dots, y_{n-1}) = \tilde{a} + \sum_{i<n} \tilde{b}_i y_i + \sum_{i,j<n} c_{ij} y_i y_j.$$

Hence the statement follows.

We still have to investigate the case when $\partial_n^3 H$ and $\partial_n^3 K$ are both identically zero. In this case we have to prove that the statement holds with $d_n = 0$. By (4) of Lemma 6 there exists a constant $c_1 \in \mathbb{C}$ such that with $c_{nn} = c_1/2$ for all sufficiently small x', y', x_n and y_n we have

$$\partial_n^2 H(x' + x_n f_n) = 2 c_{nn} \quad \text{and} \quad \partial_n^2 K(y' + y_n f_n) = 2 c_{nn}.$$

Let us observe that $\partial_i \partial_n H(x')$ and $\partial_i \partial_n K(y')$ is the same constant for all $i < n$: If $\partial_i^3 H$ and $\partial_i^3 K$ are non-zero, then (15) holds true and from the last line of (15) we obtain $D_i = 0$, hence the statement follows from the first line of (15). If $\partial_i^3 H$ and $\partial_i^3 K$ are both identically zero, then the statement follows from case (4) of Lemma 7.

Now let us integrate the equations above. We arrive at

$$\partial_n H(x' + x_n f_n) = 2 c_{nn} x_n + B_1(x_1, \dots, x_{n-1})$$

and

$$\partial_n K(y' + y_n f_n) = 2c_{nn} y_n + \tilde{B}_1(y_1, \ldots, y_{n-1}).$$

The functions B_1 and $\tilde{B}_1$ are in $\mathcal{C}^\infty$. Differentiating these equations with respect to x_1 and y_1 and then substituting $x_n = 0$ and $y_n = 0$, respectively, we obtain that $\partial_1 B_1 \equiv \partial_1 \tilde{B}_1$ are constant. Denoting this constant by $2c_{1,n}$ we have that

$$\partial_n H(x' + x_n f_n) = 2c_{nn} x_n + 2c_{1,n} x_1 + B_2(x_2, \ldots, x_{n-1})$$

and

$$\partial_n K(y' + y_n f_n) = 2c_{nn} y_n + 2c_{1,n} y_1 + \tilde{B}_2(y_2, \ldots, y_{n-1}).$$

Continuing by induction, we finally obtain (denoting again B_n and $\tilde{B}_n$ by b_n and $\tilde{b}_n$, respectively) that

$$\partial_n H(x' + x_n f_n) = 2c_{nn} x_n + 2c_{1,n} x_1 + \cdots + 2c_{n-1,n} x_{n-1} + b_n$$

and

$$\partial_n K(y' + y_n f_n) = 2c_{nn} y_n + 2c_{1,n} y_1 + \cdots + 2c_{n-1,n} y_{n-1} + \tilde{b}_n.$$

Integrating these equations gives

$$H(x' + x_n f_n) = c_{nn} x_n^2 + \sum_{i=1}^{n-1} 2c_{i,n} x_i x_n + b_n x_n + a_1(x_1, \ldots, x_{n-1})$$

and

$$K(y' + y_n f_n) = c_{nn} y_n^2 + \sum_{i=1}^{n-1} 2c_{i,n} y_i y_n + \tilde{b}_n y_n + \tilde{a}_1(y_1, \ldots, y_{n-1}).$$

Substituting $x_n = 0$ and $y_n = 0$ and using the equations (8) and (9), respectively, leads to

$$a_1(x_1, \ldots, x_{n-1}) = a + \sum_{i<n} b_i x_i + \sum_{i,j<n} c_{i,j} x_i x_j$$
$$+ \ln\big(c\,\sigma(d_1 x_1 + d_{n-1} x_{n-1} + e; \omega_1, \omega_2)\big),$$

and

$$\tilde{a}_1(y_1, \ldots, y_{n-1}) = \tilde{a} + \sum_{i<n} \tilde{b}_i y_i + \sum_{i,j<n} c_{i,j} y_i y_j$$
$$+ \ln\big(\tilde{c}\,\sigma(d_1 y_1 + \ldots + d_{n-1} y_{n-1} + \tilde{e}; \omega_1, \omega_2)\big),$$

hence because $d_n = 0$, the statement follows. Thus the statement (4) is completely proven.

Now we prove (5). The case $n = 1$ follows from Lemma 6. Our induction hypothesis is that

$$H(x) = a + \sum_{i<n} b_i x_i$$
$$+ \sum_{i,j<n} c_{i,j} x_i x_j + \ln\big(c\,\sigma(d_1 x_1 + \ldots + d_{n-1} x_{n-1} + e; \omega, \infty)\big)$$

and

$$K(y) = \tilde{a} + \sum_{i<n} \tilde{b}_i y_i + \sum_{i,j<n} c_{i,j} y_i y_j,$$

if we are in the subspace spanned by the first $n - 1$ standard basis vectors $f_1, f_2, \ldots, f_{n-1}$ Moreover d_i is equal to zero if and only if $\partial_i^3 H$ is identically zero.

First we shall investigate the case when $\partial_n^3 H$ is never zero. From Lemma 6 we obtain that

$$\partial_n^2 H(x' + x_n f_n)$$
$$= -\mathcal{P}(x_n + e_1(x_1, \ldots, x_{n-1}), \tilde{\omega}(x_1, \ldots, x_{n-1}), \infty) + c_1 \tag{27}$$

and that

$$\partial_n^2 K(y' + y_n f_n) = c_1 \tag{28}$$

whenever x', y', x_n and y_n are sufficiently small, where x_n, y_n are real numbers and x', y' are vectors from $\mathbb{R}^n$ with last coordinate equal to zero.

The induction hypothesis implies $\partial_i^2 K(y') = 2c_{ii}$. Using this, we obtain from Lemma 7 that with some complex constants D_i, c_{in} and c_{nn}

$$\partial_i^2 K(y) = 2c_{ii}, \qquad \partial_i \partial_n K(y) = 2c_{in}, \qquad \partial_n^2 K(y) = 2c_{nn},$$
$$\partial_i \partial_n H(x) - 2c_{in} = D_i(\partial_i^2 H(x) - 2c_{ii}),$$
$$\partial_n^2 H(x) - 2c_{nn} = D_i^2(\partial_i^2 H(x) - 2c_{ii}), \tag{29}$$
$$\partial_n^3 H(x) = D_i^3 \partial_i^3 H(x)$$

whenever x, y are sufficiently small. From the last line it follows that $D_i \neq 0$ for all $i < n$. From (28) and the first line of (29) it follows that $c_1 = 2c_{nn}$, moreover that

$$\frac{(\partial_n^3 H(x))^2 - 4(\partial_n^2 H(x) - 2c_{nn})^3}{(\partial_n^2 H(x) - 2c_{nn})^2} = D_1^2 \frac{(\partial_1^3 H(x))^2 - 4(\partial_1^2 H(x) - 2c_{11})^3}{(\partial_1^2 H(x) - 2c_{11})^2}.$$

By the induction hypothesis, using the differential equation of the $\mathcal{P}$ functions we obtain that if $\omega = \infty$, then the right hand side is zero.

If $\tilde{\omega}(x_1,\dots,x_{n-1})$ would be somewhere different from infinity, then the left hand side would be $4\pi^2/\tilde{\omega}(x_1,\dots,x_{n-1})^2$, which is impossible, hence $\tilde{\omega}(x_1,\dots,x_{n-1}) \equiv \infty$. If $\omega \neq \infty$, then the right hand side is

$$4d_1^2 D_1^2 \pi^2/\omega^2 \neq 0,$$

hence the left hand side is non-zero, i. e. $\tilde{\omega}(x_1,\dots,x_{n-1}) = \infty$ is nowhere valid. Thus the left hand side is $4\pi^2/\tilde{\omega}(x_1,\dots,x_{n-1})^2$, hence we obtain that $\tilde{\omega}(x_1,\dots,x_{n-1}) = \pm\omega/(d_1 D_1)$. Because the sign is unimportant (the lattice remains the same, we may suppose that $\tilde{\omega}(x_1,\dots,x_{n-1}) = \omega/(d_1 D_1)$ everywhere. Using this together with the notation $d_n = D_1 d_1$ and

$$\mathcal{P}(z,\omega,\infty)/\alpha^2 = \mathcal{P}(\alpha z, \alpha\omega, \infty) \qquad (\alpha \in \mathbb{C} \setminus \{0\})$$

(27) goes over into

$$\partial_n^2 H(x' + x_n f_n) = -d_n^2 \mathcal{P}(d_n x_n + d_n e_1(x_1,\dots,x_{n-1}), \omega, \infty) + 2c_{nn}. \quad (30)$$

If we substitute $x_n = 0$ into (30) we obtain that

$$\partial_n^2 H(x') = -d_n^2 \mathcal{P}(d_n e_1(x_1,\dots,x_{n-1}), \omega, \infty) + 2c_{nn}.$$

Now we first differentiate (30) with respect to x_n and then we put $x_n = 0$ to get

$$\partial_n^3 H(x') = -d_n^3 \mathcal{P}'(d_n e_1(x_1,\dots,x_{n-1}), \omega, \infty).$$

Defining

$$X(x_1,\dots,x_{n-1}) = \mathcal{P}(d_n e_1(x_1,\dots,x_{n-1}), \omega, \infty)$$

and

$$Y(x_1,\dots,x_{n-1}) = \mathcal{P}'(d_n e_1(x_1,\dots,x_{n-1}), \omega, \infty)$$

the above equations show that the functions X and Y can be represented by means of $\partial_n^2 H(x')$ and $\partial_n^3 H(x')$ and hence they are in C^∞. Suppose that $\omega \neq \infty$. The functions X and Y satisfy the equation $Y^2 = 4X^3 - 4\pi^2 X^2/\omega^2$. Because Y is non-zero at the origin, it is uniquely defined by X around the origin. For the same reason the mapping $z \mapsto \mathcal{P}(z, \omega, \infty)$ has inverse on some neighbourhood of $d_n e_1(0,\dots,0)$. Let I denote the inverse of this mapping and let $e_1^*(x_1,\dots,x_{n-1}) = I\left(-(\partial_n^2 H(x') - 2c_{nn})/d_n^2\right)/d_n$. In some neighbourhood of the origin we have

$$X(x_1,\dots,x_{n-1}) = \mathcal{P}(d_n e_1^*(x_1,\dots,x_{n-1}), \omega, \infty).$$

Because Y is uniquely defined by X, on some neighbourhood of the origin it is also true that

$$Y(x_1,\dots,x_{n-1}) = \mathcal{P}'(d_n e_1^*(x_1,\dots,x_{n-1}), \omega, \infty).$$

But by Lemma 1 this means that

$$d_n\big(e_1(x_1,\dots,x_{n-1}) - e_1^*(x_1,\dots,x_{n-1})\big) \in \Omega.$$

Hence we may suppose that $e_1 = e_1^*$, i. e. that $e_1 \in C^\infty$. If $\omega = \infty$ we get $e_1 \in C^\infty$ similarly.

Let us investigate the function e_1. Substituting $x_n = 0$ into (30) and using the third line of (29) and the induction hypothesis we obtain that

$$
\begin{aligned}
d_n^2 \mathcal{P}(d_n e_1(x_1,\dots&,x_{n-1}),\omega,\infty) - 2c_{nn}\\
&= -\partial_n^2 H(x') = -D_1^2\partial_1^2 H(x') + 2D_1^2 c_{11} - 2c_{nn} \qquad (31)\\
&= D_1^2 d_1^2 \mathcal{P}(d_1 x_1 + \dots + d_{n-1}x_{n-1} + e,\omega,\infty) - 2c_{nn}.
\end{aligned}
$$

Using that $d_n = D_1 d_1$ this equation implies

$$
\begin{aligned}
\mathcal{P}(d_n e_1(x_1,\dots&,x_{n-1}),\omega,\infty)\\
&= \mathcal{P}(d_1 x_1 + \dots + d_{n-1}x_{n-1} + e,\omega,\infty).
\end{aligned}
\qquad (32)
$$

On the other hand, differentiating equation (30) with respect to x_n and then substituting $x_n = 0$, moreover using the last line of (29) and the induction hypothesis we arrive at

$$
\begin{aligned}
d_n^3 \mathcal{P}'(d_n e_1(x_1,\dots&,x_{n-1}),\omega,\infty)\\
&= -\partial_n^3 H(x') = -D_1^3 \partial_2^3 H(x') \qquad (33)\\
&= D_1^3 d_1^3 \mathcal{P}'(d_1 x_1 + \dots + d_{n-1}x_{n-1},\omega,\infty),
\end{aligned}
$$

i. e., because of $D_1 d_1 = d_n$, we get

$$
\begin{aligned}
\mathcal{P}'(d_n e_1(x_1,\dots&,x_{n-1}),\omega,\infty)\\
&= \mathcal{P}'(d_1 x_1 + \dots + d_{n-1}x_{n-1},\omega,\infty).
\end{aligned}
\qquad (34)
$$

By (32) — for any fixed $x_2,\dots,x_{n-1}$ — using the implicit function theorem we have that the function e_1 is continuously differentiable with respect to x_1. Differentiation of (32) with respect to x_1 yields

$$
\begin{aligned}
d_n \mathcal{P}'(d_n e_1(x_1,\dots&,x_{n-1}),\omega,\infty)\partial_1 e_1(x_1,\dots,x_{n-1})\\
&= d_1 \mathcal{P}'(d_1 x_1 + \dots + d_{n-1}x_{n-1},\omega,\infty).
\end{aligned}
\qquad (35)
$$

Now, (31) and (32) result in

$$\partial_1 e_1(x_1,\dots,x_{n-1}) = d_1/d_n$$

everywhere. But this means that for some continuous function

$$e_2(x_2,\dots,x_{n-1})$$

we have

$$e_1(x_1, \dots, x_{n-1}) = d_1 x_1 / d_n + e_2(x_2, \dots, x_{n-1}).$$

Putting this into equation (32) and choosing $x_2 = \cdots = x_{n-1} = 0$ from Lemma 2 we obtain also that $d_n e_2(0, \dots, 0) - e$ is in the (degenerated) lattice Ω generated by ω.

In the next step we show that (12) and (13) are valid with $i = 2$. Substituting $x_n = 0$ into (30) and using the third line of (29) and the induction hypothesis we obtain that

$$
\begin{aligned}
d_n^2 \mathcal{P}(d_1 x_1 &+ d_n e_2(x_2, \dots, x_{n-1}), \omega, \infty) - 2c_{nn} \\
&= -\partial_n^2 H(x') = -D_2^2 \partial_2^2 H(x') + 2D_2^2 c_{22} - 2c_{nn} \\
&= D_2^2 d_2^2 \mathcal{P}(d_1 x_1 + \dots + d_{n-1} x_{n-1} + e, \omega, \infty) - 2c_{nn}.
\end{aligned}
\tag{36}
$$

Using (32) from this equation we get that $D_2 d_2 = d_n$. Differentiating equation (30) with respect to x_n and then substituting $x_n = 0$, moreover using the last line of (29) and the induction hypothesis we obtain that

$$
\begin{aligned}
d_n^3 \mathcal{P}'(d_1 x_1 &+ d_n e_2(x_2, \dots, x_{n-1}), \omega, \infty) \\
&= -\partial_n^3 H(x') = -D_2^3 \partial_2^3 H(x') \\
&= D_2^3 d_2^3 \mathcal{P}'(d_1 x_1 + \dots + d_{n-1} x_{n-1}, \omega, \infty),
\end{aligned}
\tag{37}
$$

i. e., because of $D_2 d_2 = d_n$, it follows that

$$
\begin{aligned}
\mathcal{P}'(d_1 x_1 &+ d_n e_2(x_2, \dots, x_{n-1}), \omega, \infty) \\
&= \mathcal{P}'(d_1 x_1 + \dots + d_{n-1} x_{n-1}, \omega, \infty).
\end{aligned}
\tag{38}
$$

Equation (38) is exactly equation (25) with $\omega_1 = \omega$, $\omega_2 = \infty$. Following the arguments after equation (25) we arrive (with obvious changes) at (5).

We have to investigate also the case when $\partial_n^3 H$ and $\partial_n^3 K$ are both zero everywhere. In this case we have to prove that the statements holds true with $d_n = 0$. By Lemma 6 we obtain that there exists a constant $c_1 \in \mathbb{C}$ such that for all sufficiently small x', y', x_n and y_n we have

$$\partial_n^2 H(x' + x_n f_n) = c_1 \quad \text{and} \quad \partial_n^2 K(y' + y_n f_n) = c_1,$$

hence with notation $c_{nn} = c_1/2$ we get

$$\partial_n^2 H(x' + x_n f_n) = 2c_{nn} \quad \text{and} \quad \partial_n^2 K(y' + y_n f_n) = 2c_{nn}.$$

Let us note that $\partial_i \partial_n H(x')$ and $\partial_i \partial_n K(y')$ are the same constant for all $i < n$: If $\partial_i^3 H$ and $\partial_i^3 K$ are non-zero, then equations (29) are holds true and from the last line of (29) we get $D_i = 0$, hence the statement follows from the first and second line of (29). If $\partial_i^3 H$ and $\partial_i^3 K$ are both zero, then the statement follows from case (4) of Lemma 7. Now we may follow the lines of the proof of (4).

Finally, the proof of (6) is analogous to the proof of (5). □

Theorem 5. *Suppose that the functions* $f_1, f_2, g_1, g_2, h_1, h_2 : \mathbb{R}^n \to \mathbb{C}$ *satisfy the functional equation*

$$f_1(u + v)f_2(u - v) = g_1(u)h_1(v) + g_2(u)h_2(v) \qquad (u, v \in \mathbb{R}^n),$$

the functions f_1 *and* f_2 *are measurable and none of them is almost everywhere zero. Then there exist constants* $a, \tilde{a}, e, \tilde{e} \in \mathbb{C}$, *linear mappings* $L, L_1, L_2 : \mathbb{R}^n \to \mathbb{C}$ *and a quadratic mapping* $Q : \mathbb{R}^n \to \mathbb{C}$, *moreover lattice constants* ω_1, ω_2 *such that*

$$f_1(x) = \exp(a + L_1(x) + Q(x))\sigma(L(x) + e, \omega_1, \omega_2) \quad \text{if} \quad x \in \mathbb{R}^n \quad \text{and}$$
$$f_2(y) = \exp(\tilde{a} + L_2(y) + Q(y))\sigma(L(y) + \tilde{e}, \omega_1, \omega_2) \quad \text{if} \quad y \in \mathbb{R}^n$$

or

$$f_1(x) = \exp(a + L_1(x) + Q(x))\sigma(L(x) + e, \omega_1, \infty) \quad \text{if} \quad x \in \mathbb{R}^n \quad \text{and}$$
$$f_2(y) = \exp(\tilde{a} + L_2(y) + Q(y)) \quad \text{if} \quad y \in \mathbb{R}^n$$

or

$$f_1(x) = \exp(a + L_1(x) + Q(x)) \quad \text{if} \quad x \in \mathbb{R}^n \quad \text{and}$$
$$f_2(y) = \exp(\tilde{a} + L_2(y) + Q(y))\sigma(L(y) + \tilde{e}, \omega_1, \infty) \quad \text{if} \quad y \in \mathbb{R}^n.$$

The functions g_1, g_2, h_1, h_2 *can be obtained from the representation of* f_1, f_2 *using Lemma 4 and Remark 4.*

Proof. Let us first suppose that there exists a point $x_0 \in \mathbb{R}^n$ with $f_1(x_0) \neq 0$ and an index $1 \leq i \leq n$ such that with the notation $H(x) = \ln(f_1(x + x_0)/f_1(x_0))$ the derivative $\partial_i^3 H$ is not identically zero on any neighbourhood of the origin or that there exists a point $y_0 \in \mathbb{R}^n$ with $f_2(y_0) \neq 0$ and an index $1 \leq i \leq n$ such that with the notation $K(y) = \ln(f_2(y + y_0)/f_2(y_0))$ the derivative $\partial_i^3 K$ is not identically zero on any neighbourhood of the origin. We shall treat only the first case since the second case can be treated in exactly the same way. Without loss of generality we may also suppose that $i = 1$. Thus there exists an $x_0 \in \mathbb{R}^n$ such that $\partial_1^3 H$ is not identically zero on any neighbourhood of the origin. (Note that H depends on x_0.) We shall prove that $x_0, y_0 \in \mathbb{R}^n$ and $\delta > 0$ may be chosen so

(a) that $\partial_1^3 H$ is not identically zero on any neighbourhood of the origin,

(b) that the equations (6)–(9) of Remark 7 are also satisfied for all $|x|, |y| < \delta$, and

(c) that moreover for all $1 \leq j \leq n$ the following conditions are also satisfied:

(1) $\partial_j^3 H(x)$ is either zero for all $|x| < \delta$ or different from zero for all $|x| < \delta$;

(2) $\partial_j^3 K(y)$ is either zero for all $|y| < \delta$ or different from zero for all $|y| < \delta$;

(3) if $\partial_j^3 H \neq 0$ or $\partial_j^3 K \neq 0$, (or both) everywhere in a δ-neighbourhood of the origin, then $\partial_j^2 H(x) - \partial_j^2 K(y)$ is also different from zero for all $|x| < \delta$, $|y| < \delta$.

To prove this statement let us start with an $\varepsilon > 0$ and with points x_1 and y_1 with $f_1(x_1) \neq 0$ and $f_2(y_1) \neq 0$ for which the functions $H_1(x) = \ln(f_1(x + x_1)/f_1(x_1))$ and $K_1(y) = \ln(f_2(y + y_1)/f_2(y_1))$ are defined on the open ball centered at the origin and having radius ε, moreover that $\partial_1^3 H_1$ is not identically zero on any neighbourhood of the origin. Decreasing ε if necessary we may suppose that for all x_0, y_0 satisfying $|x_0 - x_1| < \varepsilon$, $|y_0 - y_1| < \varepsilon$, $f_1(x_0) \neq 0$ and $f_2(y_0) \neq 0$ the functions $H(x) = \ln(f_1(x+x_0)/f_1(x_0))$ and $K(y) = \ln(f_2(y+y_0)/f_2(y_0))$ (also depending upon x_0 and y_0, respectively) are defined on the open ball having radius $\delta = \varepsilon - \max\{|x_0 - x_1|, |y_0 - y_1|\}$ and centered at the origin. Let us investigate the connection between H_1 and H and between K_1 and K, respectively. If ε is small enough then

$$H(x) = \ln(f_1(x + x_0)/f_1(x_0))$$
$$= \ln((f_1(x + (x_0 - x_1) + x_1)/f_1(x_1)) - \ln(f_1(x_0)/f_1(x_1))$$
$$= H_1(x + (x_0 - x_1)) + \ln(f_1(x_1)/f_1(x_0)).$$

This means that the derivatives of H around the origin behave in the same way like the derivatives of H_1 around $x_0 - x_1$. Similarly we obtain that

$$K(y) = K_1(y + (y_0 - y_1)) + \ln(f_2(y_1)/f_2(y_0)),$$

i. e., that the derivatives of K around the origin behave in the same way like the derivatives of K_1 around $y_0 - y_1$.

It is now clear that decreasing ε if necessary we may suppose that for any $|x_0|, |y_0| < \varepsilon$ the functions H and K satisfy the system of partial differential equations (6)–(9) of Remark 7 on the open ball around the origin having radius δ and that $\partial_1^3 H$ is not identically zero on any neighbourhood of the origin.

Let us consider the open set

$$X = \{x \in \mathbb{R}^n : |x| < \varepsilon, \partial_1^3 H_1(x) \neq 0\}.$$

This is a Baire space. For each $2 \leq j \leq n$ the set X_j of all those points $x \in X$ which either have a neighbourhood in X on which $\partial_j^3 H_1$ is nowhere zero or have a neighbourhood in X on which $\partial_j^3 H_1$ is identically zero, is a dense open set in X (note that X_j is the boundary of $\{x \in X : \partial_j^3 H_1(x) = 0\}$). Hence the set $\cap_{j=2}^n X_j$ is a dense open subset of X. Let x_2 be an arbitrary point of this set. There is a neighbourhood of $x_1 + x_2$ such that for all points x_0 from this neighbourhood (1) is satisfied. We obtain the point y_2 similarly: Let us consider the open set $Y = \{y \in \mathbb{R}^n : |y| < \varepsilon\}$. This is a Baire space. For each $1 \leq j \leq n$ the set Y_j of all those points $y \in Y$ which either have a neighbourhood in Y such that $\partial_j^3 K_1$ is nowhere zero on this neighbourhood or have a neighbourhood in Y such that $\partial_j^3 K_1$ is identically zero on this neighbourhood, is a dense open set in Y. Hence the open set $\cap_{j=1}^n Y_j$ is dense in Y. Let y_2 be an arbitrary point of this set. There exists a neighbourhood of $y_1 + y_2$ such that for all points y_0 from this neighbourhood (2) is satisfied. Finally, we have to prove that x_0 and y_0 may be chosen so that (3) is also satisfied. This easily follows by induction. Let $x_0 = x_1 + x_2$ and $y_0 = y_1 + y_2$. If (3) is already satisfied for $1 \leq j < k$, and if $\partial_k^3 H_1(x_2) = \partial_k^3 H(0) \neq 0$, then let us replace x_2 by $x_2 + te_k$, where e_k is the k-th vector of the standard base and t is small enough so that the condition (c) remains valid, and if $\partial_k^3 H_1(x_2) = \partial_k^3 H(0) = 0$, then, using that $\partial_k^3 K_1(y_2)) = \partial_k^3 K(0) \neq 0$, let us replace y_2 by $y_2 + te_k$, where t is small enough so that the condition (c) remains valid. If with these new values x_2, y_2 we reconsider the points $x_0 = x_1 + x_2$ and $y_0 = y_1 + y_2$, then (3) can be satisfied for $1 \leq j \leq k$.

Now let us apply Lemma 8. We obtain that there exist constants $a', \tilde{a}', c, \tilde{c}, e', \tilde{e}' \in \mathbb{C}$, linear maps $L, L_1', L_2' : \mathbb{R}^n \to \mathbb{C}$ and a quadratic map $Q : \mathbb{R}^n \to \mathbb{C}$ such that for a sufficiently small neighbourhood of the origin we have

$$H(x) = a' + L_1'(x) + Q(x) + \ln \sigma\big(c(L(x) + e', \omega_1, \omega_2)\big) \qquad \text{and}$$
$$K(y) = \tilde{a}' + L_2'(y) + Q(y) + \ln\big(\tilde{c}\sigma(L(y) + \tilde{e}', \omega_1, \omega_2)\big)$$

or

$$H(x) = a' + L_1'(x) + Q(x) + \ln\big(c\sigma(L(x) + e', \omega_1, \infty)\big) \qquad \text{and}$$
$$K(y) = \tilde{a}' + L_2'(y) + Q(y)$$

or

$$H(x) = a' + L_1'(x) + Q(x) \qquad \text{and}$$
$$K(y) = \tilde{a}' + L_2'(y) + Q(y) + \ln\big(\tilde{c}\sigma(L(y) + \tilde{e}', \omega_1, \infty)\big).$$

This means that the mappings $x \mapsto f_1(x + x_0)$ and $y \mapsto f_2(y + y_0)$ have the stated representations on a small neighbourhood of the origin. But from Remark 4 we know that these functions on a small neighbourhood of the origin satisfy a functional equation having type (1) with everywhere analytic and on every neighbourhood of the origin linearly independent functions $\tilde{g}_1, \tilde{g}_2, \tilde{h}_1, \tilde{h}_2$ on the right hand side. Hence Lemma 5 can be applied and f_1 and f_2 have the representation stated in the theorem everywhere.

Only missing is the investigation of the case that for each $x_0 \in \mathbb{R}^n$, for which $f_1(x_0) \neq 0$ each partial derivate $\partial_i^3 H$ of the function $H(x) = \ln(f_1(x + x_0)/f_1(x_0))$ is identically zero on some neighbourhood of the origin, and for each $y_0 \in \mathbb{R}^n$, for which $f_2(y_0) \neq 0$ each partial derivate $\partial_i^3 K$ of the function $K(y) = \ln(f_2(y + y_0)/f_2(y_0))$ is identically zero on some neighbourhood of the origin. We shall prove that in this case the statement of the theorem is satisfied with $L \equiv 0$, i. e. we shall prove that there exist complex constants $a, \tilde{a}$, linear mappings $L_1, L_2 : \mathbb{R}^n \to \mathbb{C}$, and a quadratic map $Q : \mathbb{R}^n \to \mathbb{C}$ such that

$$
\begin{aligned}
f_1(x) &= \exp(a + L_1(x) + Q(x)) && \text{if } x \in \mathbb{R}^n && \text{and} \\
f_2(y) &= \exp(\tilde{a} + L_2(y) + Q(y)) && \text{if } y \in \mathbb{R}^n.
\end{aligned} \tag{4}
$$

Here also we first prove that the above representation is satisfied locally. By the statement (4) of Lemma 7 for any x_0 and y_0 for which $f_1(x_0) \neq 0$ and $f_2(y_0) \neq 0$, we have

$$
\begin{aligned}
H(x) &= a' + L_1'(x) + Q(x) && \text{and} \\
K(y) &= \tilde{a}' + L_2'(y) + Q(y),
\end{aligned}
$$

for appropriate complex constants $a', \tilde{a}'$, linear functions $L_1', L_2' : \mathbb{R}^n \to \mathbb{C}$ and a quadratic function $Q : \mathbb{R}^n \to \mathbb{C}$. Now we use $f_1(x + x_0) = f_1(x_0) \exp(H(x))$ and $f_2(y+y_0) = f_2(y_0) \exp(K(y))$, introduce new variables and on a neighbourhood of x_0 and y_0 arrive at

$$
\begin{aligned}
f_1(x) &= \exp(a + L_1(x) + Q(x)) && \text{and} \\
f_2(y) &= \exp(\tilde{a} + L_2(y) + Q(y)),
\end{aligned}
$$

with appropriate complex constants $a, \tilde{a}$ and appropriate linear maps $L_1, L_2 : \mathbb{R}^n \to \mathbb{C}$. Let us fix a pair x_0, y_0 and let us consider the maximal radius $0 < \delta \leq \infty$ for which f_1 is nowhere zero on the open ball centered at x_0 and having radius δ. Because f_1 is analytic on this ball, the above representation is satisfied everywhere on this ball with a, L_1 and Q belonging to the pair x_0, y_0. If we assume that $\delta < \infty$ then the mapping $x \mapsto a + L_1(x) + Q(x)$ would be bounded on the ball, the function f_1

cannot be continuous because $|f_1|$ would have a positive lower bound. This means that $\delta = \infty$, i. e. that the above representation is satisfied on the whole $\mathbb{R}^n$ with a, L_1 and Q belonging to the pair x_0, y_0. Similarly, the function f_2 also can be represented in form (4) with $\tilde{a}$, L_2 and Q belonging to x_0, y_0. □

Remark 8. Besides the regularity results the main idea for solving the functional equation in Theorem 5 was to find an appropriate partial differential equation (or a system of partial differential equations) the solutions of which lead to the solutions of the functional equation. In Remark 5 to Remark 7 we have presented appropriate partial differential equations for a big class of corresponding functional equations. Specializing the function F in $F(u, v) = \sum_{i=1}^{k} g_i(u)h_i(v)$ (see Remark 5) many interesting functional equations can be solved as shown in this paper, where the special case $F(u, v) = f_1(u+v)f_2(u-v)$ and $k = 2$ was solved.

References

[1] J. Aczél, *The state of the second part of Hilbert's fifth problem*, Bull. Amer. Math. Soc. (N. S.) **20** (1989), 153–163.

[2] J. A. Baker, *On the functional equation $f(x)g(y) = \prod_{i=1}^{n} h_i(a_i x + b_i y)$*, Aequationes Math. **11** (1974), 154–162.

[3] M. Bonk, *The Characterization of Theta Functions by Functional Equations*, Abh. Math. Sem. Univ. Hamburg **65** (1995), 29–55.

[4] M. Bonk, *The addition theorem of Weierstraß's sigma function*, Math. Ann. **298** (1994), 591–601.

[5] M. Bonk, *The addition formula for theta functions*, Aequationes Math. **53** (1997), 54–72.

[6] J. Dieudonné, *Grundzüge der modernen Analysis I–IX*, VEB Deutscher Verlag der Wissenschaften, Berlin, 1971–1988.

[7] H. Gauchman and L. A. Rubel, *Sums of products of functions of x times functions of y*, Linear Alg. Appl. **125** (1989), 19–63.

[8] H. Haruki, *Studies on certain functional equations from the standpoint of analytic function theory*, Sci. Rep. Osaka Univ. **14** (1965), 1–40.

[9] D. Hilbert, *Gesammelte Abhandlungen Band III*, Springer Verlag, Berlin–Heidelberg–New York, 1970.

[10] A. Járai, *On regular solutions of functional equations*, Aequationes Math. **30** (1986), 21–54.

[11] A. Járai, *On Lipschitz property of continuous solutions of functional equations*, Aequationes Math. **47** (1994), 69–78.

[12] A. Járai and W. Sander, *A regularity theorem in information theory*, Publ. Math. Debrecen **50** (1997), 339–357.

[13] A. Járai and L. Székelyhidi, *Regularization and general methods in the theory of functional equations*, Survey paper. Aequationes Math. **52** (1966), 10–29.

[14] K. Lajkó, *On the functional equation* $f(x)g(y) = h(ax + by)k(cx + dy)$, Periodica Math. Hung. **11** (1980), 187–195.

[15] T. M. Rassias and J. Šimša, *Finite Sums Decompositions in Mathematical Analysis*, John Wiley & Sons, Chichester, 1995.

[16] R. Rochberg and L. A. Rubel, *A functional equation*, Indiana Univ. Math. J. **41** (1992), 363–376.

[17] W. Sander, *Problem 19*, Aequationes Math. **46** (1993), 294.

[18] S. Saks and A. Zygmund, *Analytic Functions*, Warszawa–Wrocław 1952.

ON A MIKUSIŃSKI–JENSEN FUNCTIONAL EQUATION

Károly Lajkó and Zsolt Páles

Institute of Mathematics and Informatics, University of Debrecen,
H-4010 Debrecen, Pf. 12, Hungary

lajko@math.klte.hu, pales@math.klte.hu

Dedicated to the 70th birthday of Professor Mátyás Arató

Abstract   The following Mikusiński-Jensen type functional equation

$$f\left(\frac{x+y}{2}\right)\left[f(x)+f(y)-2f\left(\frac{x+y}{2}\right)\right]=0 \qquad (x,y\in I)$$

is investigated. The main result states that if I is an open real interval, or more generally, if I is a convex subset of a linear space whose intersection with straight lines is an open segment, then the above equation is equivalent to the so called Jensen functional equation.

Keywords: Jensen functional equation, conditional functional equation

Mathematics Subject Classification (2000): 39B22, 39B52

1.   INTRODUCTION

During the 31th International Symposium on Functional Equations (Debrecen, Hungary, 1993) C. Alsina [3] asked about the continuous solutions of the functional equation (in the case $I = \mathbb{R}$)

$$
\begin{aligned}
f(x)f(px + (1-p)y) &+ f(y)f((1-p)x + py) \\
&= f(px + (1-p)y)^2 + f((1-p)x + py)^2 \qquad (x,y\in I),
\end{aligned}
\tag{1}
$$

---

[0]This research has been supported by the Hungarian Scientific Research Fund (OTKA) Grant T-030082 and by the Higher Education, Research, and Development Fund (FKFP) Grant 0310/1997.

Z. Daróczy and Z. Páles (eds.),
Functional Equations - Results and Advances, 81–87.
© 2002 *Kluwer Academic Publishers*.

where I is a real interval, $p \in]0, 1[$ is a fixed parameter, and f is a real valued function on I. An answer to the above problem was already given during the meeting by A. Járai and Gy. Maksa [7] supposing that $I = \mathbb{R}$. The measurable nowhere zero solutions of (1) were also found by A. Járai and Gy. Maksa in [8]. They proved the following result.

Theorem 1. *Let $p \in]0, 1[$ be fixed, $I \subset \mathbb{R}$ be an interval of positive length, and $f : I \to \mathbb{R} \setminus \{0\}$ be measurable on a subset of I of positive Lebesgue measure. Then f satisfies the functional equation (1) if and only if*

- *either f is constant on I and p is arbitrary,*
- *or f is of the form $f(x) = ae^{bx}$ and $p = 1/3$,*
- *or f is of the form $f(x) = a(x + c)$ and $p = 1/2$,*

where a, b, c are arbitrary real constants with $a \neq 0$, $-c \notin I$.

It is interesting that, under the conditions of the theorem, (1) has nonconstant solutions if and only if either $p = 1/3$ or $p = 1/2$. In the case $p = 1/2$, (1) reduces to the following Mikusiński–Jensen type functional equation:

$$f\left(\frac{x+y}{2}\right)\left[f(x) + f(y) - 2f\left(\frac{x+y}{2}\right)\right] = 0 \qquad (x, y \in I). \qquad \text{(M-J)}$$

The aim of the present note is to show that a function $f : I \to \mathbb{R}$, where I is an open interval, is a solutions of (M-J) if and only if f satisfies the Jensen functional equation

$$f(x) + f(y) = 2f\left(\frac{x+y}{2}\right) \qquad (x, y \in I). \qquad \text{(J)}$$

Thus, the solutions of (M-J) can completely be described. The results will also be extended to the case when I is a convex subset of a real vector space intersecting any straight line in an open interval.

We point out that the openness of I is essential, that is, the equivalence of the two functional equations (M-J) and (J) is valid if and only if the underlying real interval I is open. The general solutions of (M-J) over arbitrary interval will also be obtained.

The equation (M-J) is formally a special case of the so called Mikusiński–Pexider equation, that is,

$$f(x + y)[g(x + y) - h(x) - k(y)] = 0$$

that was studied by Baron and Ger [4] in the semigroup setting. Unfortunately, the results obtained in [4] cannot be applied directly to (M-J) since the natural domains for (M-J) are convex sets.

2. SOLUTION OF (M-J) ON INTERVALS

In order to determine the solutions of (M-J), we need to recall the following result that describes the general solution of the Jensen equation over arbitrary convex sets of a linear space. This lemma is a special case of [6, Theorem] (cf. also [9]).

Lemma 1. *Let X be a real linear space, $I \subset X$ be a nonempty convex set, and let $f : I \to \mathbb{R}$. Then f satisfies the Jensen equation (J) if and only if there exist an additive function $A : X \to \mathbb{R}$ and a constant $b \in \mathbb{R}$ such that*

$$f(x) = A(x) + b \qquad (x \in I). \tag{2}$$

In particular, if I is a nonempty interval, then it is also convex. Thus, the general solutions of (J) over real intervals are also of the form (2) (see [1], [5] and [10]).

In the rest of this section let I denote an open real interval. Below, we introduce the notion of locally Jensen functions. Our idea is to prove that the solutions of (M-J) belong to this class of functions.

Definition 1. A function $f : I \to \mathbb{R}$ is called *locally Jensen at* $p \in I$ if there exists $r_p > 0$ such that f satisfies the Jensen equation on the open interval $I_p =]p - r_p, p + r_p[\subset I$. $f : I \to \mathbb{R}$ is called *locally Jensen on I* if it is locally Jensen at each point $p \in I$.

First we prove that each solution f of the functional equation (M-J) is locally Jensen at those points $p \in I$ where f does not vanish. The result also shows that the positive number r_p can be chosen to be maximal such that $]p - r_p, p + r_p[\subset I$. In the next step, we will prove that the solutions of (M-J) are locally Jensen on I.

Lemma 2. *Let I be an open interval, $p \in I$, and $f : I \to \mathbb{R}$ be a solution of the functional equation (M-J) such that $f(p) \neq 0$. Then f is Jensen on the interval $I_p := I \cap (2p - I)$.*

Proof. The interval I_p is symmetric with respect to p. Therefore, if $x \in I_p$, then $2p - x \in I_p$. Substituting $x \in I_p$, $y := 2p - x$ into (M-J), we get

$$f(p)[f(x) + f(2p - x) - 2f(p)] = 0 \qquad (x \in I_p),$$

whence, by the assumption $f(p) \neq 0$, we obtain the identity

$$f(2p - x) = 2f(p) - f(x) \qquad (x \in I_p). \tag{3}$$

Replacing x by $2p - x$ and y by $2p - y$, equation (M-J) yields

$$f\left(2p - \frac{x+y}{2}\right)\left[f(2p-x) + f(2p-y) - 2f\left(2p - \frac{x+y}{2}\right)\right] = 0 \qquad (x, y \in I_p).$$

Now using (3), we get

$$\left[2f(p) - f\left(\frac{x+y}{2}\right)\right]\left[f(x) + f(y) - 2f\left(\frac{x+y}{2}\right)\right] = 0 \qquad (x, y \in I_p).$$

Adding this equation to (M-J) and using again that $f(p) \neq 0$, we obtain

$$f(x) + f(y) - 2f\left(\frac{x+y}{2}\right) = 0 \qquad (x, y \in I_p),$$

which means that f is Jensen on I_p.   □

Lemma 3. *Let I be an open interval and $f : I \to \mathbb{R}$ be an arbitrary solution of the functional equation (M-J). Then f is a locally Jensen function on I.*

Proof. Let $p \in I$ be fixed. We distinguish three cases.

If $f(p) \neq 0$, then, by Lemma 2, f is locally Jensen at p.

If $f(p) = 0$ and there exits $r_p > 0$ such that f is identically zero on $I_p :=]p - r_p, p + r_p[$ then f is clearly Jensen on I_p, and hence it is locally Jensen at p.

The last case is when $f(p) = 0$ and there exists a sequence $p_n \to p$ such that $f(p_n) \neq 0$. Let $r_p > 0$ be such that

$$]p - 3r_p, p + 3r_p[\subset I$$

and choose $n \in \mathbb{N}$ so that $p_n \in I_p :=]p - r_p, p + r_p[$. Then

$$I_p \subset J_n :=]p_n - 2r_p, p_n + 2r_p[\subset]p - 3r_p, p + 3r_p[\subset I.$$

Applying Lemma 2, we obtain that f is Jensen on J_n. The inclusion $I_p \subset J_n$ yields that f is locally Jensen at p.

Thus, f is locally Jensen at p in each of the above three cases.   □

Now we are in the position to state and prove the main result of this paper.

Theorem 2. *Let I be an open interval. Then $f : I \to \mathbb{R}$ is a solution of the Mikusiński–Jensen functional equation (M-J) if and only if there exist an additive function $A : \mathbb{R} \to \mathbb{R}$ and a constant $b \in \mathbb{R}$ such that f is of the form (2).*

Proof. If f is of the form (2), then (M-J) holds trivially. To prove the necessity of this representation, assume that f is a solution of (M-J).

Let $p_0 \in I$ be fixed. Then, by Lemma 3, f is Jensen on an open interval I_0 containing p_0. By Lemma 1, there exist an additive function $A : \mathbb{R} \to \mathbb{R}$ and a constant $b \in \mathbb{R}$ such that, for all $x \in I_0$,

$$f(x) = A(x) + b. \qquad (4)$$

We are going to show that (4) holds on I instead of I_0. The union of all open intervals $J \subset I$ such that $p_0 \in J$ and f is of the form (4) on J is the maximal open subinterval H of I with the above properties. It suffices to show that $H = I$.

Assume, on the contrary, that $p = \sup H \in I$. By Lemma 3, f is locally Jensen at p, hence there exists an open interval I_p around p such that f is Jensen on I_p. Again, by Lemma 1, there exist an additive function $A_p : \mathbb{R} \to \mathbb{R}$ and a constant $b_p \in \mathbb{R}$ such that

$$f(x) = A_p(x) + b_p \qquad (p \in I_p). \tag{5}$$

Thus we have that

$$A(x) + b = f(x) = A_p(x) + b_p \qquad (p \in H \cap I_p).$$

The intersection $H \cap I_p$ is a nonempty open interval, and hence, one can easily verify that $A_p \equiv A$ and $b_p = b$. Therefore, we obtain that (4) is valid for $x \in H \cup I_p = H \cup [p, p + r_p[$, which contradicts the maximality of H.

The case when $\inf H \in I$ leads to an analogous contradiction. Due to these contradictions, we have that $H = I$, that is, f is represented by (4) on the whole interval I. □

It is interesting to observe that the result of Theorem 2 is not valid if the underlying interval I is not open. That is the two functional equations (M-J) and (J) are equivalent to each other if and only if I is open.

Let $I := [0, 1[$ and define $f : I \to \mathbb{R}$ by

$$f(x) := \begin{cases} 0, & \text{if } x \in]0, 1[, \\ c, & \text{if } x = 0, \end{cases}$$

where $c \neq 0$ is an arbitrary constant. If $x, y \in [0, 1[$ and $x + y > 0$, then $f\left(\frac{x+y}{2}\right) = 0$, hence (M-J) is valid. If $x = y = 0$, then the second factor of the left hand side of (M-J) vanishes. Thus f is a solution of (M-J) but it is not a Jensen function.

Motivated by this example and using the result of Theorem 2, we can describe the solutions of the functional equation (M-J) over arbitrary real intervals.

Theorem 3. *Let I be a proper real interval. Then $f : I \to \mathbb{R}$ is a solution of the Mikusiński–Jensen functional equation (M-J) if and only if either*

- there exist an additive function $A : \mathbb{R} \to \mathbb{R}$ and a constant $b \in \mathbb{R}$ such that f is of the form (2), or
- f is identically zero on the interior of I and is arbitrary at the endpoints of I.

Proof. The sufficiency can be checked easily.

Assume that f is a solution of (M-J). Then it also satisfies this equation on the interior of I, therefore, by Theorem 2, there exist an additive function $A : \mathbb{R} \to \mathbb{R}$ and $b \in \mathbb{R}$ such that (4) holds for all interior points x of I. To complete the proof, it suffices to show that, if (4) is not valid at one of the endpoints of I then $A \equiv 0$ and $b = 0$.

Let $q \in I$ be an endpoint of I such that $f(q) \neq A(q) + b$. Without loss of generality, we may assume that q is the left endpoint of I. If $f(x) = A(x) + b$ is not identically zero on the interior of I, then there exist $p \in I^\circ$ such that $f(p) = A(p) + b \neq 0$ and $2p - q \in I^\circ$. Substituting $x = q$, $y = 2p - q$ into the functional equation (M-J), and using that (4) is valid for $x = p$ and $x = 2q - p$, we get that

$$0 = f(p)[f(q) + f(2p - q) - 2f(p)] = f(p)[f(q) - A(q) - b] \neq 0,$$

which is an obvious contradiction. Hence f must be zero on I°.   □

3.   SOLUTION OF (M-J) ON CONVEX SETS

A convex set I of a real linear space will be called *intrinsically convex* if, for all $u, v \in I$, the set

$$I_{u,v} := \{\, t \in \mathbb{R} \mid tu + (1 - t)v \in I \,\}$$

is an open real interval.

If X is a topological real linear space and C is an open convex set then it is intrinsically convex, however the converse of this statement is obviously not valid in general.

Theorem 4. *Let I be an intrinsically convex subset of a real linear space X. Then $f : I \to \mathbb{R}$ is a solution of the Mikusiński–Jensen functional equation (M-J) if and only if there exist an additive function $A : X \to \mathbb{R}$ and a constant $b \in \mathbb{R}$ such that f is of the form (2).*

Proof. Let $u, v \in I$ be fixed and introduce the function $g : I_{u,v} \to \mathbb{R}$ by

$$g(t) = f(tu + (1 - t)v) \qquad (t \in I_{u,v}).$$

Substituting $x = tu + (1 - t)v$, $y = su + (1 - s)v$ into (M-J), we get that the function g satisfies the functional equation (M-J) over $I_{u,v}$. This interval being open, Theorem 2 yields that g satisfies the Jensen equation on $I_{u,v}$. Hence $g(0) + g(1) = 2g(1/2)$, i.e.,

$$f(v) + f(u) = f\left(\frac{u+v}{2}\right)$$

for arbitrary $u, v \in I$. Thus, f satisfies the Jensen equation on I. Using Lemma 1, the statement follows. $\qquad\square$

References

[1] J. Aczél, *Lectures on Functional Equations and Ttheir Applications*, Academic Press, New York–London, 1966.

[2] C. Alsina and J. L. García-Roig, *On two functional equations related to a result of Gregorie de Saint Vincent*, Results in Math. **28** (1995), 33–39.

[3] C. Alsina, *1. Problem (in Report of Meeting)*, Aequationes Math. **47** (1994), 302.

[4] K. Baron and R. Ger, *On Mikusiński–Pexider functional equation*, Colloq. Math. **28** (1973), 307–312.

[5] Z. Daróczy, K. Lajkó, and L. Székelyhidi, *Functional equations on ordered fields*, Publ. Math. Debrecen **24** (1977), 173–179.

[6] R. Ger, *On extension of polinomial functions*, Results Math. **26** (1994), 281–289.

[7] A. Járai and Gy. Maksa, *2. Remark (in Report of Meeting)*, Aequationes Math. **47** (1994), 302.

[8] A. Járai and Gy. Maksa, *The measurable solutions of a functional equation of C. Alsina and J. L. Garcia-Roig*, C. R. Math. Rep. Acad. Sci. Canada **17** (1995), 7–10.

[9] M. Kuczma, *An Introduction to the Theory of Functional Equations and Inequalities*, Państwowe Wydawnictwo Naukowe and Uniwersytet Śląnski, Warszawa–Kraków–Katowice, 1985.

[10] K. Lajkó, *Applications of extensions of additive functions*, Aequationes Math. **11** (1974), 68–76.

II
STABILITY
OF FUNCTIONAL EQUATIONS

STABILITY OF THE MULTIPLICATIVE CAUCHY FUNCTIONAL EQUATION IN ORDERED FIELDS

Zoltán Boros

Institute of Mathematics and Informatics, University of Debrecen,

4010 Debrecen, Pf. 12, Hungary

boros@math.klte.hu

Abstract   An analogue of Baker's superstability theorem, a related example, and further stability type results concerning the multiplicative Cauchy functional equation are presented for functions with values in ordered fields.

Keywords: multiplicative Cauchy functional equation, stability, ordered field, free Abelian group

Mathematics Subject Classification (2000): 39B52, 39B72, 12J15, 20K30

1.   THE WEAK SUPERSTABILITY PHENOMENON

The stability of the exponential equation has been first proved by Baker, Lawrence, and Zorzitto [2] for real valued functions. Their result has been extended and greatly simplified by Baker [1]: If S is a semigroup, $0 < \delta \in \mathbb{R}$, and f is a real or complex valued function defined on S such that

$$|f(xy) - f(x)f(y)| \leq \delta \qquad \text{for all} \quad x, y \in S,$$

---

[0]This research has been supported in part by the Hungarian Scientific Research Fund (OTKA) Grant T-030082, by the Hungarian Higher Education Research Development Programs (FKFP) Grant 0310/1997, and by the Kereskedelmi és Hitelbank Rt. Universitas Foundation.

Z. Daróczy and Z. Páles (eds.),

Functional Equations - Results and Advances, 91–98.

© *2002 Kluwer Academic Publishers.*

then either

$$|f(x)| \leq \frac{1 + \sqrt{1 + 4\delta}}{2} \qquad \text{for all} \quad x \in S$$

or $f(xy) = f(x)f(y)$ for all $x, y \in S$.

As the semigroup operation for S is also written multiplicatively, it is natural to summarize the above theorem in the following way: if the multiplicative Cauchy difference of f is bounded, then f is either bounded or multiplicative. Since in a typical stability result one should only expect that f is a sum of a bounded and a multiplicative function, Baker's result is usually cited as the superstability of the (exponential or) multiplicative Cauchy functional equation.

In the same paper Baker noted that the main tool in his argument is the multiplicativity of the modulus (or norm) on the range of f. He also presented an example to show that his theorem does not work for functions with values in a normed algebra for which the norm is not multiplicative. Results into this direction have been obtained by Ger and Šemrl [5].

Let us now recall the usual definition of $|r|$ for $r \in \mathbb{R}$, which requires only the existence of a neutral element and an order relation: $|r| = r$ if $r \geq 0$, while $|r| = -r$ if $r < 0$. Now, if r is taken from an arbitrary ordered field, then the mapping $r \mapsto |r|$ defined above is also multiplicative and subadditive (but not real valued). This observation suggests that we could try to extend Baker's result to functions with values in an arbitrary ordered field.

The statement and the proof of the following theorem are analogous to those of [1, Theorem 1]. It is the only difference that the sharp quantitative estimation cannot be formulated.

Theorem 1. *If S is a semigroup, R is an ordered field, $f : S \rightarrow R$, $0 < \delta \in R$, and*

$$|f(xy) - f(x)f(y)| \leq \delta \qquad \text{for every} \quad x, y \in S, \tag{1}$$

then f is either bounded (with respect to the order relation in R) or multiplicative.

Proof. For every $x, y, z \in S$ we have

$$|f(xy)f(z) - f(xyz)| \leq \delta \qquad \text{and} \qquad |f(xyz) - f(x)f(yz)| \leq \delta$$

as particular cases of (1). These inequalities imply

$$|f(xy)f(z) - f(x)f(yz)| \leq 2\delta.$$

Hence

$$|f(xy) - f(x)f(y)||f(z)| = |f(xy)f(z) - f(x)f(y)f(z)|$$
$$\leq |f(xy)f(z) - f(x)f(yz)| + |f(x)f(yz) - f(x)f(y)f(z)|$$
$$\leq 2\delta + |f(x)|\delta.$$

So we have proved that

$$|f(xy) - f(x)f(y)||f(z)| \leq (2 + |f(x)|)\,\delta \qquad \text{for every} \quad x, y, z \in S. \tag{2}$$

Now let us assume that f is not multiplicative, i.e. there exist x_0, $y_0 \in S$ such that $f(x_0 y_0) \neq f(x_0)f(y_0)$. Then (2) implies the estimation

$$|f(z)| \leq \frac{(2 + |f(x_0)|)\,\delta}{|f(x_0 y_0) - f(x_0)f(y_0)|} \qquad \text{for every} \quad z \in S. \tag{3}$$

$\square$

As one can observe, the upper bound for $|f|$ in inequality (3) is not universal in the sense that it involves f as well, while the estimation in Baker's theorem does not depend on f. In the spirit of the stability concepts introduced by the author in his recent paper [3], we shall say that the multiplicative Cauchy functional equation is *weakly superstable* in (S, R), if any function $f : S \to R$ having a bounded multiplicative Cauchy difference

$$(x, y) \mapsto f(xy) - f(x)f(y) \qquad ((x, y) \in S \times S)$$

is either multiplicative or bounded (in R). In this sense, Theorem 1 states the *weak superstability* of the multiplicative Cauchy functional equation for functions that map a semigroup into an ordered field.

While it is clear that the formula $(1 + \sqrt{1 + 4\delta})/2$ cannot be interpreted in an arbitrary ordered field, one may expect that another universal estimation, which is comparable with δ, could be established for $|f|$. As the subsequent example shows, that is not the case.

Before formulating our example we must introduce some basic notions concerning ordered fields. One can easily check that the additive subsemigroup of an ordered field R generated by its multiplicative unit element 1 serves as a model for Peano's axioms. It is therefore reasonable to denote it by N. Then $\mathbb{Q}$ may denote the subfield of R generated by N. So we may consider N and $\mathbb{Q}$ as subsets of R.

Definition. *Let R denote an ordered field. We call $x \in R$ finite, if there exists $n \in$ N such that $|x| \leq n$. We say that $x \in R$ is infinitesimal,* if $|x| < 1/n$ for every $n \in \mathbb{N}$. The set of all finite elements of R will be denoted by $\mathcal{F}_R$, while $\mathcal{I}_R$ will denote the set of all infinitesimal elements of R. We say that R is Archimedean if $\mathcal{F}_R = R$. Otherwise, we say that R is non-Archimedean.

Remark. One can easily check that $\mathcal{F}_R$ is a ring and $\mathcal{I}_R$ is an ideal in $\mathcal{F}_R$. Clearly, R is Archimedean if, and only if, $\mathcal{I}_R = \{0\}$.

For construction and further study of non-Archimedean fields the reader is referred to the book of Lightstone and Robinson [6, Chapter 1] or other monographs on this subject.

Example. Let us consider a non-Archimedean ordered field R, take $0 < \delta \in \mathcal{F}_R$, $0 < h \in \mathcal{I}_R$, and define $f : \mathbb{Q} \to R$ by

$$f(r) = r + h\delta \qquad (r \in \mathbb{Q}).$$

Then $f(\mathbb{Q}) \subset \mathcal{F}_R$ (and thus any positive element of $R \setminus \mathcal{F}_R$ is an upper bound for $|f|$), but $f(n) > n$ for every $n \in \mathbb{N}$, so $|f|$ cannot have any finite upper bound. On the other hand, if $r, q \in \mathbb{Q}$, then

$$1 - r - q \in \mathbb{Q} \subset (\mathcal{F}_R \setminus \mathcal{I}_R) \cup \{0\} \qquad \text{and} \qquad h\delta \in \mathcal{I}_R \setminus \{0\},$$

which yields $1 - r - q - h\delta \in \mathcal{F}_R \setminus \{0\}$, and thus

$$\frac{f(rq) - f(r)f(q)}{\delta} = h(1 - r - q - h\delta) \in \mathcal{I}_R \setminus \{0\} \qquad \text{for all} \quad r, q \in \mathbb{Q}.$$

Hence, f is not multiplicative, but

$$|f(rq) - f(r)f(q)| < \frac{\delta}{n} \qquad \text{for all} \quad r, q \in \mathbb{Q}, n \in \mathbb{N}.$$

In fact, it is enough to consider the last inequality for $n = 1$. While δ could be arbitrarily small (infinitesimal, if we wish), the upper bounds for $|f|$ are always larger than any natural number.

2.   STABILITY WITH RESPECT TO THE MULTIPLICATIVE STRUCTURE

A different approach to the stability of the multiplicative (or, with an additive notation, exponential) Cauchy functional equation is suggested by Roman Ger [4], [5]. He claims that the superstability phenomenon is caused by mixing the additive and the multiplicative structures of the range. According to Ger's proposal, we should assume that the ratio $f(xy)/[f(x)f(y)]$ is close to 1. In order to avoid zeros in the denominator, this assumption should be formulated as

$$f(xy) \in f(x)f(y)U, \tag{4}$$

where U is a fixed neighbourhood of 1. However, supposing $f(y) = 0$ and (4), we obtain $f(xy) = 0$. Therefore, in the interesting case when f is defined on a group, (4) holds identically, and $f(y) = 0$ for some y, it follows that f is identically equal to zero. In the rest of this section we shall assume that f has no zeros (see also [4, Proposition 1]). On the other hand, one cannot completely get rid of the additive structure if it is involved in the metric. Since in non-Archimedean ordered fields the mapping $(x, y) \mapsto |x - y|$ is not a metric, it is natural to formulate our problem in a purely multiplicative setting. Let us assume that G is a groupoid, R is an ordered field, $1 < k \in R$, and $f : G \to R \setminus \{0\}$ satisfies

$$\frac{1}{k} \le \frac{f(xy)}{f(x)f(y)} \le k \qquad \text{for all} \quad x, y \in G. \tag{5}$$

Now the question arises whether there exist a multiplicative function $g : G \to R \setminus \{0\}$ and $1 < K \in R$ such that

$$\frac{1}{K} \le \frac{f(x)}{g(x)} \le K \qquad \text{for all} \quad x \in G. \tag{6}$$

It is also interesting to compare K with k.

In what follows we treat only the special case when the domain is a commutative group equipped with a base. As we wish to make use of the algebraic structure of the multiplicative group $(R \setminus \{0\}, \cdot)$, we begin with a stability type theorem fitting to group theory.

Theorem 2. *Let $(G, \cdot)$ be a free Abelian group, $(A, \cdot)$ be an Abelian group, and B be a subgroup of A. If $f : G \to A$ satisfies*

$$f(xy)\,(f(x)f(y))^{-1} \in B \qquad \text{for every} \quad x, y \in G, \tag{7}$$

then there exists a homomorphism $g : G \to A$ such that

$$f(x)\,(g(x))^{-1} \in B \qquad \text{for every} \quad x \in G. \tag{8}$$

Proof. From (7) we obtain

$$f(x) \in B^{-1}f(xy)\,(f(y))^{-1} = Bf(xy)\,(f(y))^{-1} \qquad \text{for all} \quad x, y \in G$$

and hence

$$f(x) \in \bigcap_{y \in G} Bf(xy)\,(f(y))^{-1} \qquad \text{for all} \quad x \in G.$$

This yields that the intersection is non-empty for every $x \in G$.

The hypothesis that $(G, \cdot)$ is free means that there exists a subset $H \subset G$ and for every $x \in G$ there is a unique function $k_x : H \to \mathbb{Z}$ such that $k_x(h) = 0$ for all but finitely many $h \in H$ and

$$x = \prod_{h \in H} h^{k_x(h)}$$

(the set H is called a *base*). From these assumptions one can easily obtain the identity

$$k_{xy}(h) = k_x(h) + k_y(h) \qquad \text{for every} \quad x, y \in G, \ h \in H. \tag{9}$$

For every $h \in H$ choose

$$g_0(h) \in \bigcap_{y \in G} B f(hy) \, (f(y))^{-1},$$

and define

$$g(x) = \prod_{h \in H} (g_0(h))^{k_x(h)} \qquad (x \in G).$$

Then $g \, |_H = g_0$ and, applying equation (9), we immediately obtain that $g : G \to A$ is a homomorphism.

We also state that

$$g(x) \in \bigcap_{y \in G} B f(xy) \, (f(y))^{-1} \tag{10}$$

for every $x \in G$. In order to prove this statement, let

$$T = \{ x \in G \mid x \text{ satisfies (10)} \}.$$

Since g is an extension of g_0, the definition of g_0 yields $H \subset T$. Let us now take $x, z \in T$ and $u \in G$. Applying (10) for x and z (in place of x) and specifying $y = z^{-1}u$ (in both cases), we obtain

$$g(xz^{-1}) = g(x) \, (g(z))^{-1}$$

$$\in B f(xz^{-1}u) \, (f(z^{-1}u))^{-1} \left(B f(zz^{-1}u) \, (f(z^{-1}u))^{-1} \right)^{-1}$$

$$= B f(xz^{-1}u) \, (f(z^{-1}u))^{-1} \, f(z^{-1}u) \, (f(u))^{-1} \, B^{-1}$$

$$= B f(xz^{-1}u) \, (f(u))^{-1}.$$

Thus $xz^{-1} \in T$, so T is a subgroup of G. Moreover, as it is already encountered, T contains H and H generates G, therefore $T = G$.

Finally, applying the inclusions (7) and (10) for every $x \in G$ and for arbitrary $y \in G$, we conclude that there exist b_1, $b_2 \in B$ such that

$$f(x)\,(g(x))^{-1} = f(x)\left(b_1 f(xy)\,(f(y))^{-1}\right)^{-1} = f(x)f(y)\,(f(xy))^{-1}\,b_1^{-1}$$
$$= \left(f(xy)\,(f(x)f(y))^{-1}\right)^{-1} b_1^{-1} = b_2^{-1}b_1^{-1} \in B.$$

$\square$

Now we can apply this theorem to the subgroup $\mathcal{F}_R \setminus \mathcal{I}_R$ of the multiplicative group $(R \setminus \{0\}, \cdot)$ in an ordered field $(R, +, \cdot, \leq)$.

Corollary. *Let G be a free Abelian group and R denote an ordered field. If $f : G \to R \setminus \{0\}$ satisfies*

$$\frac{f(xy)}{f(x)f(y)} \in \mathcal{F}_R \setminus \mathcal{I}_R \qquad \text{for every} \quad x, y \in G, \qquad (11)$$

then there exists $g : G \to R \setminus \{0\}$ such that

$$g(xy) = g(x)g(y) \qquad \text{for every} \quad x, y \in G \qquad (12)$$

and

$$\frac{f(x)}{g(x)} \in \mathcal{F}_R \setminus \mathcal{I}_R \qquad \text{for every} \quad x \in G. \qquad (13)$$

Let us note that the multiplicative function g in Corollary (and thus in Theorem 2) is not necessarily unique. For instance, in the case $G = \{2^m \mid m \in \mathbb{Z}\}$ (which can be considered as a multiplicative subgroup of $R \setminus \{0\}$) g could be replaced by $x \mapsto xg(x)$.

Now let us assume that R is a non-Archimedean ordered field, G is a free Abelian group, and $f : G \to R \setminus \{0\}$ satisfies (5) for some $k \in \mathbb{N}$. Then (11) also holds. Therefore, choosing $0 < K \in R \setminus \mathcal{F}_R$ and applying Corollary, we obtain that there exists $g : G \to R \setminus \{0\}$ such that (12) and (6) hold.

References

[1] J. A. Baker, *The stability of the cosine equation*, Proc. Amer. Math. Soc. **80** (1980), 411–416.

[2] J. A. Baker, J. Lawrence, and F. Zorzitto, *The stability of the equation $f(x + y) = f(x)f(y)$*, Proc. Amer. Math. Soc. **74** (1979), 242–246.

[3] Z. Boros, *Stability of the Cauchy equation in ordered fields*, Math. Pannonica **11** (2000), 191–197.

[4] R. Ger, *Superstability is not natural*, Rocznik Naukowo-Dydaktyczny WSP w Krakowie, Prace Mat. **159** (1993), 109–123.

[5] R. Ger and P. Šemrl, *The stability of the exponential equation*, Proc. Amer. Math. Soc. **124** (1996), 779–787.

[6] A. H. Lightstone and A. Robinson, *Nonarchimedean Fields and Asymptotic Expansions*, North-Holland, Amsterdam, 1975.

ON APPROXIMATELY MONOMIAL FUNCTIONS

Attila Gilányi

Institute of Mathematics and Informatics, University of Debrecen,
4010 Debrecen, Pf. 12, Hungary
gil@math.klte.hu

Abstract   In this paper we prove that a function f mapping from the set of the reals into a Banach space satisfying

$$\lim \frac{\|\Delta_y^n f(x) - n! f(y)\|}{|y|^\alpha} = 0$$

for a positive integer n, a real number $\alpha < n$ and as (x, y) tends to $(-\infty, \infty)$, $(\infty, -\infty)$, (∞, ∞) or $(-\infty, -\infty)$ can be approximated by a monomial function of degree n.

Keywords: functional equation, monomial function, stability

Mathematics Subject Classification (2000): 39B82, 39A11

In the present paper we investigate functions which satisfy the so called monomial functional equation only approximately and only on a certain domain. We consider, for a function f mapping from the set of reals $\mathbb{R}$ into a linear normed space X, the well-known difference-operator Δ which is defined by $\Delta_y^1 f(x) = f(x+y) - f(x)$ and, for a positive integer n, by $\Delta_y^{n+1} f(x) = \Delta_y^1 \Delta_y^n f(x)$, where x are y are real numbers. We call f a monomial function of degree n if $\Delta_y^n f(x) - n! f(y) = 0$ for all $x, y \in \mathbb{R}$. In the papers [6], [7], [8] and [11] monomial functions were studied with

---

[0]Research supported by the Higher Education, Research and Development Fund (FKFP), Grant Nr. 0215/2001 and by the Hungarian Scientific Research Fund (OTKA), Grant Nr. T-030082. The author is also grateful for the Bolyai János Kutatási Ösztöndíj of the Hungarian Academy of Sciences.

Z. Daróczy and Z. Páles (eds.),
Functional Equations - Results and Advances, 99–111.
© 2002 *Kluwer Academic Publishers.*

the help of the operator

$$\tilde{D}^{n,\alpha} f(\xi) = \lim_{\substack{(x,y)\to(\xi,0) \\ x\le\xi\le x+ny}} \frac{\|\Delta_y^n f(x) - n! f(y)\|}{y^\alpha} \qquad (\xi \in \mathbb{R})$$

derived from the Dinghas intervall-derivative. In the works above a global characterization and local stability results were given, if (x,y) tends to $(0,0)$. (For similar theorems concerning polynomial functions and the Dinghas interval derivative we refer to [2], [16] and [17]). Motivated by these results, we investigate the case when x and y tend to (plus or minus) infinity in the limit above and we show stability theorems for monomial functional equations in the sense used by S. Ulam [22] (cf. also [3] and [5]).

We prove that if X is a Banach space, n is a positive integer, $\varepsilon, \delta, A \ge 0$ and $\alpha < n$ are real numbers and a function $f : \mathbb{R} \to X$ satisfies the inequality

$$\|\Delta_y^n f(x) - n! f(y)\| \le \varepsilon |x|^\alpha + \delta y^\alpha \qquad (x \le -A, \, y \ge A)$$

then there exists a uniquely determined monomial function $g : \mathbb{R} \to X$ of degree n such that

$$\|f(y) - g(y)\| \le (c\varepsilon + d\delta)|y|^\alpha \qquad (|y| \ge A),$$

where c and d are real constants depending on n and α. (Here and in the sequel let $0^0 = 1$ and $0^\alpha = 0$ for $\alpha < 0$.) As a consequence of this theorem we obtain that, for $\alpha < n$, the property

$$\lim_{(x,y)\to(-\infty,\infty)} \frac{\|\Delta_y^n f(x) - n! f(y)\|}{y^\alpha} = 0$$

implies the existence of a (unique) monomial function $g : \mathbb{R} \to X$ of degree n for which

$$\lim_{|y|\to\infty} \frac{\|f(y) - g(y)\|}{|y|^\alpha} = 0.$$

We also show that analogous statements are valid if (x,y) tends to (∞,∞), $(-\infty,-\infty)$ or $(\infty,-\infty)$. These stability results have the special character that we make assumptions only on a certain subset of $\mathbb{R} \times \mathbb{R}$. Similar stability problems for monomial functions of degree 1 and 2 were considered by F. Skof in [18] and [19], and by F. Skof and S. Terracini in [20]. Finally, we remark that our theorem with $A = \alpha = 0$ and $n = 1$ yields the stability of the Cauchy equation investigated by

D. H. Hyers [12]; in the case when $A = \alpha = 0$, it implies the so-called Hyers-Ulam stability of monomial functional equations (cf. eg. [1], [21], [9]) and it gives a stability theorem for the Cauchy equation studied in [4] and [15] for $\alpha < 1$ with $A = 0$ and $n = 1$ (cf. [10] for monomial equations).

Lemma 1. *Let X be a linear normed space, n be a positive integer, α be an arbitrary, ε, δ and A nonnegative real numbers and let $f : \mathbb{R} \to X$ be a function satisfying*

$$\|\Delta_y^n f(x) - n! f(y)\| \le \varepsilon |x|^\alpha + \delta y^\alpha \qquad (x \le -A, \, y \ge A). \qquad (1)$$

For an arbitrary integer $l \ne 0$, we have

$$\|f(ly) - l^n f(y)\| \le (c_l \varepsilon + d_l \delta)|y|^\alpha \qquad (|y| \ge A), \qquad (2)$$

where c_l and d_l are real constants depending on l, n and α.

Proof. Let ε, δ, α, A, and n be given and let $f : \mathbb{R} \to X$ satisfy (1). In the first part of the proof we show that, for any positive integer l, there exist real constants $c_l = c(l, n, \alpha)$, and $d_l = d(l, n, \alpha)$ such that

$$\|f(lz) - l^n f(z)\| \le (c_l \varepsilon + d_l \delta) z^\alpha \qquad (z \ge A). \qquad (3)$$

Inequality (3) holds for $l = 1$ trivially. Let $l \ge 2$ and define the functions $F_i : [A, \infty) \to X$ by

$$F_i(z) = \Delta_z^n f(-iz) - n! f(z) \qquad (z \ge A) \qquad (4)$$

for $i = 1, \ldots, (l-1)n + 1$ and $G : [A, \infty) \to X$ by

$$G(z) = \Delta_{lz}^n f(-[(l-1)n + 1]z) - n! f(lz) \qquad (z \ge A).$$

Using the well-known formula

$$\Delta_y^n f(x) = \sum_{j=0}^{n} (-1)^{n-j} \binom{n}{j} f(x + jy) \qquad (x, y \in \mathbb{R}), \qquad (5)$$

we can write these functions in the form

$$F_i(z) = \sum_{j=1}^{ln+1} \alpha_{i-1}^{(j-1)} f((n-j)z) - n! f(z) \qquad (z \ge A) \qquad (6)$$

and

$$G(z) = \sum_{j=1}^{ln+1} \beta^{(j-1)} f((n-j)z) - n! f(lz) \qquad (z \ge A), \qquad (7)$$

where

$$\alpha_i^{(i+j)} = \begin{cases} (-1)^j \binom{n}{n-j}, & \text{if } 0 \leq j \leq n, \\ 0, & \text{otherwise} \end{cases} \qquad (8)$$

for $i = 0, \ldots, (l-1)n$, $j = -i, \ldots, ln - i$ and

$$\beta^{(j)} = \begin{cases} (-1)^{\frac{j}{l}} \binom{n}{n-\frac{j}{l}}, & \text{if } l \mid j, \\ 0, & \text{if } l \nmid j \end{cases} \qquad (9)$$

for $j = 0, \ldots, ln$. By Lemma 1 in [8] (cf. also Hilfssatz 2.2.2 in [6] or Lemma 2.2 in [7]), there exist positive integers $K_0, \ldots, K_{(l-1)n}$ such that

$$K_0 + \cdots + K_{(l-1)n} = l^n \qquad (10)$$

and

$$K_0 \alpha_0^{(j)} + \cdots + K_{(l-1)n} \alpha_{(l-1)n}^{(j)} - \beta^{(j)} = 0$$

for $j = 0, \ldots, ln$. Taking the linear combination of the functions $F_1, \ldots, F_{(l-1)n+1}$ with the coefficients $K_0, \ldots, K_{(l-1)n}$, and using (6), (7) and the last equation, we obtain

$$K_0 F_1(z) + \cdots + K_{(l-1)n} F_{(l-1)n+1}(z) + (K_0 + \cdots + K_{(l-1)n})n! f(z)$$
$$= G(z) + n! f(lz) \qquad (z \geq A),$$

therefore, by (10), we have

$$K_0 F_1(z) + \cdots + K_{(l-1)n} F_{(l-1)n+1}(z) + l^n n! f(z)$$
$$= G(z) + n! f(lz) \qquad (z \geq A)$$

that is,

$$f(lz) - l^n f(z)$$
$$= \tfrac{1}{n!} \left(K_0 F_1(z) + \cdots + K_{(l-1)n} F_{(l-1)n+1}(z) - G(z) \right) \qquad (z \geq A). \quad (11)$$

If we replace (x, y) by $(-z, z), (-2z, z), \ldots, (-[(l-1)n+1]z, z)$ for $z \geq A$ in (1) we get

$$\|F_i(z)\| \leq (i^\alpha \varepsilon + \delta) z^\alpha \qquad (z \geq A) \qquad (12)$$

for $i = 1, \ldots, (l-1)n + 1$. Replacing (x, y) by $(-[(l-1)n+1]z, lz)$ for $z \geq A$ in (1), we obtain

$$\|G(z)\| \leq l^\alpha \delta z^\alpha \qquad (z \geq A).$$

The last two inequalities with (10) and (11) imply

$$\|f(lz) - l^n f(z)\| \leq \left(\frac{\sum_{i=1}^{(l-1)n+1} K_{i-1} i^\alpha}{n!} \varepsilon + \frac{(l^\alpha + l^n)}{n!} \delta \right) z^\alpha \qquad (z \geq A).$$

$$(13)$$

Therefore, (3) holds for positive integers l.

Now we show the existence of real numbers c_{-1}, d_{-1} for which

$$\|f(-z) - (-1)^n f(z)\| \leq (c_{-1}\varepsilon + d_{-1}\delta)|z|^\alpha \qquad (|z| \geq A). \qquad (14)$$

If we replace x by $-z$ and y by $2z$ for $z \geq A$ in (1) we obtain that the function $F : [A, \infty) \to X$ defined by

$$F(z) = \Delta_{2z}^n f(-z) - n! f(2z) \qquad (z \geq A)$$

satisfies

$$\|F(z)\| \leq (\varepsilon + 2^\alpha \delta)z^\alpha \qquad (z \geq A). \qquad (15)$$

Using the simple formula

$$\sum_{j=0}^n (-1)^{n-j} \binom{n}{j} \left(\frac{2j-1}{2} \right)^n = n!$$

and (5), this function can be written in the form

$$F(z) = \sum_{j=0}^n (-1)^{n-j} \binom{n}{j} \left[f((2j-1)z) - \left(\frac{2j-1}{2} \right)^n f(2z) \right]$$

$$(16)$$

$$(z \geq A).$$

According to the first part of the proof, there exist real numbers c_1, $c_3, \ldots, c_{2n-1}$, $d_1, d_3, \ldots, d_{2n-1}$ and c_2, d_2 such that

$$\|f((2j-1)z) - (2j-1)^n f(z)\| \leq (c_{2j-1}\varepsilon + d_{2j-1}\delta)z^\alpha \qquad (z \geq A)$$

for $j = 1, \ldots, n$ and

$$\|f(2z) - 2^n f(z)\| \leq (c_2 \varepsilon + d_2 \delta)z^\alpha \qquad (z \geq A).$$

The combination of these inequalities gives

$$\left\| f((2j-1)z) - \left(\frac{2j-1}{2} \right)^n f(z) \right\|$$

$$\leq \left[\left(\left(\frac{2j-1}{2} \right)^n c_2 + c_{2j-1} \right) \varepsilon + \left(\left(\frac{2j-1}{2} \right)^n d_2 + d_{2j-1} \right) \delta \right] z^\alpha \qquad (z \geq A)$$

for $j = 1, \dots, n$. This property with (15) and (16) implies the existence of constants c_{-1}, d_{-1} satisfying

$$\|f(-z) - (-1)^n f(z)\| \leq (c_{-1}\varepsilon + d_{-1}\delta)z^\alpha \qquad (z \geq A),$$

thus, (14) holds, which implies our statement. $\qquad\square$

Lemma 2. *Let X be a linear normed space, n be a positive integer and A be a nonnegative real number. If a function $f : \mathbb{R} \to X$ satisfies*

$$\Delta_y^n f(x) - n! f(y) = 0 \qquad (x \leq -A, \, y \geq A) \tag{17}$$

then f is a monomial function of degree n, that is,

$$\Delta_y^n f(x) - n! f(y) = 0 \tag{18}$$

for $x, y \in \mathbb{R}$.

Proof. Let A and n be given and let $f : \mathbb{R} \to X$ satisfy (17). We prove (18) for $y \geq A$. Let $\bar{x}, \bar{y} \in \mathbb{R}$, $\bar{y} \geq A$ be fixed. Assumption (17) implies (18) for $\bar{x} \leq -A$, therefore, we may suppose that $\bar{x} > -A$. Let p denote the smallest (positive) integer for which $\bar{x} - p\bar{y} \leq -A$. We consider the real numbers

$$m_k = \Delta_{\bar{y}}^n f(x_k) - n! f(\bar{y}), \tag{19}$$

where $x_k = \bar{x} - (p - k)\bar{y}$, and we prove by induction that $m_k = 0$ for $k = 0, \dots, p$. Property (17) gives $m_0 = 0$. Let $k \in \{1, \dots, p\}$ and suppose that $m_s = 0$ for $s = 0, \dots, k-1$, that is,

$$\Delta_{\bar{y}}^n f(x_k - i\bar{y}) - n! f(\bar{y}) = 0 \qquad (i = 1, \dots, k). \tag{20}$$

Let $l \geq 2$ be an integer for which $x_k - (l-1)n\bar{y} \leq -A$. If we replace (x, y) by $(x_k - i\bar{y}, \bar{y})$ in (17) for $i = k, \dots, (l-1)n$, we get

$$\Delta_{\bar{y}}^n f(x_k - i\bar{y}) - n! f(\bar{y}) = 0 \qquad (i = k, \dots, (l-1)n).$$

These equations and (20) yield

$$\Delta_{\bar{y}}^n f(x_k - i\bar{y}) - n! f(\bar{y}) = 0 \qquad (i = 1, \dots, (l-1)n).$$

Writing $x = x_k - (l-1)n\bar{y}$ and $y = l\bar{y}$ in (17), we obtain

$$\Delta_{l\bar{y}}^n f(x_k - (l-1)n\bar{y}) - n! f(l\bar{y}) = 0.$$

Using the coefficients defined in (8) and (9), the equations above can be written in the form

$$\sum_{j=0}^{ln} \alpha_i^{(j)} f(x_k + (n - j)\bar{y}) - n! f(\bar{y}) = 0 \qquad (i = 1, \dots, (l-1)n)$$

and

$$\sum_{j=0}^{ln} \beta^{(j)} f(x_k + (n-j)\bar{y}) - n! f(l\bar{y}) = 0,$$

respectively. Furthermore, (19) has the form

$$m_k = \sum_{j=0}^{ln} \alpha_0^{(j)} f(x_k + (n-j)\bar{y}) - n! f(\bar{y}).$$

By Lemma 1 in [8], there exist positive integers $K_0, \ldots, K_{(l-1)n}$ such that

$$K_0 + \cdots + K_{(l-1)n} = l^n$$

and

$$K_0 \alpha_0^{(j)} + \cdots + K_{(l-1)n} \alpha_{(l-1)n}^{(j)} - \beta^{(j)} = 0$$

for $j = 0, \ldots, ln$. The combination of the last five equations yields $K_0 m_k = n! f(l\bar{y}) - l^n n! f(\bar{y})$. Since $\bar{y} \geq A$, Lemma 1 with $\varepsilon = \delta = 0$ gives $f(l\bar{y}) - l^n f(\bar{y}) = 0$, that is, $m_k = 0$. Thus, we have

$$\Delta_y^n f(x) - n! f(y) = 0 \qquad (x, y \in \mathbb{R}, \ y \geq A), \qquad (21)$$

that is, (18) is valid for $x, y \in \mathbb{R}, \ y \geq A$.

According to Zs. Páles' extension theorem for linear functional equations (cf. Tétel 3.12 and Megjegyzés 3.13 in [13] or Theorem 4 and Remark 4 in [14]), if Y is a linear normed space, S is a subring with identity of the set of the real numbers, H is a subset of the reals closed under addition and under multiplication by positive elements of S, furthermore, p is a positive integer, $a_1, \ldots, a_p \in [0, \infty) \cap S$, $b_1, \ldots, b_p \in (0, \infty) \cap S$ and $c_0, \ldots, c_p$ are arbitrary real numbers and the function $h : H \to Y$ satisfies the functional equation

$$h(x) = c_0 + \sum_{j=1}^{p} c_j h(a_j x + b_j y) \qquad (x, y \in H)$$

then there exists a uniquely determined $h^* : H^* \to Y$ such that

$$h^*(x) = c_0 + \sum_{j=1}^{p} c_j h^*(a_j x + b_j y) \qquad (x, y \in H^*)$$

and $h^*(x) = h(x)$ for $x \in H$, where

$$H^* = \bigcap_{j=1}^{p} \{x \in \mathbb{R} \mid \exists \nu \in \mathbb{N} : a_j^\nu x \in H - H\}.$$

Writing the equation in (21) in the form

$$f(x) = \sum_{j=1}^{n} (-1)^{j+1} \binom{n}{j} f(x + jy) - (-1)^{n+1} n! f(y)$$

(cf. (5)) and applying Páles' result in the case when $Y = X$, $S = \mathbb{Z}$, $H = [A, \infty)$, $p = n + 1$ and $a_j = 1$, $b_j = j$, $c_j = (-1)^{j+1} \binom{n}{j}$ for $j = 1, \ldots, n$, $a_{n+1} = 0$, $b_{n+1} = 1$, $c_{n+1} = -(-1)^{n+1} n!$, we easily obtain our statement. $\square$

Theorem 1. *Let X be a Banach space, n be a positive integer, $\varepsilon \geq 0$, $\delta \geq 0$, $A \geq 0$ and $\alpha < n$ be real numbers and suppose that $f : \mathbb{R} \to X$ is a function fulfilling*

$$\|\Delta_y^n f(x) - n! f(y)\| \leq \varepsilon |x|^\alpha + \delta y^\alpha \qquad (x \leq -A,\ y \geq A). \qquad (22)$$

Then there exists a uniquely determined monomial function $g : \mathbb{R} \to X$ of degree n satisfying

$$\|f(y) - g(y)\| \leq (c\varepsilon + d\delta)|y|^\alpha \qquad (|y| \geq A), \qquad (23)$$

where c and d are real constants depending on n and α.

Proof. We prove the theorem using the so called Hyers-method (cf. [12]). Let n, ε, δ, A, α and f satisfy the assumptions above. Let, furthermore, $l \geq 2$ be a fixed integer. By Lemma 1, there exist real numbers c_l and d_l such that

$$\left\| \frac{1}{l^n} f(ly) - f(y) \right\| \leq \frac{1}{l^n} (c_l \varepsilon + d_l \delta)|y|^\alpha \qquad (|y| \geq A).$$

Using this property and the triangle inequality we obtain

$$\left\| f(y) - \frac{1}{l^{mn}} f(l^m y) \right\| \leq \left\| f(y) - \frac{1}{l^n} f(ly) \right\| + \left\| \frac{1}{l^n} f(ly) - \frac{1}{l^{2n}} f(l^2 y) \right\| +$$

$$\cdots + \left\| \frac{1}{l^{(m-1)n}} f(l^{m-1} y) - \frac{1}{l^{mn}} f(l^m y) \right\|$$

$$\leq \frac{1}{l^n} \sum_{j=0}^{m-1} l^{j(\alpha-n)} (c_l \varepsilon + d_l \delta)|y|^\alpha \qquad (|y| \geq A) \qquad (24)$$

for every positive integer m. Since $\alpha < n$ we have

$$\sum_{j=0}^{\infty} l^{j(\alpha-n)} = \frac{1}{1 - l^{\alpha-n}},$$

therefore, for the functions $g_m : (-\infty, -A] \cup [A, \infty) \to X$ defined by

$$g_m(y) = \frac{f(l^m y)}{l^{mn}} \qquad (|y| \geq A)$$

we get

$$\|g_k(y) - g_m(y)\| \leq l^{m(\alpha-n)} \frac{1}{l^n - l^\alpha}(c_l \varepsilon + d_l \delta)|y|^\alpha \qquad (|y| \geq A)$$

for positive integers $k > m$. Thus, $(g_k(y))$ is a Cauchy sequence for fixed real numbers $|y| \geq A$. Since X is complete, we can define the function $g : (-\infty, -A] \cup [A, \infty) \to X$ by

$$g(y) = \lim_{m \to \infty} g_m(y) \qquad (|y| \geq A).$$

Replacing (x, y) by $(l^m x, l^m y)$ for $x \leq -A, y \geq A$ in (22) we obtain

$$\|\Delta_{l^m y}^n f(l^m x) - n! f(l^m y)\| \leq l^{\alpha m}(\varepsilon |x|^\alpha + \delta y^\alpha) \qquad (x \leq -A, y \geq A)$$

for any positive integer m. Dividing this inequality by l^{mn} and letting $m \to \infty$ we get

$$\Delta_y^n g(x) - n! g(y) = 0 \qquad (x \leq -A, y \geq A).$$

Thus, Lemma 2 implies that g is a monomial function of degree n. Property (24) yields

$$\left\| \frac{1}{l^{mn}} f(l^m y) - f(y) \right\| \leq \frac{1}{l^n - l^\alpha}(c_l \varepsilon + d_l \delta)|y|^\alpha \qquad (|y| \geq A),$$

therefore,

$$\|g(y) - f(y)\| \leq \frac{1}{l^n - l^\alpha}(c_l \varepsilon + d_l \delta)|y|^\alpha \qquad (|y| \geq A), \qquad (25)$$

that is, (23) holds, too.

To prove uniqueness we suppose that $g, \bar{g} : \mathbb{R} \to X$ are different monomial functions of degree n such that

$$\|f(y) - g(y)\| \leq C|y|^\alpha \qquad (|y| \geq A)$$

and

$$\|f(y) - \bar{g}(y)\| \leq \bar{C}|y|^\alpha \qquad (|y| \geq A),$$

where C and $\bar{C}$ are real constants. Using the triangle inequality we get

$$\|g(y) - \bar{g}(y)\| \leq (C + \bar{C})|y|^\alpha \qquad (|y| \geq A). \tag{26}$$

The functions g and $\bar{g}$ are different monomial functions, so, there exists a $y \in \mathbb{R}$ such that $|y| \geq A$ and $g(y) \neq \bar{g}(y)$. There exists a positive integer l such that

$$l^{n-\alpha} > \frac{(C + \bar{C})|y|^\alpha}{\|g(y) - \bar{g}(y)\|}.$$

By Lemma 1 we have $g(ly) = l^n g(y)$ and $\bar{g}(ly) = l^n \bar{g}(y)$. Thus

$$\|g(ly) - \bar{g}(ly)\| > (C + \bar{C})|ly|^\alpha,$$

which contradicts (26). $\qquad\qquad\qquad\qquad\qquad\qquad\qquad\qquad\qquad$ □

Corollary 1. *Let X be a Banach space, n be a positive integer and $\alpha < n$ be a real number. If a function $f : \mathbb{R} \to X$ satisfies*

$$\lim_{(x,y)\to(-\infty,\infty)} \frac{\|\Delta_y^n f(x) - n!f(y)\|}{y^\alpha} = 0$$

then there exists a uniquely determined monomial function $g : \mathbb{R} \to X$ of degree n for which

$$\lim_{|y|\to\infty} \frac{\|f(y) - g(y)\|}{|y|^\alpha} = 0.$$

Proof. The statement is a simple consequence of Theorem 1. $\qquad\qquad$ □

Theorem 2. *Let X be a Banach space, n be a positive integer, $\varepsilon \geq 0$, $\delta \geq 0$, $A \geq 0$ and $\alpha < n$ be real numbers and $f : [A, \infty) \to X$ be a function satisfying*

$$\|\Delta_y^n f(x) - n!f(y)\| \leq \varepsilon x^\alpha + \delta y^\alpha \qquad (x, y \geq A). \tag{27}$$

There exists a uniquely determined monomial function $g : \mathbb{R} \to X$ of degree n such that

$$\|f(y) - g(y)\| \leq (c\varepsilon + d\delta)y^\alpha \qquad (y \geq A), \tag{28}$$

where c and d are real constants depending on n and α.

Proof. Let ε, δ, α, A, and n be given and let $f : \mathbb{R} \to X$ satisfy (27).

I. First we prove that, for a fixed positive integer l, there exist real numbers $c_l = c(l, n, \alpha)$ and $d_l = d(l, n, \alpha)$ such that

$$\|f(ly) - l^n f(y)\| \leq (c_l \varepsilon + d_l \delta) y \qquad (y \geq A). \qquad (29)$$

The statement is trivial for $l = 1$. If $l \geq 2$, we define the functions $F_i : [A, \infty) \to X$ by

$$F_i(z) = \Delta_z^n f(iz) - n! f(z) \qquad (z \geq A)$$

for $i = 1, \ldots, (l-1)n + 1$ and $G : [A, \infty) \to X$ by

$$G(z) = \Delta_{lz}^n f(z) - n! f(lz) \qquad (z \geq A).$$

If we replace (x, y) by $(z, z), (2z, z), \ldots, [(l-1)n + 1]z, z)$ and by (z, lz) for $z \geq A$ in (27), we obtain

$$\|F_i(z)\| \leq (i^\alpha \varepsilon + \delta) z^\alpha \qquad (z \geq A)$$

for $i = 1, \ldots, (l-1)n + 1$ and

$$\|G(z)\| \leq l^\alpha \delta z^\alpha \qquad (z \geq A).$$

By the method used in the first part of the proof of Lemma 1, we get that there exist positive integers such that

$$\|f(lz) - l^n f(z)\| \leq \left(\frac{\sum_{i=1}^{(l-1)n+1} K_{i-1} i^\alpha}{n!} \varepsilon + \frac{(l^\alpha + l^n)}{n!} \delta \right) z^\alpha \qquad (z \geq A),$$

thus, (29) holds for positive integers l.

II. Using the property above, we can show, similarly to the proof of Theorem 1, that the function $g : [A, \infty) \to \mathbb{R}$ defined by

$$g(y) := \lim_{m \to \infty} \frac{f(l^m y)}{l^{mn}} \qquad (y \geq A)$$

satisfies (28). Assumption (27) implies that

$$\left\| \Delta_{l^m y}^n f(l^m x) - n! f(l^m y) \right\| \leq l^{\alpha m} (\varepsilon x^\alpha + \delta y^\alpha) \qquad (x, y \geq A),$$

for positive integers l and m, therefore,

$$\Delta_y^n g(x) - n! g(y) = 0 \qquad (x, y > A).$$

By Zs. Páles' results in [13] and [14] (cf. the proof of Lemma 2 in this paper), we obtain that g can be extended to a monomial function of degree n defined on $\mathbb{R}$. Finally, it is easy to see, that g (and, therefore, also its extension) is uniquely determined.   □

Corollary 2. *Let X be a Banach space, n be a positive integer and $\alpha < n$ be a real number. If a function $f : \mathbb{R} \to X$ satisfies*

$$\lim_{(x,y)\to(\infty,\infty)} \frac{\|\Delta_y^n f(x) - n! f(y)\|}{y^\alpha} = 0$$

then there exists a uniquely determined monomial function $g : \mathbb{R} \to X$ of degree n such that

$$\lim_{y\to\infty} \frac{\|f(y) - g(y)\|}{y^\alpha} = 0.$$

Proof. The statement follows from Theorem 2.   □

Remark. It is easy to see that similar results to the ones above are valid if (x,y) tends to $(-\infty, -\infty)$ or $(\infty, -\infty)$.

References

[1] M. Albert and J. A. Baker, *Functions with bounded n^{th} differences*, Ann. Polon. Math. **43** (1983), 93–103.

[2] A. Dinghas, *Zur Theorie der gewöhnlichen Differentialgleichungen*, Ann. Acad. Sci. Fennicae, Ser. A I **375** (1966).

[3] G. L. Forti, *Hyers–Ulam stability of functional equations in several variables*, Aequationes Math. **50** (1995), 142–190.

[4] Z. Gajda, *On stability of additive mappings*, Internat. J. Math. Math. Sci. **14** (1991), 431–434.

[5] R. Ger, *A survey of recent results on stability of functional equations*, Proc. of the 4th International Conference on Functional Equations and Inequalities, Pedagogical University, Cracow, 1994, 5–36.

[6] A. Gilányi, *Charakterisierung von monomialen Funktionen und Lösung von Funktionalgleichungen mit Computern*, Diss., Univ. Karlsruhe, 1995.

[7] A. Gilányi, *A characterization of monomial functions*, Aequationes Math. **54** (1997), 289–307.

[8] A. Gilányi, *On locally monomial functions*, Publ. Math. Debrecen **51** (1997), 343–361.

[9] A. Gilányi, *Hyers-Ulam stability of monomial functional equations on a general domain*, Proc. Natl. Acad. Sci. USA **96** (1999), 10588–10590.

[10] A. Gilányi, *On the stability of monomial functional equations*, Publ. Math. Debrecen **56** (2000), 201–212.

[11] A. Gilányi, *Local stability and global superstability of monomial functional equations*, in Advances in Equations and Inequalities, Hadronic Press, Palm Harbor, USA (1999), 73-95.

[12] D. H. Hyers, *On the stability of the linear functional equation*, Proc. Natl. Acad. Sci. USA **27** (1941), 222–224.

[13] Zs. Páles, *Újabb módszerek a függvényegyenletek és függvényegyenlőtlenségek elméletében*, MTA Doktori értekezés, Debrecen, 1999.

[14] Zs. Páles, *Extension theorems for functional equations with bisymmetric operations*, Aequationes Math., to appear.

[15] Th. M. Rassias, *On the stability of the linear mapping in Banach spaces*, Proc. Amer. Math. Soc. **72** (1978), 297–300.

[16] A. Simon and P. Volkmann, *Eine Charakterisierung von polynomialen Funktionen mittels der Dinghasschen Intervall-Derivierten*, Results in Math. **26** (1994), 382–384.

[17] A. Simon and P. Volkmann, *Perturbations de fonctions additives*, Ann. Math. Silesianae **11** (1997), 21–27.

[18] F. Skof, *Sull'approssimazione delle applicazioni localmente δ-additive*, Atti della Accademia delle Scienze di Torino **117** (1983), 377–389.

[19] F. Skof, *Proprietà locali e approssimazione di operatori*, Rend. Sem. Mat. Fis. Milano **53** (1983), 113–129.

[20] F. Skof and S. Terracini, *Sulla stabilità dell'equazione funzionale quadratica su un dominio ristretto*, Atti della Accademia delle Scienze di Torino **121** (1987), 153–167.

[21] L. Székelyhidi, *The stability of linear functional equations*, C. R. Math. Rep. Acad. Sci. Canada **3** (1981), 63–67.

[22] S. Ulam, *A Collection of Mathematical Problems*, Intersence Tracts in Pure and Applied Mathematics, vol. 8, Intersence Publishers, New York–London, 1960.

LES OPÉRATEURS DE HYERS

Zenon Moszner

Institut de Mathématique, Académie Pédagogique,
Podchorążych 2 30-084 Kraków, Pologne
e-mail: zmoszner@wsp.krakow.pl

Abstract   In virtue of the well – known theorem of Hyers the linear space **C** of all those mappings of a Banach space into another whose Cauchy difference is bounded is the direct sum of the spaces: **A** of additive mappings and **B** of boundes ones. Properties of the projections of **C** into **A**, resp. **B** (called Hyers operators) are studied in the case where the functions are from ℝ into ℝ.

Keywords: operators of Hyers (opérateurs de Hyers), continuity (continuité)

Mathematics Subject Classification (2000): 26A15, 39B22

En résolvant le problème de S. M. Ulam [7], D. H. Hyers a démontré le théorème suivant [3]

Soient $(X, \|.\|)$ et $(Y, \|.\|)$ des espaces de Banach. Si pour un $\varepsilon > 0$ la fonction $f : X \to Y$ remplit la condition

$$\|f(x+y) - f(x) - f(y)\| \leq \varepsilon \qquad \text{pour} \quad x, y \in X,$$

alors il existe exactement une fonction additive $a : X \to Y$ telle que

$$\|f(x) - a(x)\| \leq \varepsilon \qquad \text{pour} \quad x \in X.$$

Nous disons d'après ce théorème que l'équation fonctionnelle de Cauchy de la fonction additive

$$f(x+y) = f(x) + f(y)$$

est stable. La théorie de la stabilité des équations fonctionnelles est déjà très riche (voir la revue [1]).

Z. Daróczy and Z. Páles (eds.),
Functional Equations - Results and Advances, 113–122.

Il résulte de ce théorème que chaque fonction $f : X \to Y$ dont la différence de Cauchy $f(x+y) - f(x) - f(y)$ est bornée est la somme d'une fonction $a : X \to Y$ additive et d'une fonction $b : X \to Y$ bornée ($b := f - a$). De plus l'espace $\mathbf{C}$ des fonctions $f : X \to Y$ dont la différence de Cauchy est bornée forme un espace vectoriel, analogiquement les espaces $\mathbf{A}$ et $\mathbf{B}$ des fonctions de X à Y additives et des fonctions bornées forment des espaces vectoriels et puisque la somme de la fonction additive et de la fonction bornée appartient à $\mathbf{C}$ et $\mathbf{A} \cap \mathbf{B} = \{0\}$, nous constatons que $\mathbf{C}$ est la somme directe de ces espaces $\mathbf{A}$ et $\mathbf{B}$ ($\mathbf{C} = \mathbf{A} \oplus \mathbf{B}$). Les projections de $f \in \mathbf{C}$ sur l'espace $\mathbf{A}$ (désignée dans la suite par $a(f)$) et sur l'espace $\mathbf{B}$ (désignée par $b(f)$), sont nommées les opérateurs (linéaires, comme les projections) de Hyers. Nous allons considérer quelques propriétés de ces opérateurs et de ces opérateurs sur le domaine restreint.

Les plus intéressantes sont les continuités des opérateurs $a(f)(x)$ et $b(f)(x)$ ($f \in \mathbf{C}, x \in X$) comme les fonctions de x ou de f ou de la paire (f, x). Nous considérons dans la suite comme X et Y l'espace $\mathbb{R}$ avec la norme simple ($\|.\| = |.|$).

Les projections de la nature analogue sont considérées aussi dans [2] dans le cas plus générale et pour l'équation des fonctions exponentielles, mais sans des recherches des propriétés de ces projections.

Partie I

1.   LA CONTINUITÉ PAR RAPPORT À X

Nous allons chercher des conditions suffisantes et nécessaires (ou seulement suffisantes ou seulement nécessaires) qui rempliées par la fonction f donnent la continuité de la fonction $a(f)(x)$. Puisque $a(f) \in \mathbf{C}$ pour chaque $f \in \mathbf{C}$ et $a(a(f)) = a(f)$, donc chaque propriété P de cette sorte doit impliquer la continuité de la fonction additive jouissant de la propriété P. Nous allons considérer dans la suite seulement les propriétés de cette façon.

Ils existent des propriétés de cette sorte qui ne sont ni suffisantes ni nécessaires pour la continuité de $a(f)$. Par exemple telle est la propriété suivante: $f(x) = \alpha x$ pour un $\alpha \in \mathbb{R}$ sur une base H de Hamel. En effet la fonction additive qui a cette propriété est continue. De plus il existe la fonction $f \in \mathbf{C}$ et ayant cette propriété pour laquelle $a(f)$ n'est pas continue. Pour la démonstration posons $f(x) = x$ sur $H, b(h) = h$ pour un h de H et $b(x) = 0$ en outre sur $\mathbb{R}$, prolongéons la fonction $f(x) - b(x)$ sur H à la fonction $a(x)$ additive sur $\mathbb{R}$ et considérons la fonction $f = a + b$ pour $x \in \mathbb{R}$. Cette fonction f appartient à $\mathbf{C}$, a la propriété en considération et $a(f)(x) = a(x)$ n'est pas continue, puisque le noyau de $a(x)$ n'est ni $\mathbb{R}$ ni $\{0\}$. Il existe aussi évidemment la fonction f pour laquelle $a(f)(x)$ a la propriété en considération, mais f ne remplit pas cette condition (puisque la fonction b peut être choisie arbitrairement).

Théorème 1.   *a) $a(f)(x)$ est continue par rapport à x si et seulement si f est bornée d'une côté sur un ensemble de la mesure intérieure de Lebesgue positive ou s'il existe la direction asymptotique de f à*

$$+\infty \text{ ou à } -\infty \ \Big(\lim_{x\to\pm\infty} \frac{f(x)}{x} \text{ est finie} \Big).$$

b) Si f est continue au moins en un point ou si elle est mesurable au sens de Lebesgue, alors $a(f)(x)$ est continue. L'implication inverse n'est pas vraie.

c) Il existe la propriété pour la fonction additive $a(f)$ qui implique que f a cette propriété, mais l'implication inverse n'est pas vraie.

Démonstration.   Ad a) S'il existe la direction asymptotique de f à $+\infty$ ou à $-\infty$, la fonction f est bornée d'une côté sur un ensemble de la mesure intérieure positive et dans ce cas la fonction $a(f)(x) = f(x) - b(x)$ est aussi bornée sur cet ensemble et comme additive, elle est continue. Inversement si $a(f)$ est continue, elle a la direction asymptotique à $+\infty$ et aussi à $-\infty$, donc la fonction $f = a(f) + b(f)$ a aussi cette direction ($b(f)$ est bornée) et de là f est bornée sur un ensemble de mesure intérieure positive.

Ad b) Si f est continue au moins en un point, elle est bornée dans un entourage de ce point, donc $a(f)$ est continue d'après a). Si f est mesurable, d'après le théorème de Lusin elle est continue sur un compact F de mesure positive, donc elle est bornée sur F et $a(f)$ est continue d'après a).

Inversement si $a(f)$ est continue, f ne doit être ni mesurable ni continue en un point puisque $b(f)$ ne doit être ni mesurable ni continue en un point dans ce cas.

Ad c) Soit la propriété consiste à l'appartenance de la fonction à l'ensemble qui se compose des fonctions de la forme $b(x) + \alpha x$, avec toutes les fonctions $b(x)$ bornées et chaque $\alpha \in \mathbb{R}$ et de la fonction $1 + a(x)$, avec la fonction $a(x)$ additive et discontinue. Cette propriété satisfait à c).

$\square$

S'il s'agit de la continuité de l'opérateur $b(f)(x)$ par rapport à x remarquons seulement que la continuité de f implique la continuité de $a(f)(x)$ (voir b) dans le théorème 1), donc aussi la continuité de $b(f) = f - a(f)$. La continuité de $b(f)$ n'implique pas naturellement la continuité de f. De plus $f(x)$ est continue si (évidemment) et seulement si $a(f)(x)$ et

$b(f)(x)$ sont continues. Si $f(x)$ est continue, alors $a(f)(x)$ est la même, donc $b(f)(x)$ est aussi continue.

2.   LA CONTINUITÉ PAR RAPPORT À F

Remarquons au commencement que dans l'espace **C** nous n'avons pas de la norme naturelle (par exemple la norm sup) par rapport à quelle nous pourrions considérer la continuité, puisque les fonctions de **C** ne doivent pas être bornées. Nous allons donc considérer la continuité par rapport à f en prenant dans **C** la convergence simple de la suite f_n vers la fonction f ($f_n(x) \to f(x)$ pour chaque $x \in \mathbb{R}$, en abrégé $f_n \to f$ ou $\lim_{n \to +\infty} f_n = f$).

Faisons les deux remarques au sujet de cette convergence.

1. La famille **C** n'est pas fermée par rapport à cette convergence simple. Il suffit de considérer la suite des fonctions définies comme il suit: $f_n(x) = x^3$ pour $|x| \leq n$, $f_n(x) = n$ pour $x > n$, $f_n(x) = -n$ pour $x < -n$. Cet exemple montre aussi que l'espace **B** n'est pas fermé par rapport à la convergence simple. Au contraire l'espace **A** est fermé en ce sens puisque la limite d'une suite des fonctions additives est aussi additive.

2. La convergence simple de $f_n \in$ **C** vers $f \in$ **C** n'implique pas de la convergence simple de $a(f_n)$ vers $a(f)$. En effet pour la suite f_n définie comme il suit: $f_n(x) = x$ pour $|x| \leq n$, $f_n(x) = n$ pour $x > n$ et $f_n(x) = -n$ pour $x < -n$, nous avons $f_n(x) \to f(x) = x$ et $a(f_n)(x) = 0 \nrightarrow a(f)(x) = x$. Analogiquement $b(f_n)(x) = f_n(x) \nrightarrow 0 = b(f)(x)$. Cela montre que les opérateurs $a(f)$ et $b(f)$ ne sont pas continues relativement à f par rapport à la convergence simple. Puisque notre suite f_n converge presque uniformément (c. à d. converge uniformément sur tout compact de $\mathbb{R}$), donc aussi converge quasi-uniformément ([6] p. 143) et converge de façon continue au sens étroit ([4] p. 97), par conséquent elle tend de même de façon continue ([4] p. 95 et 97), toutes ces convergences aussi ne suffisent pas. Il faut donc ajouter quelques conditions complémentaires à la convergence simple pour avoir cette continuité.

La convergence uniforme suffit ici mais elle est trop forte, comme cela montre le théorème suivant.

Théorème 2. *Si* **C** $\ni f_n \to f$ *et*

$$\exists M \, \exists k \, \forall x \in \mathbb{R} \, \forall n \geq k \quad |f_n(x) - f_k(x)| \leq M,$$

alors $f \in$ **C** *et* $a(f) = a(f_n)$ *pour n suffisement grand.*

Démonstration. Nous avons d'après $f_n = b(f_n) + a(f_n)$

$$|b(f_n)(x) + a(f_n)(x) - b(f_k)(x) - a(f_k)(x)| \leq M \qquad \text{pour} \quad n \geq k.$$

Fixons n, puisque $b(f_m)(x)$ est une fonction bornée il existe M_{nk} tel que

$$\forall x \in \mathbb{R} \quad |a(f_n)(x) - a(f_k)(x)| \leq M + M_{nk},$$

d'où pour chaque m entier positif

$$\forall m \quad m|a(f_n)(x) - a(f_k)(x)| \leq M + M_{nk},$$

alors $a(f_n)(x) = a(f_k)(x)$ pour $n \geq k$ et pour chaque $x \in \mathbb{R}$.

Nous avons pour $n \to +\infty$ d'après la supposition faite que

$$\forall x \in \mathbb{R} \quad |f(x) - a(f_k)(x) - b(f_k)(x)| \leq M,$$

d'où $f - a(f_k)$ est bornée, puisque $b(f_k)$ est bornée. Il en résulte que $f = (f - a(f_k)) + a(f_k) \in \mathbf{C}$ et de plus $a(f_n) = a(f_k) = a(f)$ pour $n \geq k$. $\qquad\qquad\square$

La condition au sujet de f_n dans le théorème 2 est naturellement plus faible que la convergence uniforme de f_n sur $\mathbb{R}$. Elle n'est pas nécessaire pour la convergence de $a(f_n)$ vers $a(f)$ si f_n tend vers f. Exemple: $f_n(x) = \alpha_n x$ pour $\alpha_n \in \mathbb{R}, \alpha_n \to \alpha, \alpha_n \neq \alpha$.

Il suit du théorème de Hyers, cité au début de cette note, que $a(f) = \lim\limits_{\mathbb{N} \ni m \to +\infty} \dfrac{f(mx)}{m}$. On peut noter la convergence $a(f_n)$ vers $a(f)$ pour $f_n \to f$ comme il suit

$$\lim_{n \to +\infty} \left(\lim_{m \to +\infty} \frac{f_n(mx)}{m} \right) = \lim_{m \to +\infty} \left(\lim_{n \to +\infty} \frac{f_n(mx)}{m} \right), \qquad (1)$$

donc le problème de la continuité de $a(f)$ c'est le problème de la commutativité des limites ci-dessus par rapport à m et n. Pour donner une condition nécessaire et suffisante pour cette continuité nous donnerons au commencement les deux lemmes au sujet de la commutativité des limites d'une suite double $s(m, n)$.

Lemme 1. [5].
Nous avons

$$\lim_{n \to +\infty} \left(\lim_{m \to +\infty} s(m, n) \right) = \lim_{m \to +\infty} \left(\lim_{n \to +\infty} s(m, n) \right) \qquad (2)$$

si et seulement s'ils existent $\lim\limits_{m \to +\infty} \left(\lim\limits_{n \to +\infty} s(m, n) \right)$ *et* $\lim\limits_{m \to +\infty} s(m, n)$
pour chaque $n \in \mathbb{N}$ *et de plus*

$$\forall \varepsilon > 0 \, \forall \eta > 0 \, \exists \delta > 0 \, \exists m > \varepsilon \, \forall n > \delta \quad \left[\left| s(m, n) - \lim_{m \to +\infty} s(m, n) \right| \leq \eta \right].$$

$$(3)$$

Démonstration de la suffisance. Désignons par

$$g := \lim_{m \to +\infty} (\lim_{n \to +\infty} s(m,n)),$$

$\phi(n) := \lim_{m \to +\infty} s(m,n)$ et $\psi(m) := \lim_{n \to +\infty} s(m,n)$ et soit $\eta > 0$ arbitraire. Il existe $\varepsilon > 0$ tel que

$$\forall m > \varepsilon \quad [|\psi(m) - g| < \eta/3].$$

Il existe d'après la condition (3) un $\overline{m}$ et $\delta_1 > 0$ tels que

$$\overline{m} > \varepsilon \quad \text{et} \quad \forall n > \delta_1 \quad [|s(\overline{m},n) - \phi(n)| < \eta/3].$$

Il existe aussi $\delta_2 > 0$ pour lequel

$$\forall n > \delta_2 \quad [|s(\overline{m},n) - \psi(\overline{m})| < \eta/3].$$

Il en résulte que pour chaque $n > \delta := \max(\delta_1, \delta_2)$ nous avons

$$|\phi(n) - g| \leq |\phi(n) - s(\overline{m},n)| + |s(\overline{m},n) - \psi(\overline{m})| + |\psi(\overline{m}) - g| < \eta,$$

donc la démonstration de la suffisance est finie.

Démonstration de la nécessité. Ils existent dans ce cas toutes les limites dans (2). Soient $\varepsilon > 0$ et $\eta > 0$ arbitraires et soient $g, \phi(n)$ et $\psi(m)$ les mêmes que plus haut. Puisque $\lim_{m \to +\infty} \psi(m) = g$ il existe $\varepsilon_1 > 0$ tel que

$$\forall m > \varepsilon_1 \quad [|\psi(m) - g| < \eta/3],$$

et puisque $\lim_{n \to +\infty} \phi(n) = g$ il existe $\delta_1 > 0$ pour lequel

$$\forall n > \delta_1 \quad [|\phi(n) - g| < \eta/3].$$

Soit $\overline{m} > \max(\varepsilon_1, \varepsilon)$. Puisque $\lim_{n \to +\infty} s(\overline{m},n) = \psi(\overline{m})$ il existe $\delta_2 > 0$ tel que

$$\forall n > \delta_2 \quad [s(\overline{m},n) - \psi(\overline{m})| < \eta/3].$$

Il en résulte que pour chaque $n > \max(\delta_1, \delta_2)$ nous avons

$$|s(\overline{m},n) - \phi(n)| \leq |s(\overline{m},n) - \psi(\overline{m})| + |\psi(\overline{m}) - g| + |g - \phi(n)| < \eta/3,$$

donc la condition (3) a lieu.   □

Lemme 2. *La condition* (3) *est équivalente à la suivante*

$$\liminf_{m \to +\infty} \limsup_{n \to +\infty} \left| s(m,n) - \lim_{m \to +\infty} s(m,n) \right| = 0. \tag{4}$$

Démonstration. Désignons $w(m,n) := \left| s(m,n) - \lim_{m \to +\infty} s(m,n) \right|$.

Si (3) a lieu, pour chaque $\eta > 0$ et pour chaque N il existe un $m > N$ tel que $\limsup_{n \to +\infty} w(m,n) \leq \eta$. Pour $\eta = 1/k$ il existe donc une suite $m_k \to +\infty$ pour laquelle $\limsup_{n \to +\infty} w(m_k,n) \leq 1/k$, d'où (4) a lieu. Inversement si (4) a lieu, il existe une suite $m_k \to +\infty$ telle que $\lim_{k \to +\infty} \limsup_{n \to +\infty} w(m_k,n) = 0$.

Soient $\varepsilon > 0$ et $\eta > 0$ arbitraires.

Il existe $m_k > \varepsilon$ tel que $\limsup_{n \to +\infty} w(m_k,n) \leq \eta/2$ et de là $w(m_k,n) \leq \eta$ pour n suffisement grand, donc nous avons (3). $\qquad\square$

Nous pouvons démontrer d'après ces lemmes le théorème suivant.

Théorème 3. *Supposons que f_n, $f \in C$ et $f_n \to f$. Dans ce cas $a(f_n) \to a(f)$ si et seulement si pour chaque $x \in \mathbb{R}$*

$$\liminf_{m \to +\infty} \limsup_{n \to +\infty} \left| \frac{f_n(mx)}{m} - a(f_n)(x) \right| = 0. \tag{5}$$

Démonstration. Fixons $x \in \mathbb{R}$. On sait d'après les considérations plus haut que la convergence $a(f_n) \to a(f)$ pour $f_n \to f$ est équivalente à la condition (1) et cette condition est équivalente à la condition (4), où $s(m,n) = \dfrac{f_n(mx)}{m}$ et $\lim_{m \to +\infty} \dfrac{f_n(mx)}{m} = a(f_n)(x)$. $\qquad\square$

Remarques. 1. La condition (5) sous la forme

$$\liminf_{m \to +\infty} \limsup_{n \to +\infty} \left| \frac{f_n(mx)}{m} - \lim_{m \to +\infty} \frac{f_n(mx)}{m} \right| = 0, \tag{6}$$

ayant lieu pour chaque $x \in \mathbb{R}$, est en effet une condition complémentaire qui avec la convergence simple $f_n \to f$ nous donne une convergence par rapport à quelle l'opérateur $a(f)$ est continue, puisque (6) avec $a(f_n)$ au lieu de f_n est évidemment remplie. Cette convergence étant aussi nécessaire pour la continuité de $a(f)$, n'est pas commode, puisque elle est compliquée. La condition (6) est la même que la condition

$$\forall \varepsilon > 0 \, \forall \eta > 0 \, \exists \delta > 0 \, \exists m > \varepsilon \, \forall n > \delta \quad \left| \frac{f_n(mx)}{m} - \lim_{m \to +\infty} \frac{f_n(mx)}{m} \right| \leq \eta.$$

2. La condition (5) est équivalente à la suivante

$$\liminf_{m \to +\infty} \limsup_{n \to +\infty} \left| \frac{b(f_n)(mx)}{m} \right| = 0, \tag{7}$$

puisque $f_n(mx) = a(f_n)(mx) + b(f_n)(mx) = ma(f_n)(x) + b(f_n)(mx)$. Cette condition est remplie évidemment si la suite $b(f_n)$ est uniformément bornée.

3. Il suffit supposer la condition (5) seulement pour $x \in H$, où H est une base de Hamel arbitrairement fixée, puisque la convergence $a(f_n) \to a(f)$ pour $x \in H$ implique $a(f_n) \to a(f)$ sur $\mathbb{R}$ tout entier ($a(f_n)$ sont des fonctions additives).

4. S'il existe $\lim\limits_{n \to +\infty} b(f_n)(x) =: g(x)$, dans ce cas (7) est équivalente à la condition

$$\liminf_{m \to +\infty} \left| \frac{g(mx)}{m} \right| = 0,$$

qui est remplie évidement si g est bornée.

Il est vrai le théorème suivant:

Théorème 4. *Soit* $\mathbf{C} \ni f_n \to f \in \mathbf{C}$. $a(f_n) \to a(f)$ *si et seulement s'il existe* $\lim\limits_{n \to +\infty} b(f_n)$ *bornée.*

Démonstration. Pour la démonstration de "seulement si" remarquons que $b(f_n) = f_n - a(f_n) \to f - a(f) = b(f)$ bornée.

La démonstration de "si" est évidente d'après la considération au commencement du point 4 et d'après le théorème 3. On peut donner aussi la démonstration directe. En effet sous nos suppositions si $g = \lim\limits_{n \to +\infty} b(f_n)$, alors $a(f_n) = f_n - b(f_n) \to f - g$ et puisque l'espace $\mathbf{A}$ est fermé, $f - g$ est additive et $f = g + (f - g)$ nous donne $a(f) = f - g = \lim\limits_{n \to +\infty} a(f_n)$. $\qquad\square$

Le raisonnement dernier montre aussi que s'il existe $\lim\limits_{n \to +\infty} b(f_n)$ bornée pour une suite $f_n \in \mathbf{C}$ telle que $f_n \to f$, alors $f \in \mathbf{C}$.

Les exemples considérés plus haut montrent que l'existence $\lim\limits_{n \to +\infty} a(f_n)$ (additive dans ce cas) pour $f_n \to f$ ne suffit pas pour que $a(f_n) \to a(f)$, puisque on peut se passer que $a(f)$ n'existe pas ($f \notin \mathbf{C}$) ou si elle existe on peut avoir lieu $a(f_n) \nrightarrow a(f)$.

5. Les continuités des opérateurs $a(f)$ et $b(f)$ sont équivalentes puisque $f = b(f) + a(f)$.

6. La continuité de l'opérateur $a(f)(x)$ par rapport à x au point x_0 est équivalente à la condition

$$\lim_{x \to x_0} \lim_{m \to +\infty} \frac{f(mx)}{m} = \lim_{m \to +\infty} \lim_{x \to x_0} \frac{f(mx)}{m},$$

(donc à la commutativité des limites) mais seulement sous la supposition que $f(mx)$ est continue au point mx_0 pour chaque $m \in \mathbb{N}$ et déjà la continuité de f au moins en un point nous donne la continuité de $a(f)(x)$ par rapport à x sur $\mathbb{R}$ tout entier (voir le théorème 1b).

3.   LA CONTINUITÉ PAR RAPPORT À (F, X)

La continuité relativement à x et à f est nécessaire pour cette continuité. L'implication inverse est aussi vraie. En effet si $a(f)(x)$ est continue par rapport à x, nous avons $a(f)(x) = \alpha(f)x$ pour un $\alpha(f) \in \mathbb{R}$. La continuité de $a(f)(x)$ par rapport à f (relativement à la convergence de f_n vers f définie par $f_n \to f$ et par la condition (6)) nous donne $a(f_n)(x) \to a(f)(x)$, alors $\alpha(f_n)x \to \alpha(f)x$, d'où $\alpha(f_n) \to \alpha(f)$. Il en résulte que $a(f_n)(x_n) = \alpha(f_n)x_n \to \alpha(f)x = a(f)(x)$, si $x_n \to x$ dans $\mathbb{R}$ et f_n converge vers f au sens plus haut.

La même situation a lieu pour $b(f)(x)$ si nous ajoutons à la convergence de f_n vers f considérée plus haut qu'elle est continue, c.à d. que $f_n(x_n) \to f(x)$ si $x_n \to x$. En effet si $b(f)(x)$ est continue par rapport à x et par rapport à f relativement à cette convergence dans ce cas les fonctions $c(f)(x) = f(x)$ et $a(f)(x)$ sont continues par rapport à (f, x), donc $b(f)(x) = f(x) - a(f)(x)$ est la même.

Partie II

Soit $\varepsilon > 0$. Désignons par

 - $\mathbf{C}(\varepsilon)$ la famille des fonctions dont la valeur absolue de la différence de Cauchy est bornée par ε,
 - $\mathbf{B}(\varepsilon)$ la famille des fonctions qui sont bornée en valeur absolue par ε.

Nous avons d'après le théorème de Hyers la décomposition unique $f = a(f) + b(f)$, où $f \in \mathbf{C}(\varepsilon), a(f) \in \mathbf{A}$ et $b(f) \in \mathbf{B}(\varepsilon)$. Remarquons inversement que la somme $a + b$, où $a \in \mathbf{A}$ et $b \in \mathbf{B}(\varepsilon)$, ne doit pas appartenir à $\mathbf{C}(\varepsilon)$. Par exemple la fonction définie comme il suit: $b(1) = \varepsilon$, $b(x) = 0$ en outre, n'appartient pas à $\mathbf{C}(\varepsilon)$. Mais il suffit pour $a + b \in \mathbf{C}(\varepsilon)$ que $b \in \mathbf{C}(\varepsilon)$. Il résulte de nos considérations que $\mathbf{C}(\varepsilon)$ est la somme directe de $\mathbf{A}$ et $\mathbf{B}(\varepsilon) \cap \mathbf{C}(\varepsilon)$ $(\mathbf{C}(\varepsilon) = \mathbf{A} \oplus [\mathbf{B}(\varepsilon) \cap \mathbf{C}(\varepsilon)])$. Les espaces $\mathbf{C}(\varepsilon), \mathbf{A}$ et $\mathbf{B}(\varepsilon) \cap \mathbf{C}(\varepsilon)$ sont fermés par rapport à la convergence simple. Puisque les opérateurs $a(f)$ et $b(f)$ considérés maintenant sont les restrictions au domaine $\mathbf{C}(\varepsilon)$ des opérateurs dans la partie I, les résultats de la partie I au sujet de la continuité par rapport à x et par rapport à f sont valables aussi pour ces opérateurs restreints. En particulier $a(f)$ est continu par rapport à la convergence simple $f_n \to f$ puisque la suite $b(f_n)$ est uniformément bornée par ε (voir la remarque 2 plus haut). De là aussi $b(f)$ est continu.

S'il s'agit de la continuité de $a(f)(x)$ et de $b(f)(x)$ par rapport à (f, x) la situation est la même que dans la Partie I.3.

Problème.

La théorie de la stabilité des équations fonctionnelles donne les généralisations du théorème de Hyers et les théorèmes analogues pour les autres équations fonctionnelles et cela permet considérer les problèmes similaires à ces qui sont l'objet de cette note.

References

[1] G. L. Forti, *Hyers-Ulam stability of functional equations in several variables*, Aequationes Math. **50** (1995), 143–190.

[2] R. Ger and P. Šemrl, *The stability of the exponential equation*, Proceedings of the Amer. Math. Soc. **124** (1996), 779–787.

[3] D. H. Hyers, *On the stability of the linear functional equation*, Proc. Nat. Acad. Sci. U.S.A. **27** (1941), 222–224.

[4] C. Kuratowski, *Topologie*, vol. I, Warszawa 1952.

[5] Z. Moszner, *O przemienności przejść granicznych*, (Sur la commutativité des limites), Wyż. Szkoła Ped. Kraków. Rocznik Nauk.-Dydakt. No 31 Prac. Mat. **5** (1968), 81–87.

[6] R. Sikorski, *Funkcje rzeczywiste*, vol. I, Warszawa 1958.

[7] S. M. Ulam, *A collection of the mathematical problems*, Interscience Publ., New York 1960.

GEOMETRICAL ASPECTS OF STABILITY

Jacek Tabor
Institute of Mathematics, Jagiellonian University,
Reymonta 4 st., 30-059 Kraków, Poland
tabor@im.uj.edu.pl

Józef Tabor
Institute of Mathematics, Pedagogical University,
Rejtana 16A st., 35-310 Rzeszów, Poland
tabor@univ.rzeszow.pl

Abstract   We investigate the problem when a function has the best approximation in a given class of functions. We give a sufficient, relatively weak
condition, for this property to hold. As a corollary we obtain that for
various functional equations this problem has a positive solution.

We also prove that in the case of the Jensen functional equation the
problem of existence and uniqueness of best approximation is strictly
connected with the problem of existence and uniqueness of Chebyshev
center of a certain set.

Keywords: best approximation, Chebyshev center, functional equation

Mathematics Subject Classification (2000): 39B52, 39B72, 41A50

1.   INTRODUCTION

By the term "stability" we mean stability in the Hyers-Ulam sense
(cf. [5]) as well as its generalizations and modyfications (cf. [4], [6]).
Roughly speaking the problem of stability of a functional equation (E)
can be formulated as follows. Does for every function f satisfying (E)
with a given accuracy there exist a solution to (E) which is "near" f?
If the answer to this question is positive we say that the equation (E) is
stable. In such a case there arises naturally the next question, whether

Z. Daróczy and Z. Páles (eds.),
Functional Equations - Results and Advances, 123–132.
© 2002 *Kluwer Academic Publishers.*

for each function satisfying f approximately there exists a solution to (E) which is nearest to f.

We will consider a little more general question, namely if for a function f there exists a solution to a given equation or inequality which is nearest to f. To investigate the problem we need to measure the distance between f and the set of solutions of a given equation (inequality). For this purpose it is convenient to use the notion of generalized metric, introduced by H. Covitz and S. D. Nadler in [1]. Instead of "generalized metric" we will use the term "extended metric", since it corresponds with the common use of terms extended reals or extended real functions. Let $\overline{\mathbb{R}}_+ := [0, +\infty]$.

Definition 1. Let X be a nonempty set. A function $d : X \times X \to \overline{\mathbb{R}}_+$ is called an extended metric if it satisfies the following conditions:

(i) $d(x, y) = 0 \Leftrightarrow x = y$,

(ii) $d(x, y) = d(y, x)$,

(iii) $d(x, z) \leq d(x, y) + d(y, z)$.

The pair (X, d) is called an extended metric space.

Notice that an extended metric space is metrizable. Namely, one can easily check that if d is an extended metric then the function

$$d^*(x, y) := \frac{d(x, y)}{1 + d(x, y)} \qquad (\tfrac{\infty}{\infty} = 1),$$

is a metric which gives the same topology as d.

In an analogous way we define extended normed spaces.

Definition 2. Let X be a real or complex vector space. A function $\| \, \| : X \to \overline{\mathbb{R}}_+$ is called an extended norm if it satisfies the following conditions:

(i) $\|x\| = 0 \Leftrightarrow x = 0$,

(ii) $\|x + y\| \leq \|x\| + \|y\|$,

(iii) $\|\alpha x\| = |\alpha| \|x\|$.

The pair $(X, \| \, \|)$ is called an extended normed space.

It is clear that an extended normed space becomes an extended metric space if we put $d(x, y) := \|x - y\|$. It is easy to verify that an extended normed space is a Hausdorff topological group. However, in general, it is not a topological vector space, as the multiplication by scalars need not be continuous.

The question of existence of the best approximation is usually considered in normed spaces (cf. [2]). We will consider this problem in extended normed spaces since the "distance" of an approximate solution of a functional equation from the set of all solutions may equal infinity.

Let (X, d) be an extended metric space, let K be a nonempty subset of X, and let $x \in X$. An element $k \in K$ is called the best approximation of x from K if

$$d(x, k) = d(x, K) := \inf\{d(x, \tilde{k}) : \tilde{k} \in K\}.$$

Let $r \in \mathbb{R}_+$. By $B(x, r)$ we denote the closed ball centered at x and with radius r.

2. BEST APPROXIMATION IN SUPREMUM NORM

In this section we investigate when a function has the best approximation in the given class of functions (we also study the case when this class is given as the set of all solutions to a functional equation or inequality).

Let S be a nonempty set and X be a normed space. For a function $f : S \to X$ by $\|f\|_{\sup}$ we denote its extended supremum norm, that is

$$\|f\|_{\sup} := \sup_{s \in S} \|f(s)\|.$$

For $f, g : S \to X$ we define $d_{\sup}(f, g) := \|f - g\|_{\sup}$.

Theorem 1. *Let S be a nonempty set and X be a dual Banach space. In $\prod_{s \in S} X$ (which can be understood as the set of all functions from S to X) we take the product topology with respect to the weak* topology in X.*

Let $\mathcal{G} \subset \prod_{s \in S} X$ be a nonempty closed set. Then every function $f : S \to X$ has the best approximation in $\mathcal{G}$, that is there exists a $g \in \mathcal{G}$ such that

$$\|f - g\|_{\sup} = d_{\sup}(f, \mathcal{G}).$$

Proof. The case $K := d_{\sup}(f, \mathcal{G}) = \infty$ is trivial.

Assume that $K < \infty$. For arbitrary $\varepsilon \geq 0$ we put

$$V_\varepsilon := \left(\prod_{s \in S} B(f(s), K + \varepsilon) \right) \cap \mathcal{G} = \{g \in \mathcal{G} : \|f - g\|_{\sup} \leq K + \varepsilon\}.$$

We prove that $V_0 \neq \emptyset$. Clearly $V_\varepsilon \neq \emptyset$ for every $\varepsilon > 0$. Since X is a dual Banach space, closed balls are compact in X with respect to the weak*

topology, and therefore $\prod_{s \in S} B(f(d), K + \varepsilon)$ is compact for every $\varepsilon \geq 0$. By Cantor's Theorem $V_0 = \bigcap_{\varepsilon > 0} V_\varepsilon$ is nonempty as an intersection of a descending family of nonempty and compact sets.

It is clear that every $g \in V_0$ satisfies the assertion. $\square$

The following general result shows that for various classes of functional equations and inequalites we have the existence of best approximation.

Theorem 2. *Let S be a nonempty set, let X be a dual Banach space and let $f_1, \ldots, f_n : S^m \to S$.*

By τ we denote the weak* topology in X. Let $F : (X, \tau)^n \to (X, \tau)$ be continuous.

Assume that a family

$$\mathcal{G} := \{g : S \to X \mid F(g(f_1(z)), \ldots, g(f_n(z))) = 0 \text{ for } z \in S^m\}$$

is nonempty.

Then every function $f : S \to X$ has the best approximation in $\mathcal{G}$, that is there exists a function $g \in \mathcal{G}$ such that

$$\|f - g\|_{\sup} = d_{\sup}(f, \mathcal{G}).$$

Proof. By Theorem 1 it is sufficient to prove that $\mathcal{G}$ is closed. So let $\{g_\lambda\}_{\lambda \in \Lambda} \subset \mathcal{G}$ be a generalized sequence which converges to some g_0. We show that $g_0 \in \mathcal{G}$.

Let $z \in S^m$ be arbitrarily chosen. Since the generalized sequence $\{g_\lambda\}_{\lambda \in \Lambda}$ is convergent to g_0, $\{g_\lambda(f_i(z))\}_{\lambda \in \Lambda}$ is convergent to $g_0(f_i(z))$ for $i = 1, \ldots, n$. Because $F : X^n \to X$ is continuous and

$$F(g_\lambda(f_1(z)), \ldots, g_\lambda(f_n(z))) = 0,$$

we obtain that $F(g_0(f_1(z)), \ldots, g_0(f_n(z))) = 0$. Hence $g_0 \in \mathcal{G}$, and therefore $\mathcal{G}$ is closed. $\square$

As a consequence of Theorem 2 we obtain the existence of best approximation for solutions of some classical functional equations and inequalities.

Corollary 1. *Let S be a nonempty set with a binary operation $\circ$ and let $f : S \to \mathbb{C}$.*

Then f has the best approximation in the class $\mathcal{M}(S, \mathbb{C}) := \{g : S \to \mathbb{C} \mid g(x \circ y) = g(x)g(y) \text{ for } x, y \in S\}$ (the class of complex-valued multiplicative functions on S), that is there exists a function $g \in \mathcal{M}(S, \mathbb{C})$ such that

$$\|f - g\|_{\sup} = d_{\sup}(f, \mathcal{M}).$$

Proof. We define $f_1, f_2, f_3 : S^2 \to S$, and $F : \mathbb{C}^3 \to \mathbb{C}$

$$f_1(x_1, x_2) := x_1, \qquad f_2(x_1, x_2) := x_2, \qquad f_3(x_1, x_2) := x_1 \circ x_2,$$

$$F(a_1, a_2, a_3) := a_1 a_2 - a_3,$$

and apply Theorem 2. $\qquad\qquad\qquad\qquad\qquad\qquad\qquad\qquad\qquad\qquad$ □

Let S be a uniquely 2-divisible semigroup. We say that $D \subset S$ is a Jensen convex set if $\frac{x+y}{2} \in D$ for all $x, y \in D$. A map $f : D \to X$ on a Jensen convex domain D is called a Jensen function whenever

$$f\left(\frac{x+y}{2}\right) = \frac{f(x) + f(y)}{2} \qquad \text{for} \quad x, y \in D.$$

The set of all Jensen convex functions is denoted by $\mathcal{J}(D, X)$.

Corollary 2. *Let S be a uniquely 2-divisible semigroup, let D be a nonempty Jensen convex subset of S, and let X be a dual Banach space.*

Then every function $f : D \to X$ has the best approximation in the class $\mathcal{J}(D, X)$, that is there exists a Jensen function $g : D \to X$ such that

$$\|f - g\|_{\sup} = d_{\sup}(f, \mathcal{J}(D, X)).$$

Proof. We define functions $f_1, f_2, f_3 : S^2 \to S$ and $F : X^3 \to X$ as follows

$$f_1(x_1, x_2) := x_1, \qquad f_2(x_1, x_2) := x_2, \qquad f_3(x_1, x_2) := \frac{x_1 + x_2}{2},$$

$$F(a_1, a_2, a_3) := \frac{a_1 + a_2}{2} - a_3,$$

and apply Theorem 2. $\qquad\qquad\qquad\qquad\qquad\qquad\qquad\qquad\qquad\qquad$ □

Let D be a Jensen convex subset of a uniquely 2-divisible semigroup. We say that $f : D \to \mathbb{R}$ is Jensen convex if

$$f\left(\frac{x+y}{2}\right) \le \frac{f(x) + f(y)}{2} \qquad \text{for} \quad x, y \in D.$$

The set of all such functions will be denoted by $\mathcal{J}_{\mathrm{con}}(D, \mathbb{R})$.

Corollary 3. *Let S be a uniquely 2-divisible semigroup, let D be a Jensen convex subset of S, and let $f : D \to \mathbb{R}$.*

Then f has the best approximation in the class $\mathcal{J}_{\mathrm{con}}(D, \mathbb{R})$, that is there exists a Jensen convex function $g : D \to X$ such that

$$\|f - g\|_{\sup} = d_{\sup}(f, \mathcal{J}_{\mathrm{con}}(D, \mathbb{R})).$$

Proof. We define functions $f_1, f_2, f_3 : S^2 \to S$, $F : \mathbb{R}^3 \to \mathbb{R}$ by the formulae

$$f_1(x_1, x_2) := x_1, \qquad f_2(x_1, x_2) := x_2, \qquad f_3(x_1, x_2) := \frac{x_1 + x_2}{2},$$

$$F(a_1, a_2, a_3) := \max\{a_3 - \frac{a_1 + a_2}{2}, 0\},$$

and apply Theorem 2. □

As we know (cf. [5]) every approximately additive function f from a commutative semigroup into a Banach space has a unique additive approximation. However in general (particularly when we deal with a functional equation on a restricted domain) we have no uniqueness. It often happens that even in this case the best approximation may exist.

Proposition 1. *Let E be a Hausdorff topological vector space, let D be a subset of E with nonempty interior, and X be a dual Banach space.*

Then every function $f : D \to X$ has the best approximation in the class $\mathcal{A}(D, X) = \{a|_D : a : E \to X, a \text{ additive}\}$, that is there exists an additive function $a : E \to X$ such that

$$\|f - a|_D\|_{\sup} = d_{\sup}(f, \mathcal{A}(D, X)).$$

Proof. Clearly $\mathcal{A}(D, X)$ is nonempty. In virtue of Theorem 2 it is enough to prove that $\mathcal{A}(D, X)$ is closed in $\prod_{d \in D} X$ (with the weak* topology in X).

Let $\{a_\lambda|_D\}_{\lambda \in \Lambda}$ be a convergent generalized sequence of elements of $\mathcal{A}(D, X)$. We need to prove that its limit is a restriction of an additive function. For this purpose we will prove that $\{a_\lambda\}_{\lambda \in \Lambda}$ is a convergent generalized sequence. Taking into consideration the topology in $\prod_{e \in E} X$ (which is the product topology with respect to the weak* topology in X) it is sufficient to prove that for every $x \in E$ the sequence $\{a_\lambda(x)\}_{\lambda \in \Lambda}$ is convergent.

Let us choose $d \in \operatorname{int} D$. Let $x \in E$ be arbitrary. Then there exists an $l \in \mathbb{N}$ such that

$$\frac{x + ld}{l + 1} \in D.$$

Since a_λ are additive functions, to show that $\{a_\lambda(x)\}_{\lambda \in \Lambda}$ is convergent, it is enough to verify that $\{a_\lambda(\frac{x+ld}{l+1})\}_{\lambda \in \Lambda}$ is convergent. However, this is the case as $\frac{x+ld}{l+1} \in D$, and therefore $\{a_\lambda(\frac{x+ld}{l+1})\}_{\lambda \in \Lambda}\}$ is convergent by the assumption.

Thus the sequence $\{a_\lambda\}_{\lambda \in \Lambda}$ is convergent. Obviously its limit is an additive function. □

We show the existence of best approximation in the case of convex functions.

Proposition 2. *Let E be a vector space, let D be a convex subset of E, and let $f : D \to \mathbb{R}$.*

Then f has the best approximation in the classes $\mathcal{G}_1, \mathcal{G}_2$ of real-valued convex, quasiconvex functions on D, that is there exist convex function $g_1 \in \mathcal{G}_1$ and quasiconvex function $g_2 \in \mathcal{G}_2$ such that

$$\|f - g_1\|_{\sup} = d_{\sup}(f, \mathcal{G}_1), \qquad \|f - g_2\|_{\sup} = d_{\sup}(f, \mathcal{G}_2).$$

Proof. We present the proof for the class $\mathcal{G}_1$ of convex functions (the proof for the class $\mathcal{G}_2$ is analogous).

By Theorem 1 we have to prove that the class $\mathcal{G}_1$ is nonempty and closed in the product topology.

It is obviously nonempty. So let $\{g_\lambda\}_{\lambda \in \Lambda}$ be a generalized sequence of elements of $\mathcal{G}_1$ which converges to some g_0. We have to show that g_0 is convex.

Let $x, y \in D$ and $\alpha \in [0, 1]$. Clearly

$$g_\lambda(\alpha x + (1 - \alpha)y) \le \alpha g_\lambda(x) + (1 - \alpha)g_\lambda(y) \qquad \text{for} \quad \lambda \in \Lambda. \quad (1)$$

Since $\{g_\lambda\}_{\lambda \in \Lambda}$ is convergent in $\prod_{d \in D} \mathbb{R}$, it is convergent pointwise, so (1) implies that

$$g_0((\alpha x + (1 - \alpha)y) \le \alpha g_0(x) + (1 - \alpha)g_0(y).$$

This means that g_0 is convex. $\square$

3. BEST APPROXIMATION IN LIPSCHITZ NORM

Up to now we have investigated approximation in the supremum norm. Since in the stability theory some other norms are also considered (cf. [6], [7]), we present a result concerning approximation in the Lipschitz norm.

Let (S, d) be a metric space and let X be a Banach space. We define its extended Lipschitz norm

$$\|f\|_{\text{lip}} := \max\{\|f\|_{\sup}, \text{lip}(f)\} \qquad \text{for} \quad f : S \to X,$$

where $\text{lip}(f)$ denotes the smallest Lipschitz constant of f if f is Lipschitz and ∞ otherwise.

Theorem 3. *Let (S, d) be a metric space, let X be a dual Banach space. In $\prod_{s \in S} X$ we take the product topology with respect to the weak* topology in X.*

Let $\mathcal{G} \subset \prod_{s \in S} X$ be a nonempty closed set. Then every function $f : S \to X$ has the best approximation in $\mathcal{G}$, i.e. there exists a $g \in \mathcal{G}$ such that

$$\|f - g\|_{\text{lip}} = d_{\text{lip}}(f, \mathcal{G}).$$

Proof. The proof is essentially a modification of that of Theorem 1.

As the case $K := d_{\text{lip}}(f, \mathcal{G}) = \infty$ is trivial, we assume that $K < \infty$. For arbitrary $\varepsilon \geq 0$ we put

$$V_\varepsilon := \{g : S \to X : \|f - g\|_{\text{sup}} \leq K + \varepsilon\},$$
$$U_\varepsilon := \{g : S \to X : \text{lip}(f - g) \leq K + \varepsilon\},$$
$$W_\varepsilon := V_\varepsilon \cap U_\varepsilon.$$

Clearly $W_\varepsilon \neq \emptyset$ for every $\varepsilon > 0$. We are going to prove that $W_0 \neq \emptyset$.

Let us notice that V_ε is a compact set, since $V_\varepsilon = \prod_{s \in S} B(f(s), K + \varepsilon)$. We now show that U_ε is closed. One can easily check that

$$U_\varepsilon = \{g : D \to X \mid \forall x, y \in S : g(x) - g(y) \in$$
$$B(f(x) - f(y), (K + \varepsilon)d(x, y)\}.$$

Suppose that $\{g_\lambda\}_{\lambda \in \Lambda} \subset U_\varepsilon$ is a generalized sequence which converges to some g_0. We need to show that $g_0 \in U_\varepsilon$. Let $x, y \in S$ be arbitrarily fixed. As $\{g_\lambda(x)\}_{\lambda \in \Lambda}$ converges to $g_0(x)$ and $\{g_\lambda(y)\}_{\lambda \in \Lambda}$ converges to $g_0(y)$ we obtain that $\{g_\lambda(x) - g_\lambda(y)\}_{\lambda \in \Lambda}$ converges to $g_0(x) - g_0(y)$. Because $g_\lambda(x) - g_\lambda(y) \in B(g(x) - g(y), (K + \varepsilon)d(x, y))$ for every $\lambda \in \Lambda$, we get that $f_0(x) - f_0(y) \in B(g(x) - g(y), (K + \varepsilon)d(x, y))$, i.e. $g_0 \in U_\varepsilon$. It means that U_ε is closed.

Therefore $W_\varepsilon = V_\varepsilon \cap U_\varepsilon$ is nonempty and compact for every $\varepsilon > 0$, and consequently $W_0 = \bigcap_{\varepsilon > 0} W_\varepsilon$ is nonempty. $\square$

4.   UNIQUENESS OF BEST APPROXIMATION

Results from the previous sections give us no information when the best approximation is uniquely determined. We are going to present some results in this direction concerning Jensen equation.

We need the following definition.

Definition 3. (cf. [3]) Let X be a normed space and $D \subset X$ a nonempty bounded set. An element $x_0 \in X$ is called a Chebyshev center of D if

$$\sup_{x \in D} \|x - x_0\| = \inf_{x \in X} (\sup_{y \in D} \|y - x\|) =: r(D).$$

The number $r(D)$ is called Chebyshev radius of D.

We would like to mention that by the classical theorem of A. Garkavi from [3] every bounded set in the dual Banach space has a Chebyshev center.

We prove that the existence and uniqueness of best approximation of a function in the class of Jensen equations is equivalent to the existence and uniqueness of Chebyshev center of a given set. This shows that there is a strong relation between approximation theory and theory of stability of functional equations.

Let S be a uniquely 2-divisible commutative semigroup. As before, by $\mathcal{J}(S, X)$, where X is a vector space, we denote the space of all Jensen functions $j : S \to X$. We say that a function $f : S \to X$ is approximately Jensen if $\sup_{s,t \in S} \|f(\frac{s+t}{2}) - \frac{f(s)+f(t)}{2}\| < \infty$.

We need also the following simple lemma.

Lemma. *Let S be a uniquely 2-divisible commutative semigroup and let X be a Banach space. Then for every approximately Jensen function $f : S \to X$ there exists a unique additive function $A_f : S \to X$ such that*

$$\|f - A_f\|_{\sup} < \infty.$$

Proof. As it is well-known, for every approximately Jensen function $f : S \to X$ there exists a Jensen function $j : S \to X$ such that $\|f - j\|_{\sup} < \infty$. Then j is of the form $j(x) = A_f(x) + b$, where A_f is an additive function. Consequently we have $\|f - A_f\|_{\sup} < \infty$. The proof of the uniqueness of A_f is routine. □

Theorem 4. *Let S be a uniquely 2-divisible commutative semigroup, let X be a Banach space, and let f be an approximately Jensen function.*

Then there exists the best Jensen approximation of f, i.e. a function $j \in \mathcal{J}(S, X)$ satisfying

$$\|f - j\|_{\sup} = d_{\sup}(f, \mathcal{J}(S, X)) < \infty, \tag{2}$$

if and only if the set $(f - A_f)(S)$ has a Chebyshev center in X.

Moreover, the best Jensen approximation of f is unique if and only if the set $(f - A_f)(S)$ has a unique Chebyshev center.

Proof. Suppose that we have a Jensen function j satisfying (2). Then

$$\|(f - A_f) - (j - A_f)\|_{\sup} = d_{\sup}(f - A_f, \mathcal{J}(S, X)) < \infty. \tag{3}$$

Since $j - A_f$ is a bounded Jensen function, it is a constant function, i.e. there exists a $c \in X$ such that $(j - A_f)(s) \equiv c$ (we show that c is a Chebyshev center of the set $(f - A_f)(S)$). In the same manner we obtain that if $\tilde{j}$ is a Jensen function such that $(f - A_f) - \tilde{j}$ is bounded then $\tilde{j} \equiv \tilde{c}$ for some $\tilde{c} \in X$. Thus (3) is equivalent to

$$\sup_{s \in S} \|(f - A_f)(s) - c\| = \inf_{\tilde{c} \in X} \sup_{s \in S} \|(f - A_f)(s) - \tilde{c}\|,$$

which means that c is a Chebyshev center of the set $(f - A_f)(S)$.

One can easily verify that if c is a Chebyshev center of $(f - A_f)(S)$ then the Jensen function $A_f + c$ is the best Jensen approximation of f.

In the analogous way one can prove the uniqueness part of the Theorem. □

Remark. A. Garkavi in [3]) proved that every bounded subset of a Banach space has at most one Chebyshev center if and only if the space is uniformly convex in every direction. This implies in particular that if X is a uniformly convex Banach space then every approximately Jensen function has a unique best Jensen approximation.

In [3] there is also construction of a Banach space E and a finite set $M \subset E$ such that M has no Chebyshev center in E. Taking into consideration a function $f : \mathbb{R} \to E$ such that $f(\mathbb{R}) = M$ we obtain an example of an approximately Jensen function which has no best Jensen approximation.

References

[1] H. Covitz and S. D. Nadler, *Multivalued contractious mappings in generalized metric spaces*, Israel J. Math. **8** (1970), 5–11.

[2] F. Deutsch, *Existence of Best Approximation*, J. Approx. Theory **28** (1980), 132–154.

[3] A. Garkavi, *The best possible net and the best possible cross section of a set in a normed space*, Izv. Akad. Nauk SSSR Ser. Mat. **26** (1962), 87–106.

[4] R. Ger, *A survey of recent results on stability of functional equation*, in *Proc. of the 4th International Conference on Functional Equations and Inequalities*, Pedagogical University in Cracow, (1994), 5–36.

[5] D. H. Hyers, *On the stability of the linear functional equation*, Proc. Natl. Acad. Sci. U.S.A. **27** (1941), 222–224.

[6] Józef Tabor, *Stability of the Cauchy type equations in $\mathcal{L}^p$ norms*, Results Math. **32** (1997), 145–158.

[7] Jacek Tabor and Józef Tabor, *Stability of the Cauchy functional equation in the class of differentiable functions*, J. Approx. Theory **98**, 167–182 (1999).

III
FUNCTIONAL EQUATIONS
IN ONE VARIABLE
AND ITERATION THEORY

ON SEMI-CONJUGACY EQUATION FOR HOMEOMORPHISMS OF THE CIRCLE

Krzysztof Ciepliński and Marek Cezary Zdun

Institute of Mathematics, Pedagogical University of Cracow,

Podchorążych 2, 30-084 Kraków, Poland

kc@wsp.krakow.pl, mczdun@wsp.krakow.pl

Abstract   The aim of this paper is to investigate, under some assumptions on homeomorphisms F_t and G_t of the unit circle $\mathbb{S}^1$, the existence of continuous as well as homeomorphic solutions of the following system of functional equations:

$$\Phi(F_t(z)) = G_t(\Phi(z)) \qquad (z \in \mathbb{S}^1, t \in M).$$

Keywords: semi-conjugacy, conjugacy, functional equation, lift, degree, rotation number

Mathematics Subject Classification (2000): 39B72, 37E10, 37C15

1.   INTRODUCTION

Let $\{F_t : \mathbb{S}^1 \to \mathbb{S}^1, \ t \in M\}$ and $\{G_t : \mathbb{S}^1 \to \mathbb{S}^1, \ t \in M\}$, where M is an arbitrary non-empty set, be families of pairwise commuting homeomorphisms of the unit circle $\mathbb{S}^1$. In this paper we study the problems of semi-conjugacy and conjugacy of such families, that is, we investigate the existence of continuous and homeomorphic solutions of the following system of functional equations:

$$\Phi(F_t(z)) = G_t(\Phi(z)) \qquad (z \in \mathbb{S}^1, t \in M). \qquad (1)$$

We begin by recalling the basic definitions and introducing some notation.

It is well-known (see for instance [1], [9] and [3]) that, for every continuous mapping $F : \mathbb{S}^1 \to \mathbb{S}^1$, there exist a continuous function $f : \mathbb{R} \to \mathbb{R}$,

135

Z. Daróczy and Z. Páles (eds.),

Functional Equations - Results and Advances, 135–158.

© 2002 *Kluwer Academic Publishers.*

which is unique up to translation by an integer, and a unique integer k such that

$$F(e^{2\pi i x}) = e^{2\pi i f(x)} \qquad (x \in \mathbb{R})$$

and

$$f(x+1) = f(x) + k \qquad (x \in \mathbb{R}).$$

The function f is said to be the lift of F and the integer k is called the degree of F, and is denoted by $\deg F$. The following two properties of the degree of circle maps will be needed throughout the paper. For any two continuous functions F, $G : \mathbb{S}^1 \to \mathbb{S}^1$,

$$\deg(F \cdot G) = \deg F + \deg G$$

and

$$\deg(F \circ G) = \deg F \cdot \deg G.$$

If $F : \mathbb{S}^1 \to \mathbb{S}^1$ is a homeomorphism, then so is its lift. Moreover, $|\deg F| = 1$. We say that a homeomorphism $F : \mathbb{S}^1 \to \mathbb{S}^1$ preserves orientation if $\deg F = 1$, which is clearly equivalent to the fact that the lift of F is increasing. For such a homeomorphism F, the number $\rho(F) \in [0, 1)$ defined by

$$\rho(F) := \lim_{n \to \infty} \frac{f^n(x)}{n} \pmod 1 \qquad (x \in \mathbb{R})$$

is called the rotation number of F. This number always exists and does not depend on x and f. Furthermore, $\rho(F)$ is rational if and only if F has a periodic point (see for instance [6] and [8]).

Throughout this note by an open arc we mean the set

$$\{e^{2\pi i t}, \ t \in (t_1, t_2)\},$$

where t_1, $t_2 \in \mathbb{R}$ are such that $t_1 < t_2 < t_1 + 1$. The closure of the set $A \subset \mathbb{S}^1$ will be denoted by $\mathrm{cl}A$ and A^d stands for the set of all cluster points of A. We also write

$$A^l := \{z^l, \ z \in A\} \qquad (l \in \mathbb{Z})$$

and

$$\Pi(t) := e^{2\pi i t} \qquad (t \in [0, 1)).$$

For every continuous mapping $F : I \to J$, where $I = \{e^{2\pi i t}, \ t \in (a, b)\}$ and $J = \{e^{2\pi i t}, \ t \in (c, d)\}$ are open arcs, there exists a unique continuous function $f : (a, b) \to (c, d)$ with

$$F(e^{2\pi i t}) = e^{2\pi i f(t)} \qquad (t \in (a, b)).$$

We call f the lift of F.

2. PRELIMINARIES

It is well-known (see for instance [8]) that if for given orientation-preserving homeomorphisms F, $G : \mathbb{S}^1 \to \mathbb{S}^1$ there exists a homeomorphism $\Phi : \mathbb{S}^1 \to \mathbb{S}^1$ such that $\Phi \circ F = G \circ \Phi$, then either $\rho(G) = \rho(F)(\mathrm{mod}1)$ or $\rho(G) = -\rho(F)(\mathrm{mod}1)$. Our first theorem generalizes this fact as well as corrects the false result given in [6] (Chapter 3, § 3).

Theorem 1. *Let F, $G : \mathbb{S}^1 \to \mathbb{S}^1$ be orientation-preserving homeomorphisms and suppose that there exists a continuous function $\Phi : \mathbb{S}^1 \to \mathbb{S}^1$ such that*

$$\Phi(F(z)) = G(\Phi(z)) \qquad (z \in \mathbb{S}^1). \tag{2}$$

Then

$$\rho(G) = \rho(F) \deg \Phi (\mathrm{mod}1). \tag{3}$$

Proof. Denote by φ, f and g the lifts of Φ, F and G, respectively, and observe that from (2) we have

$$e^{2\pi i \varphi(f(x))} = e^{2\pi i g(\varphi(x))} \qquad (x \in \mathbb{R}).$$

Since the mappings φ, f and g are continuous, there exists a $k \in \mathbb{Z}$ for which

$$\varphi(f(x)) = g(\varphi(x)) + k \qquad (x \in \mathbb{R}).$$

From this it follows by induction that

$$\varphi(f^n(x)) = g^n(\varphi(x)) + nk \qquad (n \in \mathbb{N}, x \in \mathbb{R}). \tag{4}$$

For all $x \in \mathbb{R}$ and $n \in \mathbb{N}$ let $k_n(x) \in \mathbb{Z}$ and $r_n(x) \in [0,1)$ be such that

$$f^n(x) = k_n(x) + r_n(x). \tag{5}$$

Since $\varphi(x+1) = \varphi(x) + \deg \Phi$ for $x \in \mathbb{R}$, (5) shows that

$$\varphi(f^n(x)) = \deg \Phi k_n(x) + \varphi(r_n(x)) \qquad (n \in \mathbb{N}, x \in \mathbb{R}). \tag{6}$$

Fix an $x \in \mathbb{R}$. By (5) and the definition of $\rho(F)$ we have

$$\lim_{n \to \infty} \frac{k_n(x)}{n} = \lim_{n \to \infty} \frac{f^n(x)}{n} = \rho(F) + p$$

for a $p \in \mathbb{Z}$. (6) and the fact that φ is bounded in $[0,1]$ give

$$\lim_{n \to \infty} \frac{\varphi(f^n(x))}{n} = \lim_{n \to \infty} \frac{\varphi(r_n(x))}{n} + \deg \Phi \lim_{n \to \infty} \frac{k_n(x)}{n} = \deg \Phi(\rho(F) + p).$$

On the other hand, from (4) it follows that

$$\lim_{n\to\infty} \frac{\varphi(f^n(x))}{n} = \lim_{n\to\infty} \frac{g^n(\varphi(x))}{n} + k = \rho(G) + q + k$$

for a $q \in \mathbb{Z}$. Therefore, (3) is proved.   □

The following fact follows immediately from Theorem 1.

Corollary 1. *Let* $F : \mathbb{S}^1 \to \mathbb{S}^1$ *be an orientation-preserving homeomorphism and suppose that* $s \in \mathbb{S}^1$. *If* $\Psi : \mathbb{S}^1 \to \mathbb{S}^1$ *is a continuous solution of the equation*

$$\Psi(F(z)) = s\Psi(z) \qquad (z \in \mathbb{S}^1), \tag{7}$$

then

$$s = e^{2\pi i\rho(F)\deg\Psi}.$$

Let us recall (see for instance [6]) that for a given homeomorphism $F : \mathbb{S}^1 \to \mathbb{S}^1$ such that $\rho(F) \notin \mathbb{Q}$ the set

$$L_F := \{F^n(z), \ n \in \mathbb{Z}\}^{\mathrm{d}} \qquad (z \in \mathbb{S}^1)$$

does not depend on $z \in \mathbb{S}^1$, is invariant with respect to F and either $L_F = \mathbb{S}^1$ or L_F is a perfect nowhere dense subset of $\mathbb{S}^1$.

Lemma 1. *If* $F, G : \mathbb{S}^1 \to \mathbb{S}^1$ *are commuting homeomorphisms with irrational rotation numbers, then* $L_F = L_G$.

Proof. Clearly,

$$F \circ G^n = G^n \circ F \qquad (n \in \mathbb{Z}).$$

From this and the fact that F is a homeomorphism it follows that for any $z \in \mathbb{S}^1$,

$$L_G = \{G^n(F(z)), \ n \in \mathbb{Z}\}^{\mathrm{d}} = F[\{G^n(z), \ n \in \mathbb{Z}\}]^{\mathrm{d}}$$
$$= F[\{G^n(z), \ n \in \mathbb{Z}\}^{\mathrm{d}}] = F[L_G].$$

Thus by induction we obtain $F^n[L_G] = L_G$ for $n \in \mathbb{Z}$. Finally, fixing a $z \in L_G$, we see that

$$L_F = \{F^n(z), \ n \in \mathbb{Z}\}^{\mathrm{d}} \subset (\bigcup_{n\in\mathbb{Z}} F^n[L_G])^{\mathrm{d}} = L_G^{\mathrm{d}} = L_G.$$

□

Proposition 1. *Assume that $F : \mathbb{S}^1 \to \mathbb{S}^1$ is a homeomorphism for which $\rho(F) \notin \mathbb{Q}$ and write*

$$s_F := e^{2\pi i \rho(F)}.$$

Then there exists a unique continuous function $\Phi_F : \mathbb{S}^1 \to \mathbb{S}^1$ such that

$$\Phi_F(F(z)) = s_F \Phi_F(z) \qquad (z \in \mathbb{S}^1) \tag{8}$$

and $\Phi_F(1) = 1$. Moreover, $\deg \Phi_F = 1$ and Φ_F is a homeomorphism if and only if $L_F = \mathbb{S}^1$.

For the proof of the existence of a continuous solution $\Phi_F : \mathbb{S}^1 \to \mathbb{S}^1$ of equation (8) as well as the fact that Φ_F is a homeomorphism if and only if $L_F = \mathbb{S}^1$ see for instance [7] and [6]. The uniqueness of the solution is proved in [4]. The equality $\deg \Phi_F = 1$ follows from Corollary 1.

Let $F : \mathbb{S}^1 \to \mathbb{S}^1$ be a homeomorphism such that $\rho(F) \notin \mathbb{Q}$. The set

$$K_F := \Phi_F[\mathbb{S}^1 \setminus L_F]$$

is said to be the iterative kernel of F.

Let $F, G : \mathbb{S}^1 \to \mathbb{S}^1$ be homeomorphisms with irrational rotation numbers and assume that $L_F \neq \mathbb{S}^1$ and $L_G \neq \mathbb{S}^1$. Since L_F and L_G are perfect nowhere dense subsets of $\mathbb{S}^1$, the sets $\mathbb{S}^1 \setminus L_F$ and $\mathbb{S}^1 \setminus L_G$ are countable sums of pairwise disjoint open arcs. Therefore we have the following two decompositions:

$$\mathbb{S}^1 \setminus L_F = \bigcup_{p \in T_F} I_p \quad \text{and} \quad \mathbb{S}^1 \setminus L_G = \bigcup_{q \in T_G} J_q, \tag{9}$$

where $T_F, T_G \subset \mathbb{S}^1$ are countable sets and I_p for $p \in T_F$ (respectively, J_q for $q \in T_G$) are open pairwise disjoint arcs with the middle points p (respectively, q). We also define

$$L_F^\circ := \mathbb{S}^1 \setminus \bigcup_{p \in T_F} \mathrm{cl}\, I_p \quad \text{and} \quad L_G^\circ := \mathbb{S}^1 \setminus \bigcup_{q \in T_G} \mathrm{cl}\, J_q.$$

Proposition 2. *Let $F : \mathbb{S}^1 \to \mathbb{S}^1$ be a homeomorphism for which $\rho(F) \notin \mathbb{Q}$ and $L_F \neq \mathbb{S}^1$. If $\Phi_F : \mathbb{S}^1 \to \mathbb{S}^1$ is the continuous solution of (8) such that $\Phi_F(1) = 1$, then:*

   (i) *every lift of Φ_F is increasing,*
   (ii) *for every $p \in T_F$ the function Φ_F is constant on I_p and I_p is the maximal open arc of constancy of Φ_F,*
  (iii) *if $p \neq q$, then $\Phi_F[I_p] \cap \Phi_F[I_q] = \emptyset$,*

(iv) for every $p \in T_F$ there exists a $q \in T_F$ such that $F[I_p] = I_q$,
(v) $\Phi_F[L_F] = \mathbb{S}^1$.

Conditions (i)-(iv) can be found in [4] and [10]. Analysis similar to that in the proof of Proposition 1 in [2] shows that the last statement is also true.

Lemma 2. *Let $F : \mathbb{S}^1 \to \mathbb{S}^1$ be a homeomorphism for which $\rho(F) \notin \mathbb{Q}$ and $L_F \neq \mathbb{S}^1$. If $\Phi_F : \mathbb{S}^1 \to \mathbb{S}^1$ is the continuous solution of (8) such that $\Phi_F(1) = 1$, then*

$$\Phi_F[L_F^\circ] = \mathbb{S}^1 \setminus K_F.$$

Proof. Let us first observe that by the continuity of Φ_F and Proposition 2(ii) it follows that for every $p \in T_F$, Φ_F is constant on $\mathrm{cl}I_p$. Therefore

$$K_F = \bigcup_{p \in T_F} \Phi_F[\mathrm{cl}I_p]. \tag{10}$$

Using (10) and the fact that Φ_F maps $\mathbb{S}^1$ onto $\mathbb{S}^1$ we see at once that

$$\mathbb{S}^1 \setminus K_F = \Phi_F[\mathbb{S}^1] \setminus \Phi_F[\bigcup_{p \in T_F} \mathrm{cl}I_p] \subset \Phi_F[\mathbb{S}^1 \setminus \bigcup_{p \in T_F} \mathrm{cl}I_p] = \Phi_F[L_F^\circ].$$

Now fix a $z \in \Phi_F[L_F^\circ]$ and suppose, contrary to our claim, that $z \in K_F$. Then, according to (10), there is a $p \in T_F$ such that $z = \Phi_F(w)$ for $w \in \mathrm{cl}I_p$. Since we also have $z = \Phi_F(u)$ for a $u \notin \mathrm{cl}I_p$, $\deg \Phi_F = 1$ and a lift of Φ_F is increasing, the function Φ_F is constant on an open arc J such that $I_p \subsetneq J$, which contradicts Proposition 2(ii). $\square$

Lemma 3. *Let $F_t : \mathbb{S}^1 \to \mathbb{S}^1$ for $t \in M$ be orientation-preserving homeomorphisms such that $\rho(F_{t_0}) \notin \mathbb{Q}$ for a $t_0 \in M$. Suppose that*

$$F_t \circ F_{t_0} = F_{t_0} \circ F_t \qquad (t \in M) \tag{11}$$

and write

$$s_t := e^{2\pi i \rho(F_t)} \qquad (t \in M).$$

Then for every $l \in \mathbb{Z}$ and $a \in \mathbb{S}^1$ there exists a unique continuous solution $\Psi : \mathbb{S}^1 \to \mathbb{S}^1$ of the system

$$\Psi(F_t(z)) = s_t^l \Psi(z) \qquad (z \in \mathbb{S}^1, t \in M) \tag{12}$$

such that $\Psi(1) = a$. This solution is given by the formula

$$\Psi(z) = a\Phi_{F_{t_0}}(z)^l \qquad (z \in \mathbb{S}^1) \tag{13}$$

and is of degree l.

Proof. Fix $l \in \mathbb{Z}$, $a \in \mathbb{S}^1$ and put $F := F_{t_0}$,

$$\Lambda_t(z) := \frac{1}{\Phi_F(F_t(1))} \Phi_F(F_t(z)) \qquad (z \in \mathbb{S}^1, t \in M).$$

Then by (11) and (8) we get

$$\Lambda_t(F(z)) = s_F \Lambda_t(z) \qquad (z \in \mathbb{S}^1, t \in M)$$

and $\Lambda_t(1) = 1$ for $t \in M$. Since for each $t \in M$ the function Λ_t is continuous, Proposition 1 shows that $\Lambda_t = \Phi_F$, whence we obtain

$$\Phi_F(F_t(z)) = \Phi_F(F_t(1))\Phi_F(z) \qquad (z \in \mathbb{S}^1, t \in M).$$

From Corollary 1 and the fact that $\deg \Phi_F = 1$ we conclude that the function Ψ given by (13) is a continuous solution of (12) such that $\Psi(1) = a$ and $\deg \Psi = l$.

Now assume that $\Gamma : \mathbb{S}^1 \to \mathbb{S}^1$ is a continuous solution of (12) such that $\Gamma(1) = a$ and let Ψ be given by (13). Putting $\Omega := \frac{\Gamma}{\Psi}$ we have by induction

$$\Omega(F^n(z)) = \Omega(z) \qquad (z \in \mathbb{S}^1, n \in \mathbb{Z}).$$

Fix a $c \in L_F$. Since for every $z \in \mathbb{S}^1$ there exists a sequence $(n_k)_{k \in \mathbb{N}}$ of integers such that $\lim_{k \to \infty} F^{n_k}(z) = c$, the continuity of Ω leads to

$$\Omega(z) = \lim_{k \to \infty} \Omega(F^{n_k}(z)) = \Omega(c).$$

Thus Ω is constant and the equality $\Omega(1) = 1$ gives $\Gamma = \Psi$. $\qquad \square$

If $M = \{t_0\}$ and $F := F_{t_0}$, then we get

Corollary 2. *Let $F : \mathbb{S}^1 \to \mathbb{S}^1$ be a homeomorphism for which $\rho(F) \notin \mathbb{Q}$. If a continuous function $\Psi : \mathbb{S}^1 \to \mathbb{S}^1$ of degree l satisfies equation (7) for an $s \in \mathbb{S}^1$, then there exists an $a \in \mathbb{S}^1$ such that*

$$\Psi(z) = a\Phi_F(z)^l \qquad (z \in \mathbb{S}^1).$$

Next notice that the following fact follows immediately from Lemma 1 and Lemma 3.

Remark 1. Under the hypotheses of Lemma 3, for every $t \in M$ such that $\rho(F_t) \notin \mathbb{Q}$, we have

$$L_{F_t} = L_{F_{t_0}} \qquad \text{and} \qquad K_{F_t} = K_{F_{t_0}}.$$

Corollary 3. *Let* F, $G : \mathbb{S}^1 \to \mathbb{S}^1$ *be orientation-preserving homeomorphisms and suppose that* $\rho(G) \notin \mathbb{Q}$. *If a continuous function* $\Phi : \mathbb{S}^1 \to \mathbb{S}^1$ *of degree* l *satisfies equation* (2), *then* $l \neq 0$ *and there exists a* $d \in \mathbb{S}^1$ *such that*

$$\Phi_G(\Phi(z)) = d\Phi_F(z)^l \qquad (z \in \mathbb{S}^1). \tag{14}$$

Proof. Theorem 1 makes it obvious that $l \neq 0$ and $\rho(F) \notin \mathbb{Q}$. Putting $H := \Phi_G \circ \Phi$ we deduce from (2) and Proposition 1 that

$$H(F(z)) = s_G H(z) \qquad (z \in \mathbb{S}^1)$$

and $\deg H = l$. Corollary 2 now shows that there exists a $d \in \mathbb{S}^1$ such that $H(z) = d\Phi_F(z)^l$ for $z \in \mathbb{S}^1$, and (14) is proved. $\qquad \square$

Theorem 2. *Let* F, $G : \mathbb{S}^1 \to \mathbb{S}^1$ *be orientation-preserving homeomorphisms and suppose that* $\rho(G) \notin \mathbb{Q}$. *If a continuous function* $\Phi : \mathbb{S}^1 \to \mathbb{S}^1$ *of degree* l *satisfies equation* (2), *then* Φ *maps* $\mathbb{S}^1$ *onto* $\mathbb{S}^1$, *and* L_F *onto* L_G. *If, moreover,* $L_G \neq \mathbb{S}^1$ *and* $d := \Phi_G(\Phi(1))$, *then for every* $w \in \mathbb{S}^1$, $dw^l \in K_G$ *implies* $w \in K_F$.

Proof. Notice that, by Theorem 1, $\rho(F) \notin \mathbb{Q}$ and $\deg \Phi \neq 0$. Therefore every lift of Φ maps $\mathbb{R}$ onto itself, and consequently $\Phi[\mathbb{S}^1] = \mathbb{S}^1$. Observe also that (2) and induction yield

$$\Phi(F^n(z)) = G^n(\Phi(z)) \qquad (z \in \mathbb{S}^1, n \in \mathbb{Z}). \tag{15}$$

Fix a $y \in L_G$. Then there exists a sequence $(n_k)_{k \in \mathbb{N}}$ of integers for which

$$y = \lim_{k \to \infty} G^{n_k}(\Phi(1)) = \lim_{k \to \infty} \Phi(F^{n_k}(1)).$$

Since $\mathbb{S}^1$ is compact, there is a subsequence $(n_{k_\nu})_{\nu \in \mathbb{N}}$ of the sequence $(n_k)_{k \in \mathbb{N}}$ such that $\lim_{\nu \to \infty} F^{n_{k_\nu}}(1) = x$ for an $x \in L_F$. Thus

$$y = \lim_{\nu \to \infty} \Phi(F^{n_{k_\nu}}(1)) = \Phi(x) \in \Phi[L_F],$$

and consequently $L_G \subset \Phi[L_F]$.

Now fix an $x \in L_F$ and let $(n_k)_{k \in \mathbb{N}}$ be a sequence of integers such that $\lim_{k \to \infty} F^{n_k}(1) = x$. Then it follows from (15) that

$$\Phi(x) = \lim_{k \to \infty} \Phi(F^{n_k}(1)) = \lim_{k \to \infty} G^{n_k}(\Phi(1)),$$

which gives $\Phi(x) \in L_G$. We thus get $\Phi[L_F] \subset L_G$, and, in consequence, $\Phi[L_F] = L_G$.

To see our last statement assume, for instance, that $l > 0$ and observe that from what has already been proved, we obtain $L_F \neq \mathbb{S}^1$. Corollary 3 shows that equation (14) holds for $d = \Phi_G(\Phi(1))$. Fix an $u \in K_G$. The assertion will follow, if we prove that there are l elements $w_i \in K_F$ with $dw_i^l = u$. Choose a $\xi \in \mathbb{S}^1 \setminus L_G$ for which $\Phi_G(\xi) = u$ and let $\zeta \in \mathbb{S}^1$ be such that $\xi = \Phi(\zeta)$. Since $\Phi[L_F] = L_G$, $\zeta \notin L_F$. Put $a := \varphi(0)$, where $\varphi : \mathbb{R} \to \mathbb{R}$ is a lift of Φ, and let $t \in [0, 1)$ be such that $\xi = e^{2\pi i(t+a)}$. Of course, $[a, \, a + l] \subset \varphi[[0, 1]]$, and therefore there exist $s_k \in [0, 1)$ for $k \in \{0, 1, ..., l - 1\}$ such that $\varphi(s_k) = t + a + k$ and $s_0 < s_1 < ... < s_{l-1}$. Fix a $k \in \{0, 1, ..., l - 1\}$. Writing $\zeta_k := e^{2\pi i s_k}$ we have

$$\Phi(\zeta_k) = e^{2\pi i \varphi(s_k)} = e^{2\pi i(t+k+a)} = \xi,$$

and consequently $\zeta_k \notin L_F$, since $\xi \notin L_G$. Hence $w_k := \Phi_F(\zeta_k) \in K_F$. From (14) we conclude that

$$u = \Phi_G(\xi) = \Phi_G(\Phi(\zeta_k)) = d\Phi_F(\zeta_k)^l = dw_k^l.$$

The proof is completed by showing that $w_k \neq w_j$, provided $k \neq j$. According to Proposition 2(iii), it suffices to prove that if $k \neq j$ and $\zeta_k \in I_q$, $\zeta_j \in I_p$, then $q \neq p$. Suppose, contrary to our claim, that there are $k, j \in \{0, 1, ..., l-1\}$, $k < j$ such that $\zeta_k, \zeta_j \in I_q$ for a $q \in T_F$. We will show that under this assumption $\Phi[I_q] = \mathbb{S}^1$, which, in view of equation (14) and Proposition 2(ii), is impossible. To this end, consider the set $\hat{I}_q := \Pi^{-1}[I_q]$. This set is either an open interval or $\hat{I}_q = [0, \alpha) \cup (\beta, 1)$ for some $\alpha, \beta \in (0, 1)$ such that $\alpha < \beta$. If s_k and s_j belong to an interval $J \subset \hat{I}_q$, then

$$\mathbb{S}^1 = e^{2\pi i[t+k+a, \, t+j+a]} = e^{2\pi i[\varphi(s_k), \, \varphi(s_j)]} \subset e^{2\pi i \varphi[J]} \subset e^{2\pi i \varphi[\hat{I}_q]} = \Phi[I_q],$$

and therefore $\Phi[I_q] = \mathbb{S}^1$ as claimed. Similarly, if $s_k \in [0, \alpha)$ and $s_j \in (\beta, 1)$, then

$$\mathbb{S}^1 = e^{2\pi i[a, t+a]} \cup e^{2\pi i[t+a+l-1, \, a+l]} \subset e^{2\pi i[a, \, \varphi(s_k)]} \cup e^{2\pi i[\varphi(s_j), \, a+l]}$$
$$\subset e^{2\pi i \varphi[\hat{I}_q]} = \Phi[I_q],$$

and the proof is complete.   □

For a given homeomorphism $F : \mathbb{S}^1 \to \mathbb{S}^1$ such that $\rho(F) \notin \mathbb{Q}$ and $L_F \neq \mathbb{S}^1$ we set

$$\Lambda_F := \Phi_F \, |_{T_F}.$$

Proposition 2 makes it obvious that Λ_F is a bijection mapping T_F onto K_F.

Lemma 4. *Let $F, G : \mathbb{S}^1 \to \mathbb{S}^1$ be orientation-preserving homeomorphisms and suppose that $\rho(G) \notin \mathbb{Q}$. If an orientation-preserving (respectively, orientation-reversing) homeomorphism $\Phi : \mathbb{S}^1 \to \mathbb{S}^1$ satisfies equation (2), then $K_G = dK_F$ (respectively, $K_G = dK_F^{-1}$) for $d :=$ $\Phi_G(\Phi(1))$. If, moreover, $L_G \neq \mathbb{S}^1$ and*

$$\Omega(p) := \Lambda_G^{-1}(d\Lambda_F(p)) \qquad (p \in T_F)$$

(respectively,

$$\Omega(p) := \Lambda_G^{-1}(d\Lambda_F(p)^{-1}) \qquad (p \in T_F)),$$

then

$$\Phi[I_p] = J_{\Omega(p)} \qquad (p \in T_F).$$

Proof. Clearly, $\rho(F) \notin \mathbb{Q}$. If $L_F = \mathbb{S}^1$, then from Theorem 2 it follows that

$$L_G = \Phi[L_F] = \Phi[\mathbb{S}^1] = \mathbb{S}^1,$$

and therefore $K_F = K_G = \emptyset$. The same holds for the case $L_G = \mathbb{S}^1$. Now assume that $L_F \neq \mathbb{S}^1 \neq L_G$ and observe that, by Corollary 3, (14) is true for $d = \Phi_G(\Phi(1))$. Fix a $p \in T_F$. Using (14) and Theorem 2 we see that

$$dK_F^l = d\Phi_F[\mathbb{S}^1 \setminus L_F]^l = \Phi_G[\Phi[\mathbb{S}^1 \setminus L_F]] = \Phi_G[\Phi[\mathbb{S}^1] \setminus \Phi[L_F]]$$
$$= \Phi_G[\mathbb{S}^1 \setminus L_G] = K_G,$$

where $l = \deg \Phi$. Obviously, $l = 1$ if Φ preserves orientation and $l = -1$ if Φ revers orientation. Futhermore, from Proposition 2(ii) and (14) we deduce that the function Φ_G is constant on the open arc $\Phi[I_p]$ and Proposition 2(ii) now shows that there exists a $q \in T_G$ for which $\Phi[I_p] \subset J_q$. Denoting by a and b the ends of the arc I_p we have $a, b \in L_F$, and, in consequence, $\Phi(a), \Phi(b) \in L_G$. Since Φ is a homeomorphism, $\Phi(a)$ and $\Phi(b)$ are the ends of the arc $\Phi[I_p]$, and therefore $\Phi[I_p] = J_q$. Applying the facts that $p \in I_p$ and $q \in J_q$, Proposition 2(ii) and equation (14) we get

$$\{d\Lambda_F(p)^l\} = d\Phi_F[I_p]^l = \Phi_G[\Phi[I_p]] = \Phi_G[J_q] = \{\Lambda_G(q)\}.$$

This gives $q = \Omega(p)$, and the proof is complete.   □

Lemma 5. *Let $F : \mathbb{S}^1 \to \mathbb{S}^1$ be a homeomorphism for which $\rho(F) \notin \mathbb{Q}$. Suppose that $a \in \mathbb{S}^1$ and*

$$F_a(z) := a^{-1}F(az) \qquad (z \in \mathbb{S}^1).$$

Then

$$K_{F_a} = \frac{1}{\Phi_F(a)} K_F, \qquad L_{F_a} = a^{-1} L_F, \qquad and \qquad L_{F_a}^\circ = a^{-1} L_F^\circ.$$

Proof. It is evident that $F_a : \mathbb{S}^1 \to \mathbb{S}^1$ is an orientation-preserving homeomorphism for which $\rho(F_a) = \rho(F)$. Since the function $\Phi : \mathbb{S}^1 \to \mathbb{S}^1$ defined by $\Phi(z) := a^{-1}z$ for $z \in \mathbb{S}^1$ is an orientation-preserving homeomorphism such that

$$\Phi(F(z)) = F_a(\Phi(z)) \qquad (z \in \mathbb{S}^1),$$

Theorem 2 leads to $L_{F_a} = a^{-1} L_F$. Consequently, $L_{F_a}^\circ = a^{-1} L_F^\circ$. From Lemma 4 and Corollary 3 it follows that there exists a $d \in \mathbb{S}^1$ such that $K_{F_a} = d K_F$ and

$$\Phi_{F_a}(a^{-1}z) = d\Phi_F(z) \qquad (z \in \mathbb{S}^1).$$

Substituting a for z into the last equation we get $d = \frac{1}{\Phi_F(a)}$, which completes the proof. $\qquad \square$

3. MAIN RESULTS

We start with

Theorem 3. *Let $F_t, G_t : \mathbb{S}^1 \to \mathbb{S}^1$ for $t \in M$ be orientation-preserving homeomorphisms such that*

$$F_t \circ F_s = F_s \circ F_t, \qquad G_t \circ G_s = G_s \circ G_t \qquad (s, t \in M) \qquad (16)$$

and

$$\rho(G_t) = l\rho(F_t) (\mathrm{mod}\, 1) \qquad (t \in M) \qquad (17)$$

for an $l \in \mathbb{Z}$. Assume also that there is a $t_0 \in M$ for which $\rho(G_{t_0}) \notin \mathbb{Q}$.

If $L_{G_{t_0}} = \mathbb{S}^1$, then for every $a \in \mathbb{S}^1$ there exists a unique continuous solution $\Phi : \mathbb{S}^1 \to \mathbb{S}^1$ of system (1) such that $\Phi(1) = a$. This solution is of degree l. If, moreover, $L_{F_{t_0}} = \mathbb{S}^1$ and $|l| = 1$, then Φ is a homeomorphism.

If $L_{G_{t_0}} \neq \mathbb{S}^1$ and $L_{F_{t_0}} = \mathbb{S}^1$, then system (1) has no continuous solution.

Proof. Setting $F := F_{t_0}$ and $G := G_{t_0}$ we see at once that $\rho(F) \notin \mathbb{Q}$. First, assume that $L_G = \mathbb{S}^1$ and define

$$s_t := e^{2\pi i \rho(F_t)} \qquad (t \in M).$$

Since $e^{2\pi i \rho(G_t)} = e^{2\pi i l \rho(F_t)}$ for $t \in M$, Lemma 3 yields

$$\Phi_F(F_t(z)) = s_t \Phi_F(z) \qquad (z \in \mathbb{S}^1, t \in M) \tag{18}$$

and

$$\Phi_G(G_t(z)) = s_t^l \Phi_G(z) \qquad (z \in \mathbb{S}^1, t \in M). \tag{19}$$

Fix an $a \in \mathbb{S}^1$ and let $\Phi : \mathbb{S}^1 \to \mathbb{S}^1$ be a continuous solution of system (1) such that $\Phi(1) = a$. Writing $\Lambda := \Phi_G \circ \Phi$, we deduce from (1) and (19) that $\Lambda : \mathbb{S}^1 \to \mathbb{S}^1$ is a continuous function for which $\Lambda(1) = \Phi_G(a)$ and

$$\Lambda(F_t(z)) = s_t^l \Lambda(z) \qquad (z \in \mathbb{S}^1, t \in M).$$

Lemma 3 now shows that

$$\Lambda(z) = \Phi_G(a)\Phi_F(z)^l \qquad (z \in \mathbb{S}^1)$$

and $l = \deg \Lambda = \deg \Phi$. Since $L_G = \mathbb{S}^1$, it follows from Proposition 1 that Φ_G is a homeomorphism, and therefore

$$\Phi(z) = \Phi_G^{-1}(\Phi_G(a)\Phi_F(z)^l) \qquad (z \in \mathbb{S}^1). \tag{20}$$

We now verify that for each $a \in \mathbb{S}^1$ the function $\Phi : \mathbb{S}^1 \to \mathbb{S}^1$ given by (20) has the desired properties. It is easily seen that Φ is continuous and $\Phi(1) = a$. Moreover, fixing $z \in \mathbb{S}^1$, $t \in M$ and using (18) together with (19) we have

$$\Phi(F_t(z)) = \Phi_G^{-1}(\Phi_G(a)\Phi_F(F_t(z))^l) = \Phi_G^{-1}(\Phi_G(a)s_t^l\Phi_F(z)^l)$$
$$= \Phi_G^{-1}(s_t^l\Phi_G(\Phi(z))) = G_t(\Phi(z)),$$

i.e., Φ is a solution of system (1). When $L_F = \mathbb{S}^1$ and $|l| = 1$ the mapping Φ is a homeomorphism. It follows immediately from the fact that in this case so is Φ_F.

Finally, let $L_F = \mathbb{S}^1$, $L_G \neq \mathbb{S}^1$ and suppose, contrary to our claim, that there exists a continuous solution of (1). Then, by Theorem 2,

$$\mathbb{S}^1 = \Phi[\mathbb{S}^1] = \Phi[L_F] = L_G,$$

which is impossible. $\qquad \square$

The following result will be needed in the proof of our next theorem.

Remark 2. Let $F_t, G_t : \mathbb{S}^1 \to \mathbb{S}^1$ for $t \in M$ be homeomorphisms and suppose that $a, b \in \mathbb{S}^1$, $l \in \mathbb{Z}$. If

$$F_t^a(z) := a^{-1}F_t(az) \qquad (z \in \mathbb{S}^1, t \in M) \tag{21}$$

and

$$G_t^b(z) := b^{-1} G_t(bz) \qquad (z \in \mathbb{S}^1, t \in M), \qquad (22)$$

then system (1) has a continuous solution of degree l if and only if the system

$$\Psi(F_t^a(z)) = G_t^b(\Psi(z)) \qquad (z \in \mathbb{S}^1, t \in M) \qquad (23)$$

has such a solution.

Proof. It is easy to verify that the formula

$$b\Psi(z) = \Phi(az) \qquad (z \in \mathbb{S}^1)$$

establishes the relation between the desired solutions of (1) and (23). $\square$

Theorem 4. *Let $F_t, G_t : \mathbb{S}^1 \to \mathbb{S}^1$ for $t \in M$ be orientation-preserving homeomorphisms satisfying (16) such that condition (17) holds for an $l \in \mathbb{Z}$ and there exists a $t_0 \in M$ with $\rho(G_{t_0}) \notin \mathbb{Q}$. Moreover, suppose that for all $k \in \mathbb{N}$, $t_1, ..., t_k \in M$ and $n_1, ..., n_k \in \mathbb{Z}$ such that $n_1 \rho(F_{t_1}) + ... + n_k \rho(F_{t_k}) \in \mathbb{Z}$ we have*

$$F_{t_1}^{n_1} \circ ... \circ F_{t_k}^{n_k} = \mathrm{id} = G_{t_1}^{n_1} \circ ... \circ G_{t_k}^{n_k}.$$

If $L_{F_{t_0}} \neq \mathbb{S}^1 \neq L_{G_{t_0}}$ and there is a $d \in \mathbb{S}^1$ such that

$$\text{for every } w \in \mathbb{S}^1, dw^l \in K_{G_{t_0}} \text{ implies } w \in K_{F_{t_0}}, \qquad (24)$$

then system (1) has a continuous solution of degree l depending on an arbitrary function.

Proof. For convenience we set $F := F_{t_0}$ and $G := G_{t_0}$. Clearly $\rho(F) \notin \mathbb{Q}$. First, we consider the particular case when

$$1 \in L_F^\circ, \qquad 1 \in L_G^\circ, \qquad \text{and} \qquad d = 1. \qquad (25)$$

Writing

$$s_t := e^{2\pi i \rho(F_t)} \qquad (t \in M)$$

we deduce from Lemma 3 that (18) and (19) hold true. Analyses similar to that in the proofs of Proposition 2(d) and Lemma 6 in [2] show that

$$F_t[I_p] = I_{\Lambda_F^{-1}(s_t \Lambda_F(p))} \qquad (p \in T_F, t \in M) \qquad (26)$$

and

$$G_t[J_p] = J_{\Lambda_G^{-1}(s_t^l \Lambda_G(p))} \qquad (p \in T_G, t \in M). \qquad (27)$$

Define the following relation on T_F:

$p \sim q$ if and only if there exist $k \in \mathbb{N}, t_1, ..., t_k \in M$ and

$$n_1, ..., n_k \in \mathbb{Z} \text{ such that } \Phi_F(p) = \Phi_F(q) s_{t_1}^{n_1} ... s_{t_k}^{n_k}.$$

It is easy to check that "$\sim$" is an equivalence relation. From our assumptions we have

$$K_G \subset K_F^l = \Lambda_F[T_F]^l \tag{28}$$

and therefore the set

$$T_F' := \{p \in T_F : \Lambda_F(p)^l \in K_G\}$$

is non-empty. Our next claim is that

$$\text{if } p \in T_F', \ q \in T_F \text{ and } q \sim p, \text{ then } q \in T_F'. \tag{29}$$

To see this, fix a $t \in M$ and let us observe that analysis similar to that in the proof of Lemma 1 shows that $G_t[L_G] = L_G$. Hence, in view of (19), we have

$$K_G = \Phi_G[\mathbb{S}^1 \setminus L_G] = \Phi_G[G_t[\mathbb{S}^1 \setminus L_G]] = s_t^l K_G.$$

Thus by induction we obtain

$$s_t^{ln} K_G = K_G \qquad (t \in M, n \in \mathbb{Z}).$$

Fix $p \in T_F'$ and $q \in T_F$ for which $p \sim q$ and let $k \in \mathbb{N}$, $t_1, ..., t_k \in M$, $n_1, ..., n_k \in \mathbb{Z}$ be such that $\Phi_F(q) = \Phi_F(p) s_{t_1}^{n_1} ... s_{t_k}^{n_k}$, i.e. $\Lambda_F(q) = \Lambda_F(p) s_{t_1}^{n_1} ... s_{t_k}^{n_k}$. Since $p \in T_F'$, we have $\Lambda_F(p)^l \in K_G$ and, in consequence,

$$\Lambda_F(q)^l = \Lambda_F(p)^l s_{t_1}^{ln_1} ... s_{t_k}^{ln_k} \in K_G s_{t_1}^{ln_1} ... s_{t_k}^{ln_k} = K_G.$$

This clearly forces $q \in T_F'$, and (29) is proved.

Let $E \subset T_F$ be such that for every $q \in T_F$, $\text{card}(E \cap [q]_\sim) = 1$, where $[q]_\sim$ stands for the equivalence class of q with respect to the relation "$\sim$". The only element in $E \cap [q]_\sim$ will be denoted by $A(q)$. From (29) it follows that $[q]_\sim \subset T_F'$ for $q \in T_F'$, and therefore the set

$$E' := E \cap T_F'$$

is non-empty. Moreover,

$$A(q) \in E' \qquad (q \in T_F'). \tag{30}$$

Define

$$\Omega(p) := \Lambda_G^{-1}(\Lambda_F(p)^l) \qquad (p \in T_F') \tag{31}$$

and note that Ω maps T_F' onto T_G. Indeed, fixing a $q \in T_G$ we conclude from (28) that $K_G \ni \Lambda_G(q) = \Lambda_F(p)^l$ for a $p \in T_F$. Hence $p \in T_F'$ and therefore $q = \Omega(p)$.

For $l > 0$ (respectively, $l < 0$) let

$$\Gamma_p : I_p \to J_{\Omega(p)} \qquad (p \in E')$$

be increasing (respectively, decreasing) homeomorphisms, i.e. we assume that their lifts are increasing (respectively, decreasing) homeomorphisms.

We will show that the formula

$$\Phi_0(z) := (G_{t_1}^{n_1} \circ \ldots \circ G_{t_k}^{n_k} \circ \Gamma_{A(p)} \circ F_{t_1}^{-n_1} \circ \ldots \circ F_{t_k}^{-n_k})(z)$$
$$(z \in I_p, p \in T_F'), \tag{32}$$

where $k \in \mathbb{N}$, $t_1, \ldots, t_k \in M$ and $n_1, \ldots, n_k \in \mathbb{Z}$ are such that

$$\Phi_F(p) = \Phi_F(A(p))s_{t_1}^{n_1} \ldots s_{t_k}^{n_k}, \tag{33}$$

defines a function. By (33) and (30),

$$\Lambda_F^{-1}(s_{t_1}^{-n_1} \ldots s_{t_k}^{-n_k}\Lambda_F(p)) = A(p) \in E' \tag{34}$$

and, in view of (26),

$$(F_{t_1}^{-n_1} \circ \ldots \circ F_{t_k}^{-n_k})[I_p] = I_{A(p)}. \tag{35}$$

Fix a $p \in T_F'$ and choose $k \in \mathbb{N}$, $t_1, \ldots, t_k \in M$, $n_1, \ldots, n_k \in \mathbb{Z}$ and $r \in \mathbb{N}$, $t_1', \ldots, t_r' \in M$, $n_1', \ldots, n_r' \in \mathbb{Z}$ for which

$$s_{t_1}^{n_1} \ldots s_{t_k}^{n_k} = s_{t_1'}^{n_1'} \ldots s_{t_r'}^{n_r'}.$$

Then, by the assumptions of our theorem,

$$F_{t_1}^{n_1} \circ \ldots \circ F_{t_k}^{n_k} = F_{t_1'}^{n_1'} \circ \ldots \circ F_{t_r'}^{n_r'}$$

and

$$G_{t_1}^{n_1} \circ \ldots \circ G_{t_k}^{n_k} = G_{t_1'}^{n_1'} \circ \ldots \circ G_{t_r'}^{n_r'},$$

and consequently

$$G_{t_1}^{n_1} \circ \ldots \circ G_{t_k}^{n_k} \circ \Gamma_{A(p)} \circ F_{t_1}^{-n_1} \circ \ldots \circ F_{t_k}^{-n_k}$$
$$= G_{t_1'}^{n_1'} \circ \ldots \circ G_{t_r'}^{n_r'} \circ \Gamma_{A(p)} \circ F_{t_1'}^{-n_1'} \circ \ldots \circ F_{t_r'}^{-n_r'}.$$

Thus formula (32) correctly defines a function mapping $\bigcup_{p \in T'_F} I_p$ into $\mathbb{S}^1$. Fix a $p \in T'_F$ and let $k \in \mathbb{N}$, $t_1, \ldots, t_k \in M$ and $n_1, \ldots, n_k \in \mathbb{Z}$ be such that (33) holds. Then (32), (35), (31), (34) and (27) give

$$
\begin{aligned}
\Phi_0[I_p] &= (G_{t_1}^{n_1} \circ \ldots \circ G_{t_k}^{n_k} \circ \Gamma_{A(p)} \circ F_{t_1}^{-n_1} \circ \ldots \circ F_{t_k}^{-n_k})[I_p] \\
&= (G_{t_1}^{n_1} \circ \ldots \circ G_{t_k}^{n_k} \circ \Gamma_{A(p)})[I_{A(p)}] = (G_{t_1}^{n_1} \circ \ldots \circ G_{t_k}^{n_k})[J_{\Omega(A(p))}] \\
&= (G_{t_1}^{n_1} \circ \ldots \circ G_{t_k}^{n_k})[J_{\Lambda_G^{-1}(s_{t_1}^{-ln_1} \ldots s_{t_k}^{-ln_k} \Lambda_F(p)^l)}] \\
&= J_{\Lambda_G^{-1}(\Lambda_F(p)^l)} = J_{\Omega(p)}.
\end{aligned}
$$

Since Ω maps T'_F onto T_G, this shows that $\Phi_0[\bigcup_{p \in T'_F} I_p] = \mathbb{S}^1 \setminus L_G$.

Fix $t \in M$, $p \in T'_F$, $z \in I_p$ and let $k \in \mathbb{N}$, $t_1, \ldots, t_k \in M$ and $n_1, \ldots, n_k \in \mathbb{Z}$ be such that (33) holds true. As $p \sim \Lambda_F^{-1}(s_t \Lambda_F(p))$,

$$
A(p) = A(\Lambda_F^{-1}(s_t \Lambda_F(p))) \tag{36}
$$

and, moreover, from (29) it follows that $\Lambda_F^{-1}(s_t \Lambda_F(p)) \in T'_F$. By (26), $F_t(z) \in I_{\Lambda_F^{-1}(s_t \Lambda_F(p))}$ and from (33) we see that

$$
\Phi_F(\Lambda_F^{-1}(s_t \Lambda_F(p))) = s_t \Phi_F(p) = \Phi_F(A(p)) s_t s_{t_1}^{n_1} \ldots s_{t_k}^{n_k}.
$$

(32) and (36) now give

$$
\begin{aligned}
&\Phi_0(F_t(z)) \\
&= (G_t \circ G_{t_1}^{n_1} \circ \ldots \circ G_{t_k}^{n_k} \circ \Gamma_{A(p)} \circ F_t^{-1} \circ F_{t_1}^{-n_1} \circ \ldots \circ F_{t_k}^{-n_k} \circ F_t)(z) \\
&= G_t(\Phi_0(z)).
\end{aligned}
$$

Summarizing, we have

$$
G_t(\Phi_0(z)) = \Phi_0(F_t(z)) \qquad (z \in \bigcup_{p \in T'_F} I_p, t \in M). \tag{37}
$$

Our next goal is to construct a continuous extension $\Phi : \mathbb{S}^1 \to \mathbb{S}^1$ of Φ_0 such that $\deg \Phi = l$ and (1) holds true. In order to do this, it is convenient to assume, for instance, that $l > 0$ and introduce the following notation:

$$
\hat{T}_F := \Pi^{-1}[T_F], \qquad \hat{T}_G := \Pi^{-1}[T_G],
$$

$$
\hat{T}_G^\star := \hat{T}_G \cup (\hat{T}_G + 1) \cup \ldots \cup (\hat{T}_G + l - 1),
$$

$$
\hat{K}_F := \{t \in \mathbb{R} : e^{2\pi it} \in K_F\}, \qquad \hat{K}_G := \{t \in \mathbb{R} : e^{2\pi it} \in K_G\}
$$

and

$$\hat{\check{K}}_G := \hat{K}_G \cap [0,1].$$

These sets are clearly all non-empty. Moreover, from (25) it follows immediately that

$$\hat{T}_F, \hat{T}_G \subset (0,1), \qquad \hat{T}_G^\star \subset (0,l)$$

and, in view of Lemma 2, $1 \notin K_F$, $1 \notin K_G$, since $\Phi_F(1) = \Phi_G(1) = 1$. Hence

$$\hat{K}_F \cap \mathbb{Z} = \emptyset = \hat{K}_G \cap \mathbb{Z} \qquad \text{and} \qquad \hat{\check{K}}_G = \hat{K}_G \cap (0,1).$$

Let us next observe that (28) leads to $\hat{K}_G \subset \bigcup_{j=0}^{l-1}(l\hat{K}_F + j)$. Denoting by φ_F (respectively, φ_G) the lift of Φ_F (respectively, Φ_G) such that $\varphi_F(0) = 0$ (respectively, $\varphi_G(0) = 0$) we also set

$$\hat{\varphi}_F := \varphi_F \mid_{\hat{T}_F}, \qquad \hat{\varphi}_G := \varphi_G \mid_{\hat{T}_G} \qquad \text{and} \qquad \hat{\varphi}_G^\star := \varphi_G \mid_{\hat{T}_G^\star}.$$

We leave it to the reader to verify that the functions $\hat{\varphi}_F : \hat{T}_F \to \hat{K}_F \cap (0,1)$, $\hat{\varphi}_G : \hat{T}_G \to \hat{K}_G \cap (0,1)$ and $\hat{\varphi}_G^\star : \hat{T}_G^\star \to \hat{K}_G \cap (0,l)$ are strictly increasing bijections and

$$\hat{\varphi}_G^\star(t+1) = \hat{\varphi}_G^\star(t) + 1 \qquad (t, t+1 \in \hat{T}_G^\star).$$

Putting

$$\hat{T}_F' := \{t \in \hat{T}_F : l\hat{\varphi}_F(t) \in \bigcup_{j=0}^{l-1} \hat{\check{K}}_G + j\}$$

one can also check that $\Pi[\hat{T}_F'] = T_F'$. We are now in a position to show that the function δ given by

$$\delta(t) := \hat{\varphi}_G^{\star^{-1}}(l\hat{\varphi}_F(t)) \qquad (t \in \hat{T}_F')$$

maps $\hat{T}_F'$ onto $\hat{T}_G^\star$. To do this fix an $x \in \hat{T}_G^\star$ and let $k \in \{0, 1, ..., l-1\}$ and $s \in \hat{T}_G$ be such that $x = s + k$. Obviously,

$$\hat{\varphi}_G^\star(x) = \hat{\varphi}_G^\star(s+k) = \hat{\varphi}_G^\star(s) + k = \hat{\varphi}_G(s) + k$$

and $\hat{\varphi}_G(s) \in \hat{\check{K}}_G$. This clearly forces $e^{2\pi i \hat{\varphi}_G(s)} \in K_G$ and therefore (24) and (25) yield $e^{2\pi i \frac{1}{l}(\hat{\varphi}_G(s)+k)} \in K_F$. Consequently, $y := \frac{1}{l}(\hat{\varphi}_G(s) + k) \in \hat{K}_F$ and

$$\hat{\varphi}_G^\star(x) = ly. \tag{38}$$

As $\hat{\varphi}_G^\star(x) \in \hat{K}_G \cap (0, l)$, we also see that $y \in (0, 1)$. Thus $y \in \hat{K}_F \cap (0, 1)$ and therefore, since the function $\hat{\varphi}_F : \hat{T}_F \to \hat{K}_F \cap (0, 1)$ is a bijection, $y = \hat{\varphi}_F(t)$ for a $t \in \hat{T}_F$. (38) now becomes

$$\hat{\varphi}_G^\star(x) = l\hat{\varphi}_F(t). \tag{39}$$

Next, let us observe that

$$l\hat{\varphi}_F(t) = \hat{\varphi}_G^\star(x) \in \hat{K}_G \cap (0, l) \subset \bigcup_{j=0}^{l-1} \hat{K}_G + j.$$

This gives $t \in \hat{T}_F'$, which together with (39) leads to $x = \delta(t)$. We have thus proved that δ is a strictly increasing bijection mapping $\hat{T}_F'$ onto $\hat{T}_G^\star$.

Let us next observe that

$$\Omega(e^{2\pi it}) = e^{2\pi i\delta(t)} \qquad (t \in \hat{T}_F').$$

Indeed, fixing a $t \in \hat{T}_F'$ we see that $e^{2\pi i\delta(t)} \in T_G$ and

$$\Lambda_G(e^{2\pi i\delta(t)}) = \Phi_G(e^{2\pi i\delta(t)}) = e^{2\pi i\varphi_G(\delta(t))} = e^{2\pi il\hat{\varphi}_F(t)}$$
$$= (e^{2\pi i\varphi_F(t)})^l = \Phi_F(e^{2\pi it})^l = \Lambda_F(e^{2\pi it})^l.$$

Therefore

$$e^{2\pi i\delta(t)} = \Lambda_G^{-1}(\Lambda_F(e^{2\pi it})^l) = \Omega(e^{2\pi it}).$$

Now set

$$\Pi_\star(t) := e^{2\pi it} \qquad (t \in [0, l)),$$

$$\hat{I}_t := \Pi^{-1}[I_{\Pi(t)}] \qquad (t \in \hat{T}_F)$$

and

$$\hat{J}_t := \Pi^{-1}[J_{\Pi(t)}] \qquad (t \in \hat{T}_G)$$

and let for every $t \in \hat{T}_G^\star$, $\check{J}_t$ be an open interval with the middle point t such that $\check{J}_t \subset [0, l]$ and $\Pi_\star[\check{J}_t] = J_{\Pi_\star(t)}$. We see at once that the sets $\hat{I}_s$ for $s \in \hat{T}_F$ (respectively, $\hat{J}_t$ for $t \in \hat{T}_G$) are pairwise disjoint open intervals and

$$\check{J}_t = \hat{J}_t \qquad (t \in \hat{T}_G). \tag{40}$$

For every $t \in \hat{T}'_F$ denote by $\gamma_t : \hat{I}_t \to \check{J}_{\delta(t)}$ the lift of the homeomorphism $\Phi_0 |_{I_{\Pi(t)}}$ and set

$$\tilde{\gamma}(x) = \gamma_t(x) \qquad (x \in \hat{I}_t, t \in \hat{T}'_F).$$

Since the functions $\gamma_t : \hat{I}_t \to \check{J}_{\delta(t)}$ for $t \in \hat{T}'_F$ are all homeomorphisms and $\delta[\hat{T}'_F] = \hat{T}^\star_G$, we see that $\tilde{\gamma}$ maps the set

$$U := \bigcup_{t \in \hat{T}'_F} \hat{I}_t$$

onto $\bigcup_{t \in \hat{T}^\star_G} \check{J}_t$.

Now fix a $t \in \hat{T}^\star_G$ and let $s \in \hat{T}_G$ and $k \in \{0, 1, ..., l-1\}$ be such that $t = s + k$. Then $\check{J}_{s+k} = \check{J}_s + k$ and (40) shows that $\check{J}_t = \check{J}_s + k$. Therefore

$$\bigcup_{t \in \hat{T}^\star_G} \check{J}_t = \bigcup_{t \in \hat{T}_G} \check{J}_t \cup \bigcup_{t \in \hat{T}_G+1} \check{J}_t \cup ... \cup \bigcup_{t \in \hat{T}_G+l-1} \check{J}_t$$

$$= \left(\bigcup_{t \in \hat{T}_G} \hat{J}_t \right) \cup \left(\bigcup_{t \in \hat{T}_G} \hat{J}_t + 1 \right) \cup ... \cup \left(\bigcup_{t \in \hat{T}_G} \hat{J}_t + l - 1 \right)$$

$$= \Pi^{-1}[\mathbb{S}^1 \setminus L_G] \cup (\Pi^{-1}[\mathbb{S}^1 \setminus L_G] + 1) \cup ...$$

$$... \cup (\Pi^{-1}[\mathbb{S}^1 \setminus L_G] + l - 1).$$

Hence we infer that the set $\bigcup_{t \in \hat{T}^\star_G} \check{J}_t$ is dense in $[0, l]$. It is easily seen that the function $\tilde{\gamma} : U \to \bigcup_{t \in \hat{T}^\star_G} \check{J}_t$ is continuous. We shall show that it is also strictly increasing. To do this, let us first observe that for every $t \in \hat{T}'_F$, $\tilde{\gamma} |_{\hat{I}_t}$ is strictly increasing. Next, fix $t, s \in \hat{T}'_F$ and assume that $\hat{I}_t < \hat{I}_s$, i.e. $x < y$ for $x \in \hat{I}_t$, $y \in \hat{I}_s$. Since each $r \in \hat{T}'_F$ is the middle point of an interval $\hat{I}_r$, we see that $t < s$ and consequently $\delta(t) < \delta(s)$. This clearly forces

$$\tilde{\gamma}[\hat{I}_t] = \check{J}_{\delta(t)} < \check{J}_{\delta(s)} = \tilde{\gamma}[\hat{I}_s],$$

and our assertion follows.

Writing

$$D_t := \{s \in U : s < t\} \qquad (t \in [0, 1])$$

we define

$$\gamma(t) := \begin{cases} 0, & D_t = \emptyset, \\ \sup_{s \in D_t} \tilde{\gamma}(s), & D_t \neq \emptyset, \end{cases} \qquad t \in [0, 1].$$

It is easy to check that the function $\gamma : [0, 1] \to [0, l]$ is an increasing extension of $\tilde{\gamma}$ for which $\gamma(0) = 0$ and $\gamma(1) = l$. Moreover,

$$\bigcup_{t \in \hat{T}_G^*} \check{J}_t = \tilde{\gamma}[U] = \gamma[U] \subset \gamma[[0, 1]],$$

and the density of $\bigcup_{t \in \hat{T}_G^*} \check{J}_t$ in $[0, l]$ implies that of $\gamma[[0, 1]]$. Therefore γ is continuous, and, in consequence, the function $\Phi : \mathbb{S}^1 \to \mathbb{S}^1$ given by

$$\Phi(e^{2\pi i x}) := e^{2\pi i \gamma(x)} \qquad (x \in [0, 1])$$

is a continuous extension of Φ_0.

To see (1), let us first observe that

$$\Lambda_F^{-1}(s_t \Lambda_F(p)) \in T_F' \qquad (p \in T_F', t \in M), \tag{41}$$

which follows immediately from (29). Setting

$$D_1 := \bigcup_{p \in T_F'} I_p, \qquad M_F := T_F \setminus T_F', \qquad \text{and} \qquad D_2 := \bigcup_{p \in M_F} I_p$$

and

$$H_t(z) := \frac{\Phi(F_t(z))}{G_t(\Phi(z))} \qquad (z \in \mathbb{S}^1, t \in M)$$

we conclude from (37) that $H_t(z) = 1$ for $z \in D_1$ and $t \in M$. Fix $p \in M_F$, $t \in M$ and note that γ is constant on $\hat{I}_{\Pi^{-1}(p)}$. Therefore Φ is constant on I_p, which together with (26) and the observation that $p \in M_F$ implies $\Lambda_F^{-1}(s_t \Lambda_F(p)) \in M_F$ gives the constancy of Φ on $F_t[I_p]$. Consequently, there is a $c_p \in \mathbb{S}^1$ such that $H_t(z) = c_p$ for $z \in I_p$. In particular, denoting by a one of the ends of the arc I_p, we have $a \in L_F$ and $H_t(a) = c_p$. Fix a $z_0 \in D_1$ and let $(n_k)_{k \in \mathbb{N}}$ be a sequence of integers for which $\lim_{k \to \infty} F^{n_k}(z_0) = a$. Using (26), (41) and induction one can check that $F^{n_k}(z_0) \in D_1$ for $k \in \mathbb{N}$ and the continuity of H_t now yields

$$1 = \lim_{k \to \infty} H_t(F^{n_k}(z_0)) = H_t(a) = c_p.$$

Therefore $H_t(z) = 1$ for $z \in D_2$, and consequently

$$H_t(z) = 1 \qquad (z \in D_1 \cup D_2 = \mathbb{S}^1 \setminus L_F).$$

Thus, by the density of $\mathbb{S}^1 \setminus L_F$ in $\mathbb{S}^1$ and the continuity of H_t, $H_t(z) = 1$ for $z \in \mathbb{S}^1$, and (1) is proved.

To complete the proof in case (25), it suffices to use Theorem 1 together with the assumption $\rho(G) = l\rho(F)\,(\mathrm{mod}\,1)$.

We now turn to the general case. Let us first show that there are $a \in L_F^\circ$ and $b \in L_G^\circ$ for which

$$\Phi_G(b) = d\Phi_F(a)^l. \tag{42}$$

Suppose, contrary to our claim, that

$$\Phi_G[L_G^\circ] \cap d\Phi_F[L_F^\circ]^l = \emptyset.$$

Then, in view of Lemma 2,

$$d(\mathbb{S}^1 \setminus K_F)^l = d\Phi_F[L_F^\circ]^l \subset \mathbb{S}^1 \setminus \Phi_G[L_G^\circ] = K_G$$

and therefore

$$\mathrm{card}\,d(\mathbb{S}^1 \setminus K_F)^l \leq \mathrm{card}\,K_G.$$

But from Proposition 2 it follows that $\mathrm{card}\,K_F = \mathrm{card}\,K_G = \aleph_0$, a contradiction.

Given $a \in L_F^\circ$, $b \in L_G^\circ$ satisfying (42) we define F_t^a and G_t^b by (21) and (22), respectively. It is easy to check that these functions fulfil the assumptions of our theorem. Putting

$$F_a(z) := F_{t_0}^a(z) = a^{-1}F(az) \qquad (z \in \mathbb{S}^1)$$

and

$$G_b(z) := G_{t_0}^b(z) = b^{-1}G(bz) \qquad (z \in \mathbb{S}^1)$$

we deduce from Lemma 5 that

$$1 \in a^{-1}L_F^\circ = L_{F_a}^\circ \qquad \text{and} \qquad 1 \in b^{-1}L_G^\circ = L_{G_b}^\circ.$$

We next prove that

for every $w \in \mathbb{S}^1, w^l \in K_{G_b}$ implies $w \in K_{F_a}$.

To do this, fix a $w \in \mathbb{S}^1$ for which $w^l \in K_{G_b}$ and observe that from Lemma 5 we obtain $w^l \in \frac{1}{\Phi_G(b)}K_G$. Consequently, by (42),

$$d(w\Phi_F(a))^l = w^l d\Phi_F(a)^l = w^l \Phi_G(b) \in K_G,$$

and (24) together with Lemma 5 make it obvious that

$$w\Phi_F(a) \in K_F = K_{F_a}\Phi_F(a).$$

Thus $w \in K_{F_a}$, which is our claim.

From what has already been proved, it follows that system (23) has a continuous solution of degree l and to complete the proof of Theorem 4 it is enough to apply Remark 2.   □

Remark 3. Under the hypotheses of Theorem 4 if, moreover, $|l| = 1$ and $K_{G_{t_0}} = dK_{F_{t_0}}^l$ for a $d \in \mathbb{S}^1$, then system (1) has a homeomorphic solution of degree l depending on an arbitrary function and the construction given in the proof of Theorem 4 determines all such solutions of (1).

Proof. We will follow the notation used in the proof of Theorem 4. Let us first observe that the equality $K_G = K_F^l$ yields $T_F' = T_F$.

Assume that $\Phi : \mathbb{S}^1 \to \mathbb{S}^1$ is a homeomorphic solution of (1) with $\deg \Phi = 1$ (respectively, $\deg \Phi = -1$) and put

$$\Gamma_p := \Phi \mid_{I_p}, \qquad p \in E' = E \qquad \text{and} \qquad \Phi_0 := \Phi \mid_{\mathbb{S}^1 \setminus L_F}.$$

Then, by Lemma 4, Γ_p for $p \in E'$ are increasing (respectively, decreasing) homeomorphisms mapping I_p onto $J_{\Omega(p)}$. Moreover, from (1) we deduce that

$$G_{t_1}^{m_1} \circ \dots \circ G_{t_k}^{n_k} \circ \Phi = \Phi \circ F_{t_1}^{m_1} \circ \dots \circ F_{t_k}^{n_k},$$

and therefore formula (32) holds true. The fact that the set $\mathbb{S}^1 \setminus L_F$ is dense in $\mathbb{S}^1$ now shows that the extension of Φ_0 constructed in the proof of Theorem 4 equals Φ.

On the other hand, a careful verification of the proof of Theorem 4 allows the reader to check that since $|l| = 1$ and $T_F = T_F'$, the solution Φ constructed in the proof of Theorem 4 is a homeomorphism.   □

Example. Let $L \subset \mathbb{S}^1$ be a nowhere dense perfect set with $1 \in L^\circ$,

$$\rho \in (0, \frac{1}{2}) \setminus \mathbb{Q} \qquad \text{and} \qquad s := e^{2\pi i \rho}.$$

In [10] it has been shown that for every countable set $K \subset \mathbb{S}^1$ such that $s \cdot K = K$ and $1 \notin K$ there exists a homeomorphism $F : \mathbb{S}^1 \to \mathbb{S}^1$ for which

$$\rho(F) = \rho, \qquad L_F = L, \qquad \text{and} \qquad K_F = K.$$

Putting

$$K := \{-s^{2n}, \ n \in \mathbb{Z}\}, \qquad K_1 := \{i, -i\} \cdot \{s^n, \ n \in \mathbb{Z}\}$$

and

$$K_2 := \{i, -i, \frac{1}{\sqrt{2}} + \frac{i}{\sqrt{2}}, -\frac{1}{\sqrt{2}} - \frac{i}{\sqrt{2}}\} \cdot \{s^n, \ n \in \mathbb{Z}\}$$

we see at once that

$$s \cdot K_1 = K_1, \qquad s \cdot K_2 = K_2 \qquad \text{and} \qquad s^2 \cdot K = K.$$

Let F_1, F_2, $G : \mathbb{S}^1 \to \mathbb{S}^1$ be homeomorphisms such that

$$L_{F_1} = L_{F_2} = L_G = L, \qquad \rho(F_1) = \rho(F_2) = \rho, \qquad \text{and} \qquad \rho(G) = 2\rho$$

and

$$K_{F_1} = K_1, \qquad K_{F_2} = K_2, \qquad \text{and} \qquad K_G = K.$$

It is easy to check that F_1, G and F_2, G satisfy the assumptions of Theorem 4 with $l = 2$, $d = 1$. However, since $K = K_1^2$ and $K \neq K_2^2$, we have $K_G = K_{F_1}^l$ whence $T_{F_1} = T'_{F_1}$, and $K_G \neq K_{F_2}^l$ whence $T_{F_2} \neq T'_{F_2}$.

References

[1] L. Alsedà, J. Llibre and M. Misiurewicz, *Combinatorial dynamics and entropy in dimension one*, Advanced Series in Nonlinear Dynamics, vol. 5, World Scientific Publishing Co., Inc., River Edge, NJ, 1993.

[2] M. Bajger, *On the structure of some flows on the unit circle*, Aequationes Math. **55** (1998), 106–121.

[3] L. S. Block and W. A. Coppel, *Dynamics in one dimension*, Lecture Notes in Mathematics, vol. 1513, Springer–Verlag, Berlin, 1992.

[4] K. Ciepliński, *On the embeddability of a homeomorphism of the unit circle in disjoint iteration groups*, Publ. Math. Debrecen **55** (1999), 363–383.

[5] K. Ciepliński, *On conjugacy of disjoint iteration groups on the unit circle*, European Conference on Iteration Theory (Muszyna-Złockie, 1998), Ann. Math. Sil. **13** (1999), 103–118.

[6] I. P. Cornfeld, S. V. Fomin and Ya. G. Sinai, *Ergodic theory*, Grundlehren der Mathematischen Wissenschaften, vol. 245, Springer–Verlag, New York–Berlin, 1982.

[7] W. de Melo and S. van Strein, *One-dimensional dynamics*, Ergebnisse der Mathematik und ihrer Grenzgebiete (3), vol. 25, Springer–Verlag, Berlin, 1993.

[8] Z. Nitecki, *Differentiable dynamics. An introduction to the orbit structure of diffeomorphisms*, The M.I.T. Press, Cambridge, Mass.–London, 1971.

[9] C. T. C. Wall, *A geometric introduction to topology*, Addison-Wesley Publishing Co., Reading, Mass.–London–Don Mills, Ont., 1972.

[10] M. C. Zdun, *On iterative roots of homeomorphisms of the circle*, Bull. Pol. Acad. Sci. Math. **48** (2000), 203–213.

A SURVEY OF RESULTS AND OPEN PROBLEMS ON THE SCHILLING EQUATION

Roland Girgensohn

GSF-Forschungszentrum, Institut für Biomathematik und Biometrie,

Postfach 1129, D-85758 Neuherberg, Germany

girgen@gsf.de

Abstract   The Schilling equation is the functional equation

$$4q\,f(qx) \;=\; f(x+1) + 2f(x) + f(x-1) \qquad (x \in \mathbb{R})$$

with a parameter $q \in (0,1)$. It has its origin in Physics, and although it has been studied intensively in recent years, there are still many open questions connected with it. Some questions, however, have been answered recently. In this paper, we will survey known results about Schilling's equation and point out problems which are still open.

Keywords: Schilling equation, Bernoulli convolution, Fourier transform, Pisot number, Salem number

Mathematics Subject Classification (2000): 39B22

1.   WHERE DOES IT COME FROM?

The motivation for Rolf Schilling to introduce the functional equation named after him was his study of spatially chaotic structures in amorphous (glassy) materials, see [23] or [8]. Consider a system of countably infinitely many particles on the real line (with positions $r_1, r_2, \ldots$) which interact with each other, so that a potential $V(r_1, r_2, \ldots)$ is associated with the system. One is interested in certain physical quantities E (such as energy) which depend on the positions of the particles, $E = E(r_1, r_2, \ldots)$. Of special interest here are stationary configurations: those where the partial derivatives of the potential V with respect to the coordinates $r_1, r_2, \ldots$ vanish. Schilling showed that for certain classes of

159

Z. Daróczy and Z. Páles (eds.),

Functional Equations - Results and Advances, 159–174.

© 2002 *Kluwer Academic Publishers.*

potentials V the values of E associated with those stationary configurations can be described by the random series

$$S_q = \sum_{n=0}^{\infty} B_n\, q^n,$$

where $0 < q < 1$ is a parameter and the B_n are independent random variables with

$$P(B_n = -1) = \frac{1}{4}, \qquad P(B_n = 0) = \frac{1}{2}, \qquad \text{and} \qquad P(B_n = 1) = \frac{1}{4}.$$

This random series can be visualized as shown in Figure 1. There, the base-4 digits of a $t \in [0,1]$ correspond to the random variables B_n: If the nth digit is 0, then $B_n = -1$, if the nth digit is 1 or 2, then $B_n = 0$; if the nth digit is 3, then $B_n = 1$. Thus for almost every $t \in [0,1]$, the B_n's have the desired distribution, and we can compute and plot $S_q = S_q(t)$. It is an integrable function, discontinuous precisely at the base-4 rationals.

Figure 1   Visualization of the random series

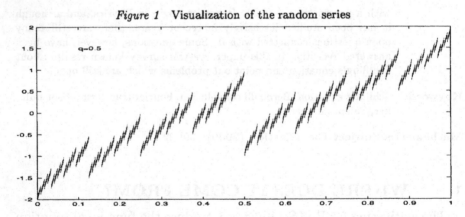

Now we are interested in the distribution of those physical quantities, and thus in the distribution of the random series. Therefore we define the distribution function F_q of S_q by

$$F_q(x) := m\{t \in [0,1] : S_q(t) \le x\}.$$

Then it is easy to see that (after rescaling) the distribution function F_q satisfies the functional equation

$$F(qx) = \frac{1}{4}\big(F(x+1) + 2F(x) + F(x-1)\big) \qquad (x \in \mathbb{R}),$$

and is the only bounded solution satisfying $F(x) = 0$ for $x < -q/(1-q)$ and $F(x) = 1$ for $x > q/(1-q)$. Moreover, it is also known and quite standard that the distribution function F_q is either absolutely continuous (having a derivative in $L^1(\mathbb{R})$ and being reconstructable from it) or strictly singular (the derivative is 0 a.e.). This follows from Kolmogorov's 0-1-law, see Theorem 35 in [13]. To decide which of the two possibilities applies for a given q, one can look at the functional equation for the density f_q of F_q, namely

$$4q\,f(qx) \;=\; f(x+1) + 2f(x) + f(x-1) \qquad (x \in \mathbb{R}). \qquad (1)$$

Here we are interested in L^1-solutions, i.e., functions $f \in L^1(\mathbb{R})$ satisfying the functional equation a.e. on $\mathbb{R}$. If (1) has a non-trivial L^1-solution f_q, then F_q is absolutely continuous with density f_q; if the only L^1-solution of (1) is the zero function, then F_q is singular. (To see this, one needs the statements about uniqueness of solutions of (1) given in the next section.) Now, the functional equation (1) is precisely the Schilling equation with parameter q. Thus we are interested in this equation and its L^1-solutions because we are interested in the distribution of the physical quantities $E(r_1, r_2, \dots)$ of Schilling's model.

In the case $q = \frac{1}{2}$, a solution for (1) can be given explicitly: $f_{0.5}(x) = \max\{1 - |x|, 0\}$. Thus this simple function is the density of the distribution function for the complicated random series shown in the picture above.

2.   CLASSIFICATION

The Schilling equation is a special case of a much more general class of equations, namely two-scale difference equations. Those are functional equations of the type

$$f(x) = \sum_{n=0}^{N} c_n\, f(\alpha x - \beta_n) \qquad (x \in \mathbb{R}) \qquad (2)$$

with $c_n \in \mathbb{C}$, $\beta_n \in \mathbb{R}$ and $\alpha > 1$. They were first discussed by Ingrid Daubechies and Jeffrey C. Lagarias in [7], who proved existence and uniqueness theorems and derived some properties of L^1-solutions. Their existence/uniqueness theorem reads as follows.

Theorem 1. *Denote* $\Delta := \alpha^{-1} \sum_{n=0}^{N} c_n$.

(a) *If* $|\Delta| \leq 1$ *and* $\Delta \neq 1$, *then Equation (2) has no non-trivial L^1-solution.*

(b) If $|\Delta| > 1$, then the vector space of L^1-solutions can be zero-, one- or infinite-dimensional.

(c) If $\Delta = 1$, then the vector space of L^1-solutions is at most one-dimensional.

The interesting case here is the case $\Delta = 1$, where, depending on the parameters, either a non-trivial L^1-solution of (2) exists which is then unique up to scaling, or the only L^1-solution is the zero function. The Schilling equation is an example of this case, because here we have $\alpha = \frac{1}{q}$, $c_0 = c_2 = \frac{1}{4q}$ and $c_1 = \frac{1}{2q}$ (and the betas are $-1, 0, 1$).

Daubechies and Lagarias moreover derived some properties of the L^1-solution of (2) in the case $\Delta = 1$. We state those results in the special case of the Schilling equation.

Theorem 2. *If an L^1-solution f_q of (1) exists, then it has the following properties.*

(a) *Its Fourier transform, defined as*

$$\widehat{f_q}(t) := \int_{-\infty}^{\infty} f_q(x)\, e^{ixt}\, dx,$$

is given by the everywhere convergent infinite product

$$\widehat{f_q}(t) = \widehat{f_q}(0) \prod_{j=1}^{\infty} \cos^2\left(q^j \frac{t}{2}\right). \tag{3}$$

(b) *If $f_q \not\equiv 0$, then $\widehat{f_q}(0) \neq 0$.*

(c) *The support of f_q satisfies* $\operatorname{supp} f_q \subseteq \left[-\dfrac{q}{1-q}, \dfrac{q}{1-q}\right] =: I_q.$

The first of these statements has a converse: If the product (3) is the Fourier transform of an L^1-function, then that function is a solution of the Schilling equation. Because of the second statement, we can always assume that a non-trivial L^1-solution of the Schilling equation is normalized by $\widehat{f_q}(0) = 1$, so that we in this sense then have uniqueness. Therefore we will from now on by $(S)_q$ denote the following problem: Find a function f_q satisfying the conditions

$$4qf(qx) = f(x+1) + 2f(x) + f(x-1) \qquad \text{for a.e. } x \in \mathbb{R}, \left.\begin{array}{r}\\ \\ \end{array}\right\} (S)_q$$

$$f \in L^1(\mathbb{R}), \qquad \int_{-\infty}^{\infty} f(x)\, dx = 1.$$

Because of the third statement, we can look for solutions in $L^1(I_q)$ instead of $L^1(\mathbb{R})$; this simplifies matters.

Theorems 1 and 2 were also proved in the special setting of the Schilling equation in [3], [2] and [10].

Now that we know that there exists at most one non-trivial L^1-solution of the Schilling equation up to scaling, the main question to be answered is: For which values of the parameter q does there exist such a solution?

3.   THE MAIN ANSWERS

The question about L^1-solutions of the Schilling equation has been open for several years; it is due to the seminal work of Boris Solomyak (Cited below) that we can now give a fairly complete answer. It turns out that there are two critical values: $\frac{1}{2}$ and $\frac{1}{2\sqrt{2}}$.

Theorem 3.

(a) For no $q < \frac{1}{2}$ does there exist a bounded L^1-solution of $(S)_q$.

(b) For almost every $q > \frac{1}{2}$ a continuous solution of $(S)_q$ exists.

(c) For no $q < \frac{1}{2\sqrt{2}}$ does there exist an L^1-solution of $(S)_q$.

(d) For almost every $q > \frac{1}{2\sqrt{2}}$ an L^1-solution of $(S)_q$ exists.

The words "almost everywhere" can not be omitted from the above statements; there are a few explicitly known values for q between $\frac{1}{2}$ and 1 where no non-trivial L^1-solution exists, as we will see below. Part (a) of Theorem 3 was proved by Karol Baron, Alice Simon and Peter Volkmann in [2]. Part (b) was proved by Gregory Derfel and Rolf Schilling in [8]. Their proof was based on earlier work by Boris Solomyak ([24] and [21]) on Bernoulli convolutions: this is a problem much older than the Schilling equation, on which a lot of work has been done (the most recent and very strong advances are those due to Boris Solomyak), and which has some relation to the Schilling equation as we will see below. Parts (c) and (d) were proved by Yuval Peres and Boris Solomyak in [22].

The situation can thus be pictured as shown in Figure 2. It shows that the "regularity" of the Schilling equation increases when q increases: No L^1-solutions exist for small q, unbounded L^1-solutions exist for a.e. q between $\frac{1}{2\sqrt{2}}$ and $\frac{1}{2}$, and continuous solutions exist for a.e. q between $\frac{1}{2}$ and 1. It can also be proved, cf. [8], that for q closer to 1, most solutions will be differentiable, twice differentiable, etc.

The answers obtained in this section are of a global character. Now we will start to look at specific values of q, hoping to obtain more information about the Schilling equation for these parameters. There are three known ways to obtain such information: direct methods, Fourier transforms and Bernoulli convolutions.

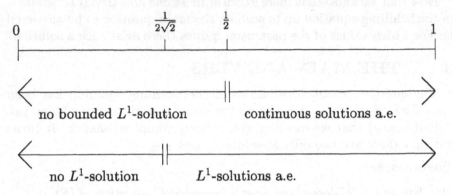

Figure 2   Existence and non-existence of solutions

4.   DIRECT METHODS

For the purposes of this section only, we are interested in pointwise defined functions which satisfy the Schilling equation everywhere and vanish outside of I_q (instead of normalized L^1-functions satisfying $(S)_q$ almost everywhere). We write $(S)_q^*$ for this problem. This setting is more appropriate for the application of "direct methods", which means methods involving iteration of the equation and drawing conclusions from that. In fact, the Schilling equation becomes simpler when q decreases (because the interval of support I_q becomes smaller), and the direct methods also work best for small q. When $q < \frac{1}{3}$, then the interval I_q is so small that at most one of the three points $x+1, x, x-1$ will be inside this interval. This means that the Schilling equation can be separated into a system of three simpler equations which can then be treated by iteration.

Theorem 4. (a) For $q \le \frac{1}{3}$, the equation (1) is equivalent to the system

$$f(x) = 2q\, f(qx) \qquad \text{for} \quad x \in I_q,$$
$$f(x) = 4q\, f(qx + q) \qquad \text{for} \quad x \in I_q,$$
$$f(x) = 4q\, f(qx - q) \qquad \text{for} \quad x \in I_q,$$
$$f(x) = 0 \qquad \text{for} \quad \frac{q^2}{1-q} < |x| < q\frac{1-2q}{1-q} \quad \text{and} \quad |x| > \frac{q}{1-q}.$$

(b) For $q < \frac{1}{3}$, every solution of $(S)_q^*$ vanishes a.e.

(c) For $q = \frac{1}{3}$, every Lebesgue-measurable solution of $(S)_q^*$ vanishes a.e.

This was proved by Wolfgang Förg-Rob in [10]. For $q < \frac{1}{3}$, this result is in a way stronger than the one by Perez and Solomyak Cited in the previous section, since Lebesgue-integrability is not assumed. I would conjecture that $q = \frac{1}{3}$ is the supremum of all numbers with this property.

Regarding solutions bounded at the origin, it was proved by Karol Baron in [1] that for any $q \leq \sqrt{2} - 1$, the zero function is the only solution bounded at the origin. This was improved later by Janusz Morawiec, in [18], [19] and [20], concerning solutions bounded at the origin and other points. We state only the first of those three results here (the others concern smaller values of q, but larger sets of points where to assume boundedness).

Theorem 5. *For $q \leq \frac{1}{3}(1 - \sqrt[3]{2} + \sqrt[3]{4})$, the zero function is the only solution of $(S)_q^*$ bounded in a neighbourhood of one of the points $\pm \sum_{i=1}^{m} q^i$, where $m \in \mathbb{N} \cup \{0, \infty\}$.*

The question if these are the largest possible values with these properties is open.

Problem 1. *(a) Determine the supremum of the set (or, even better, the set itself),*

$$M := \{q : \text{any solution of } (S)_q^* \text{ vanishes a.e.}\}.$$

At the moment we only know that $\sup M \geq \frac{1}{3}$.

(b) Determine the supremum of the set (or, even better, the set itself),

$$M(x) := \{q > \tfrac{|x|}{|x|+1} : \text{any solution of } (S)_q^* \text{ bounded}$$
$$\text{in a neighbourhood of } x \text{ vanishes}\}.$$

For example, we know that $\sup M(0) \geq \frac{1}{3}(1 - \sqrt[3]{2} + \sqrt[3]{4})$.

5. FOURIER TRANSFORMS

If an L^1-solution f_q of the problem $(S)_q$ exists, then, as stated in Section 2, its Fourier transform is given by the infinite product (3), now in the form

$$\widehat{f_q}(t) = \prod_{j=1}^{\infty} \cos^2\left(q^j \frac{t}{2}\right). \tag{4}$$

Many properties of the Schilling equation and its solutions can be derived by investigating this Fourier transform. It is possible to find specific values of q where $(S)_q$ has solutions; it is also possible to identify certain exceptional values for q between $\frac{1}{2}$ and 1 where L^1- or continuous solutions do not exist. Our first statement (unpublished so far) however deals with any value of q where an L^1-solution exists.

In fact, the first thing we notice about the function (4) is that it is nonnegative. Now we remember a theorem by Komaravolu Chandrasekharan which says that if an L^1-function has a nonnegative Fourier transform and is moreover bounded at 0, then the Fourier transform must be in $L^1(\mathbb{R})$, cf. Theorem 12 in [6]. Moreover, L^1-functions with a Fourier transform in $L^1(\mathbb{R})$ are necessarily continuous, so that we have proved the following theorem.

Theorem 6. *If an L^1-solution of $(S)_q$ exists and is bounded around the origin, then it must be continuous on $\mathbb{R}$.*

In Chandrasekharan's theorem, the origin cannot be easily replaced by other points. Therefore the question must be asked why the origin plays such a special role in Theorem 6.

Problem 2. *Determine the set $\{x \in \mathbb{R} : \text{for any } q \text{ with } x \in I_q, \text{ any } L^1\text{-solution of } (S)_q \text{ which is bounded at } x \text{ must be continuous on } \mathbb{R}\}$.*

Now we turn to an explicit construction of solutions of $(S)_q$ for certain values of q between $\frac{1}{2}$ and 1, namely for all $q = \left(\frac{1}{2}\right)^{1/n}$ with $n \in \mathbb{N}$. This construction is given via a convolution product and thus does not explicitly involve the Fourier transform. However, the Fourier transform was instrumental in finding the construction in the first place. As we have already seen, for $q = 1/2$ the Schilling equation has an explicitly given solution $f_{0.5}$ whose Fourier transform is $\widehat{f}_{0.5}(t) = \left(\frac{\sin(t/2)}{t/2}\right)^2$ (this is the same function as in (4), by Vieta's product). Now, as was first noted by Karol Baron, Alice Simon and Peter Volkmann in [2], for $q = \left(\frac{1}{2}\right)^{1/n}$ the product (4) can be written as a finite product of $\widehat{f}_{0.5}$ with scaled versions of itself:

$$\widehat{f}_q(t) = \widehat{f}_{0.5}(t) \cdot \widehat{f}_{0.5}(t/q) \cdot \ldots \cdot \widehat{f}_{0.5}(t/q^{n-1}).$$

Since this finite product is in L^1, we can apply the inverse Fourier transform to obtain an L^1-solution f_q for $q = \left(\frac{1}{2}\right)^{1/n}$. It can be written as a finite convolution product of $f_{0.5}$ with scaled versions of itself.

Theorem 7. *For $n \in \mathbb{N}$ and $q = \left(\frac{1}{2}\right)^{1/n}$, the L^1-solution f_q of $(S)_q$ exists and can be written as*

$$f_q(x) = f_{0.5}(x) * f_{0.5}(q\,x) * \cdots * f_{0.5}(q^{n-1}\,x),$$

where the convolution is defined as

$$(f * g)(t) := \int_{-\infty}^{\infty} f(u)\,g(t-u)\,du.$$

Now we turn to exceptional values of q: We will explicitly identify countably infinitely many values of q between $\frac{1}{2}$ and 1 where an L^1- (resp. continuous) solution does not exist, thereby justifying the words "almost everywhere" in the statement of Theorem 3. These exceptional values were identified by Paul Erdős and by Jean-Paul Kahane in connection with Bernoulli convolutions (see next section), long before Schilling's equation became interesting.

A *Pisot number* is defined to be an algebraic integer greater than 1 all of whose algebraic conjugates lie inside the unit disk. The best known example of a Pisot number is the golden mean $\varphi = (\sqrt{5} + 1)/2$. The characteristic property of Pisot numbers is that their powers quickly approach integers. In [9], Paul Erdős proved that if q is the inverse of a Pisot number, then the function (4) satisfies $\limsup\limits_{t \to \infty} \widehat{f_q}(t) > 0$. Since on the other hand the Riemann-Lebesgue lemma says that the Fourier transform of any L^1-function must tend to 0 if the argument tends to infinity, the function (4) cannot be the Fourier transform of an L^1-function if q is the inverse of a Pisot number.

Theorem 8. *If q is the inverse of a Pisot number, then the Schilling problem $(S)_q$ does not have an L^1-solution.*

A *Salem number* is defined to be an algebraic integer greater than 1 all of whose algebraic conjugates lie on the the unit disk with at least one on the boundary. The smallest known Salem number is the positive root of the polynomial $x^{10} + x^9 - x^7 - x^6 - x^5 - x^4 - x^3 + x + 1$, which is approximately $1.176\ldots$ In [14], Jean-Paul Kahane proved that if q is the inverse of a Salem number, then the function (4) satisfies $\lim\limits_{t \to \infty} \widehat{f_q}(t) \cdot t^\varepsilon = \infty$ for all $\varepsilon > 0$. Although this condition does not exclude the possibility that the function (4) is the Fourier transform of an L^1-function, it does mean that $\widehat{f_q}$ cannot lie in any space $L^p(\mathbb{R})$. This means in particular that $\widehat{f_q}$ is not in $L^2(\mathbb{R})$, and therefore, by Plancherel's theorem, f_q itself cannot be an L^2-function (note that $L^2(I_q) \subset L^1(I_q)$). This in turn means that f_q cannot be continuous.

Theorem 9. *If q is the inverse of a Salem number, then the Schilling problem $(S)_q$ does not have an L^2-solution.*

Pisot and Salem numbers are currently the only known exceptional values. It is known that there is a smallest Pisot number (it is the real root of the polynomial $x^3 - x - 1$, which is approximately 1.3247 ...). It is not known, however, if there is a smallest Salem number. Therefore it is also open if there is a largest exceptional value of q smaller than 1. In this context, several open problems arise.

Problem 3. *(a) Determine* sup{q : *the Schilling problem $(S)_q$ has no L^1-solution / L^2-solution / continuous solution*}.
(b) Does the Schilling problem $(S)_q$ have an L^1-solution when q is the inverse of a Salem number?
(c) Are there only countably many exceptional values?
(d) Characterize the set of all exceptional values.

More information on Pisot and Salem numbers can be found in the book [4].

6.   BERNOULLI CONVOLUTIONS

As mentioned above, those results about Pisot and Salem numbers were not obtained for the Schilling equation directly, but for Bernoulli convolutions. Consider the discrete probability density on the real line with measure $\frac{1}{2}$ at each of the two points ± 1. The corresponding measure is the so-called Bernoulli measure, denoted $b(x)$. For every $0 < q < 1$, the infinite convolution of measures

$$\mu_q(x) := b(x) * b(x/q) * b(x/q^2) * \dots$$

exists as a weak limit of the finite convolutions. The measure μ_q is supported inside the interval $\left[-\frac{1}{1-q}, \frac{1}{1-q}\right]$, it is continuous and moreover known to be either absolutely continuous or strictly singular; this is again a consequence of Theorem 35 in [13]. It can moreover be shown that the distribution function $\widetilde{F_q}$ of the measure μ_q satisfies (after rescaling) the functional equation

$$F(qx) = \frac{1}{2}\left(F(x+1) + F(x-1)\right) \qquad (x \in \mathbb{R}).$$

(This distribution function can also be introduced via a random series; see [5].) Its density (if it exists) then satisfies the functional equation

$$2q\, f(qx) = f(x+1) + f(x-1) \qquad (x \in \mathbb{R}), \tag{5}$$

and Theorem 1 and Theorem 2 (suitably modified) also apply to this functional equation. Explicitly, if (5) has a non-trivial L^1-solution $\widetilde{f}_q$, then its Fourier transform is (after the usual normalization) given by

$$\widehat{\widetilde{f}}_q(t) = \prod_{j=1}^{\infty} \cos\left(q^j\, t\right).$$

We write $(B)_q$ for the problem to determine an L^1-solution of (5) with this normalization.

Thus, up to scaling, the Fourier transform of a solution to the Schilling equation is the square of the Fourier transform of the Bernoulli density. In [5] this was used to transfer Erdős's Pisot number result on Bernoulli convolutions to the Schilling equation; in [8] it was used to transfer Kahane's Salem number result to the Schilling equation. This led to Theorems 8 and 9 of the previous section. Moreover, the following theorem suggests itself.

Theorem 10. (a) *If $\widetilde{f}_q$ is an L^1-solution of $(B)_q$, then $f_q := \widetilde{f}_q(2\,\cdot\,) *$ $\widetilde{f}_q(2\,\cdot\,)$ is an L^1-solution of $(S)_q$.*
(b) *Thus, existence of an L^1-solution of $(B)_q$ implies existence of an L^1-solution of $(S)_q$.*
(c) *Also, existence of an L^2-solution of $(B)_q$ implies existence of a continuous solution of $(S)_q$.*

Part (c) is true because the L^2-solution of $(B)_q$ has a Fourier transform in $L^2(\mathbb{R})$ whose square is in $L^1(\mathbb{R})$ and is the Fourier transform of a solution of $(S)_q$. We note that the converse implications of (b) and (c) above are *not* true. In fact, it is easy to see that for $q < \frac{1}{2}$, the support of the distribution function $\widetilde{F}_q$ is a Cantor set; therefore it cannot be absolutely continuous and $(B)_q$ cannot have a non-trivial L^1-solution. On the other hand, Solomyak's result (Theorem 3(d)) says that $(S)_q$ has L^1-solutions for almost every $q > \frac{1}{2\sqrt{2}}$. However, no value of q where the converse implication fails is currently known explicitly.

Problem 4. *Find a value of $q \in \left(\frac{1}{2\sqrt{2}}, \frac{1}{2}\right)$ such that $(S)_q$ has an L^1-solution.*

Theorem 10 has been used in [5] and in [8] to transfer the following previously obtained results on Bernoulli convolutions to the Schilling equation: In [11], Adriano M. Garsia found a class of arithmetic integers q for which $(B)_q$ has L^1-solutions; and in [24], Boris Solomyak proved that $(B)_q$ has L^2-solutions for almost every $q \in \left(\frac{1}{2}, 1\right)$.

7.   ITERATION

We have seen that a lot is known about the Schilling equation globally (for almost every q), but not many results exist for specific values of q. Does there exist, for example, an L^1-solution for $q = 0.6$, and what does it look like? To get some idea about the answer, one could try to iterate the functional equation and in this way hopefully approximate a solution. Therefore it is interesting to ask under what conditions the iterates converge to a solution if one exists.

More precisely, we define the *Schilling operator* as the linear operator $S_q : I_q \to I_q$, satisfying

$$(S_q f)(x) := \frac{1}{4q}\left[f\left(\frac{x}{q}+1\right) + 2f\left(\frac{x}{q}\right) + f\left(\frac{x}{q}-1\right) \right],$$

where the value of f is to be taken to be 0 if the argument is outside of I_q. Then S_q is a bounded linear operator with norm 1. If it can therefore be proved that for some $f_0 \in L^1(I_q)$, the sequence of iterates $\{S_q^n f_0\}$ converges in $L_1(I_q)$, then the limit will be an L^1-solution of the Schilling equation. However, even more can be said about behaviour of the iterates, because S_q is a Markov operator. Thus, iteration theory for Markov operators can be applied.

In general, if $(X, \mathcal{A}, \mu)$ is a measure space, then a *Markov operator* P is a linear mapping $P : L^1(X) \to L^1(X)$ with

(1) $Pf \geq 0$ for $f \geq 0, f \in L^1(X)$;

(2) $\|Pf\|_1 = \|f\|_1$ for $f \geq 0, f \in L^1(X)$.

A *density* is a non-negative L^1-function f with $\|f\|_1 = 1$. Since we have chosen the normalization of the solutions f_q of the Schilling equation such that they are derivatives of the increasing distribution functions F_q, they are densities (if they exist at all). A *stationary density* of a Markov operator is a density f_* with $Pf_* = f_*$. A solution of the Schilling problem $(S)_q$ is therefore a stationary density for the operator S_q. For Markov operators there are general theorems about behaviour of the sequence of iterates $\{P^n f\}$.

Theorem 11. *(a) There exists a density f_* with*

$$\lim_{n \to \infty} \|P^n f - f_*\|_1 = 0 \quad \text{for every density } f$$

if and only if $\{P^n f\}$ is precompact in $L^1(X)$ for every density f and for all densities f, g there exists $n \in \mathbb{N}$ such that

$$m(\operatorname{supp} P^n f \cap \operatorname{supp} P^n g) > 0.$$

(b) If there is a unique stationary density f_* which is moreover positive a.e., then for every density f the iterates $\{P^n f\}$ converge in the mean to f_*, i.e.,

$$\lim_{n\to\infty}\left\|\frac{1}{n}\sum_{k=0}^{n-1} P^k f - f_*\right\|_1 = 0.$$

Part a) of this theorem can be found as Theorem 3.3 in [15]; Part b) is Theorem 5.2.2 in [16].

The interesting part of this theorem is of course Part a). Unfortunately, in the case of the Schilling operator currently no proof is known for the two conditions given there. In fact, these conditions will be difficult to prove for general q, because the theorem asserts existence of a stationary density, i.e., of an L^1-solution to the Schilling equation. Thus at least for the exceptional values of q (Pisot numbers) one of the two conditions must fail.

Part b) of the theorem is easier. In fact, we already know that if there is a stationary density for the Schilling operator, then it is unique and non-negative a.e. Moreover, it was proved recently by Janusz Morawiec and myself (see [12]) that the L^1-solution of Schilling's equation (if it exists) is in fact positive a.e.

Theorem 12. *If the Schilling problem $(S)_q$ has an L^1-solution, then it is positive a.e.*

Our proof was based on the methods given in [17], where a similar statement for Bernoulli convolutions was proved.

Problem 5. *Prove or disprove: If an L^1-solution f_q of $(S)_q$ exists, then for every density f, the iterates $S_q^n f$ converge in the L^1-norm to f_q.*

Although we do not know if the iterates $\{S_q^n f\}$ converge for any density f, we can still compute some iterates to see what happens. In the following figures, the 12th iterate $S_q^{12} f_0$ is plotted for several values of q; the starting function f_0 is in every example $f_0 = \frac{1-q}{2q} \cdot \chi_{I_q}$, such that $\|f_0\|_1 = 1$. We see that for $q = \frac{1}{2}$ and $q = \frac{1}{\sqrt{2}}$, where we know that a stationary density exists, the iterates indeed seem to converge to this solution. For $q = 0.6$, existence of a solution is not known; the iterates however seem to converge to a continuous (possibly nowhere differentiable) function which would then be a solution. Finally, if q is the inverse of the golden mean, then the iterates do not converge to any meaningful function.

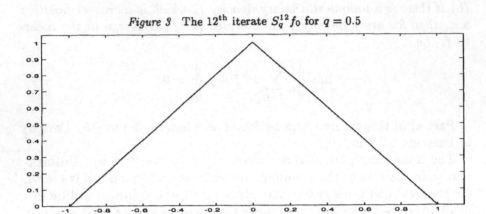

Figure 3   The 12th iterate $S_q^{12} f_0$ for $q = 0.5$

Figure 4   The 12th iterate $S_q^{12} f_0$ for $q = 0.7071067812$

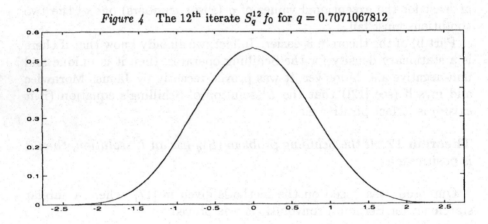

Figure 5   The 12th iterate $S_q^{12} f_0$ for $q = 0.6$

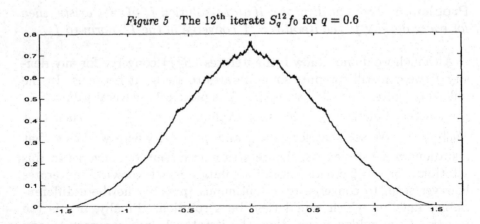

Figure 6   The 12th iterate $S_q^{12} f_0$ for $q = 0.618033988$

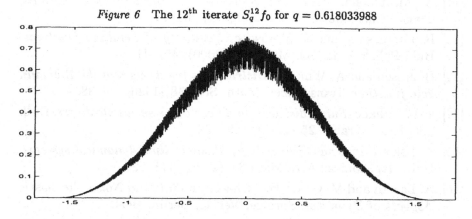

References

[1] K. Baron, *On a problem of R. Schilling*, in Selected Topics in Functional Equations (Graz, 1986), Ber. No. 286, Ber. Math.-Statist. Sekt. Forschungsgesellsch. Joanneum, Forschungszentrum Graz, Graz 1988.

[2] K. Baron, A. Simon, and P. Volkmann, *Solution d'une équation fonctionnelle dans l'espace des distributions tempérées*, C. R. Acad. Sci. Paris Ser. I Math. **319** (1994), 1249–1252.

[3] K. Baron and P. Volkmann, *Unicité pour une équation fonctionelle*, Rocznik Nauk.-Dydakt. Prace Mat. **13** (1993), 53–56.

[4] M. J. Bertin, A. Decomps-Guillox, M. Grandet-Hugot, M. Pathiaux-Delefosse, and J. P. Schreiber, *Pisot and Salem numbers*, Birkhäuser, Basel 1992.

[5] J. Borwein and R. Girgensohn, *Functional equations and distribution functions*, Results Math. **26** (1994), 229–237.

[6] K. Chandrasekharan, *Classical Fourier Transforms*, Springer-Verlag, Berlin-Heidelberg, 1989.

[7] I. Daubechies and J. C. Lagarias, *Two-scale difference equations I. Existence and global regularity of solutions*, SIAM J. Math. Anal. **22** (1991), 1388–1410.

[8] G. Derfel and R. Schilling, *Spatially chaotic configurations and functional equations*, J. Phys. A. **29** (1996), 4537–4547.

[9] P. Erdős, *On a family of symmetric Bernoulli convolutions*, Trans. Amer. Math. Soc. **61** (1939), 974–976.

[10] W. Förg-Rob, *On a problem of R. Schilling I*, Math. Pannon. **5** (1994), 29–65; *II*, ibid., 145–168.

[11] A. M. Garsia, *Arithmetic properties of Bernoulli convolutions*, Trans. Amer. Math. Soc. **102** (1962), 409–432.

[12] R. Girgensohn and J. Morawiec, *Positivity of Schilling functions*, Bull. Polish Acad. Sci. Math. **48** (2000), 407–412.

[13] B. Jessen and A. Wintner, *Distribution functions and the Riemann zeta function*, Trans. Amer. Math. Soc. **38** (1935), 48–88.

[14] J.-P. Kahane, *Sur la distribution de certaines series aléatoires*, Bull. Soc. Math. France **25** (1971), 119–122.

[15] A. Lasota, *Invariant principle for discrete time dynamical systems*, Univ. Jagellonicae Acta Math **31** (1994), 111–127.

[16] A. Lasota and M. C. Mackey, *Chaos, Fractals and Noise. Stochastic Aspects of Dynamics*, Springer Verlag, Berlin, 1994.

[17] R. D. Mauldin and K. Simon, *The equivalence of some Bernoulli convolutions to Lebesgue measure*, Proc. Amer. Math. Soc. **126** (1998), 2733–2736.

[18] J. Morawiec, *On bounded solutions of a problem of R. Schilling*, Ann. Math. Sil. **8** (1994), 97–101.

[19] J. Morawiec, *Bounded solutions of Schilling's problem*, Math. Pannon. **7** (1996), 223–232.

[20] J. Morawiec, *On locally bounded solutions of Schilling's problem*, preprint.

[21] Y. Peres and B. Solomyak, *Absolute continuity of Bernoulli convolutions, a simple proof*, Math. Research Letters **3:2** (1996), 231–239.

[22] Y. Peres and B. Solomyak, *Self-similar measures and intersections of Cantor sets*, Trans. Amer. Math. Soc. **350** (1998), 4065–4087.

[23] R. Schilling, *Spatially-chaotic structures*, in Nonlinear dynamics of solids (ed. by E. Thomas), Springer-Verlag, Berlin, 1992, pp. 213–241.

[24] B. Solomyak, *On the random series $\sum \pm \lambda^i$ (an Erdős problem)*, Annals of Math. **142** (1995), 611–625.

PROPERTIES OF AN OPERATOR ACTING ON THE SPACE OF BOUNDED REAL FUNCTIONS AND CERTAIN SUBSPACES

Hans-Heinrich Kairies

Institut für Mathematik, Technische Universität Clausthal,
Erzstraße 1, D-38678 Clausthal-Zellerfeld, Germany
kairies@math.tu-clausthal.de

Abstract   Let $\varphi : \mathbb{R} \to \mathbb{R}$ be bounded and $\alpha \in (0,1)$, $\beta \in (0,\infty)$. Then

$$F[\varphi] : \mathbb{R} \to \mathbb{R}, \qquad F[\varphi](x) := \sum_{k=0}^{\infty} \alpha^k \varphi(\beta^k x)$$

defines a two parameter family of operators $F = F(\alpha, \beta)$ acting on the Banach space of bounded real functions. It turns out that F is a continuous Banach space automorphism.

   F and F^{-1} are closely related to a de Rham type functional equation and the eigenspaces of F and F^{-1} as well as their restrictions to certain subspaces are solution spaces of systems of iterative functional equations.

Keywords: sum type Banach space operator, iterative functional equation

Mathematics Subject Classification (2000): 47B38, 39B22, 26A27

1.   INTRODUCTION. HISTORICAL BACKGROUND

Let $\varphi : \mathbb{R} \to \mathbb{R}$ be bounded and $\alpha \in (0,1)$, $\beta \in (0,\infty)$. Then

$$F[\varphi] : \mathbb{R} \to \mathbb{R}, \qquad F[\varphi](x) := \sum_{k=0}^{\infty} \alpha^k \varphi(\beta^k x)$$

is bounded as well. Functions of the type $F[\varphi]$ have been investigated by many authors who used the operator F to generate continu-

175

Z. Daróczy and Z. Páles (eds.),
Functional Equations - Results and Advances, 175–186.
© 2002 *Kluwer Academic Publishers.*

ous nowhere differentiable (cnd) and other peculiar functions from very simple ones: Weierstrass [13] 1872, Takagi [11] 1903, Hardy [5] 1916, Knopp [7] 1918, v. d. Waerden [12] 1930, Behrend [1] 1949, Mikolás [9] 1956, de Rham [10] 1957, Gajda [2] 1991, Girgensohn [3], [4] 1993/94.

We give some examples and use this to introduce some notations for functions which will be kept throughout this paper.

Weierstrass:

$$\varphi = c, \qquad c(x) = \cos 2\pi x, \qquad F[c](x) = w_{\alpha,\beta}(x) = \sum_{k=0}^{\infty} \alpha^k \cos(2\pi \beta^k x).$$

By a well known result of Hardy [5], $w_{\alpha,\beta}$ is cnd if $\alpha\beta \geq 1$.

Takagi:

$$\varphi = d, \qquad d(x) = \mathrm{dist}(x, \mathbb{Z}), \qquad F[d](x) = t_{\alpha,\beta}(x) = \sum_{k=0}^{\infty} \alpha^k d(\beta^k x);$$

$t_{1/2,2}$ is Takagi's function [11], $t_{1/10,10}$ is the famous v. der Waerden function [12]. $t_{\alpha,\beta}$ is cnd for $\alpha\beta \geq 1, \beta \in \mathbb{N}$.

Gajda:

$$\varphi = g_\mu, g_\mu(x) = -\mu \qquad \text{for} \quad x \leq -1,$$
$$g_\mu(x) = \mu x \qquad \text{for} \quad |x| < 1,$$
$$g_\mu(x) = \mu \qquad \text{for} \quad x \geq 1.$$

In [2], $F[g_\mu]$ is used as an important counterexample in the stability theory of additive functions: $f = F[g_{\varepsilon/6}]$ satisfies $|f(x+y) - f(x) - f(y)| \leq \varepsilon(|x| + |y|)$ but there is no $\delta \geq 0$ and no additive $A : \mathbb{R} \to \mathbb{R}$ such that $|f(x) - A(x)| \leq \delta|x|$ holds.

In this paper we pay attention to structural properties of the operator $F = F(\alpha, \beta)$ (its dependence on the parameters α, β will always be clear from the context and therefore not explicitly shown in the notation) which will be considered on the Banach space B of bounded real functions and on various of its subspaces. Clearly F preserves boundedness: If $\varphi : \mathbb{R} \to \mathbb{R}$ is bounded then $F[\varphi]$ is so too for every choice of $\alpha \in (0, 1)$ and $\beta \in (0, \infty)$. An analoguous statement is true if φ is bounded and continuous. However, it is no longer true if φ is, say, just continuous or differentiable or analytic. It turns out that F is a continuous Banach space automorphism. F and F^{-1} are closely connected to a de Rham type functional equation and the eigenspaces of F and F^{-1} are characterized as solution spaces of systems of iterative functional equations.

Some of the results in case $\alpha = \frac{1}{2}$, $\beta = 2$ are contained in the paper [6] of Kairies.

2.   THE OPERATOR F AND ITS CONNECTED FUNCTIONAL EQUATION

We first fix some notations.

Let $B := \{\varphi : \mathbb{R} \to \mathbb{R} ; \varphi$ bounded$\}$ and let $F : B \to B$ be given by

$$F[\varphi](x) = \sum_{k=0}^{\infty} \alpha^k \varphi(\beta^k x) \qquad (x \in \mathbb{R}) \tag{1}$$

with fixed $\alpha \in (0,1)$, $\beta \in (0, \infty)$. We shall see in a moment that a de Rham [10] type functional equation is closely related to F:

$$\psi(x) - \alpha \psi(\beta x) = \varphi(x) \qquad (x \in \mathbb{R}). \tag{2}$$

The interaction of the operator F given by (1) and the iterative functional equation (2) is exhibited by

Proposition 1. *Let $\varphi \in B$, $\alpha \in (0,1)$ and $\beta \in (0, \infty)$. Then $\psi = F[\varphi]$ if and only if ψ is a bounded solution of (2).*

Proof. (a) Let $\psi(x) = F[\varphi](x) = \sum\limits_{k=0}^{\infty} \alpha^k \varphi(\beta^k x)$. Then ψ is bounded as well:

$$|\psi(\xi)| \le (1/(1-\alpha)) \sup\{|\varphi(x)|; \ x \in \mathbb{R}\} \qquad \text{for every} \ \ \xi \in \mathbb{R}.$$

ψ satisfies (2):

$$\alpha\psi(\beta x) = \alpha[\varphi(\beta x) + \alpha\varphi(\beta^2 x) + \alpha^2\varphi(\beta^3 x) + \cdots]$$

implies

$$\varphi(x) + \alpha\psi(\beta x) = \psi(x).$$

(b) Let ψ be a bounded solution of (2). Then

$$\begin{aligned}
\psi(x) &= \alpha\psi(\beta x) + \varphi(x) \\
&= \alpha[\alpha\psi(\beta^2 x) + \varphi(\beta x)] + \varphi(x) \\
&= \alpha^2[\alpha\psi(\beta^3 x) + \varphi(\beta^2 x)] + \alpha\varphi(\beta x) + \varphi(x) \\
&\ \vdots \\
&= \alpha^m \psi(\beta^m x) + \sum_{k=0}^{m-1} \alpha^k \varphi(\beta^k x) \qquad \text{for every} \ \ x \in \mathbb{R}, \ m \in \mathbb{N}.
\end{aligned}$$

As ψ is bounded and $\alpha \in (0,1)$,

$$\psi(x) = \lim_{m \to \infty} \psi(x) = \sum_{k=0}^{\infty} \alpha^k \varphi(\beta^k x) = F[\varphi](x).$$

$\square$

3.   PROPERTIES OF $F : B \to B$

Let $\varphi \in B$ and $\|\varphi\|_B := \sup\{|\varphi(x)|;\ x \in \mathbb{R}\}$. Then $(B, \|\ldots\|_B)$ is a Banach space.

Now let $F : B \to B$ be given by (1) with fixed $\alpha \in (0,1)$, $\beta \in (0,\infty)$. The case $\beta = 1$ is rather simple and will be excluded in the sequel. However we need some facts also in this case and collect them in the following

Remark 1. Let $\alpha \in (0,1)$ and $\beta = 1$. Then

- (1) becomes $F[\varphi](x) = (1/(1-\alpha))\varphi(x)$,
- (2) becomes $(1-\alpha)\psi(x) = \varphi(x)$.
- F is a continuous Banach space automorphism with operator norm $\|F\| = 1/(1-\alpha)$.
- F has exactly one eigenvalue: $\sigma(F) = \{1/(1-\alpha)\}$ with the eigenspace $E(F, 1/(1-\alpha)) = B$.
- F^{-1} is a continuous Banach space automorphism as well, given by $F^{-1}[\psi] = (1-\alpha)\psi$ with $\|F^{-1}\| = 1-\alpha$. We have $\sigma(F^{-1}) = \{1-\alpha\}$ and $E(F^{-1}, 1-\alpha) = B$.

Theorem 1. Let $\alpha \in (0,1), \beta \in \mathbb{R}_+ \setminus \{1\}$ and F be given by (1). Then

(a) $F : B \to B$ is a continuous Banach space automorphism with $\|F\| = 1/(1-\alpha)$. We have $\sigma(F) = \{\lambda_1, \lambda_2\}$ with $\lambda_1 := 1/(1-\alpha)$, $\lambda_2 := 1/(1+\alpha)$ and
$E(F, \lambda_1) = \{\varphi \in B;\ \varphi(x) = \varphi(\beta x) \text{ for every } x \in \mathbb{R}\}$,
$E(F, \lambda_2) = \{\varphi \in B;\ \varphi(x) = -\varphi(\beta x) \text{ for every } x \in \mathbb{R}\}$.

(b) $F^{-1} : B \to B$ is given by $F^{-1}[\psi](x) = \psi(x) - \alpha\psi(\beta x)$.
F^{-1} is continuous as well with $\|F^{-1}\| = 1 + \alpha$.
We have $\sigma(F^{-1}) = \{\lambda_1^{-1}, \lambda_2^{-1}\}$ and $E(F^{-1}, \lambda_\rho^{-1}) = E(F, \lambda_\rho)$ for $\rho \in \{1,2\}$.

Proof. (a) F maps any $\varphi \in B$ to an $F[\varphi] \in B$ and clearly $F : B \to B$ is linear.

F is *injective*: Let $F[\varphi] = \mathbf{o}$. By Proposition 1, $\mathbf{o}$ has to be a solution of (2). Therefore, necessarily $\varphi = \mathbf{o}$.

F is *surjective*: Let $\psi \in B$. Then ψ is bounded.

Define $\varphi(x) := \psi(x) - \alpha\psi(\beta x)$ for $x \in \mathbb{R}$. Clearly φ is bounded, i.e., $\varphi \in B$. By Proposition 1, $\psi = F[\varphi]$ with $\varphi \in B$. For $\varphi \in B$ with $\|\varphi\|_B \leq 1$ we obtain

$$\|F[\varphi]\|_B = \sup\left\{\left|\sum_{k=0}^{\infty} \alpha^k \varphi(\beta^k x)\right|;\ x \in \mathbb{R}\right\} \leq \sum_{k=0}^{\infty} \alpha^k \cdot 1 = 1/(1-\alpha),$$

hence

$$\|F\| = \sup\{\|F[\varphi]\|_B; \ \|\varphi\|_B \le 1\} \le \frac{1}{1-\alpha}.$$

On the other hand, for the constant function $\mathbf{1}$ ($\mathbf{1}(x) = 1$ for every $x \in \mathbb{R}$) we have $\mathbf{1} \in B, \|\mathbf{1}\|_B = 1$ and $\|F[\mathbf{1}]\|_B = \sum_{k=0}^{\infty} \alpha^k \cdot 1 = 1/(1-\alpha)$, hence $\|F\| \ge 1/(1-\alpha)$.

Now we describe the set $\sigma(F)$ of eigenvalues and use to this end some properties of F^{-1} which are proved later in (b). Clearly $0 \notin \sigma(F)$. First we note that $1 \notin \sigma(F)$. This is seen by the following equivalences: $F[\varphi] = 1 \cdot \varphi$ iff $F^{-1}[\varphi] = \varphi$ iff $\varphi(x) - \alpha\varphi(\beta x) = \varphi(x)$ for every $x \in \mathbb{R}$ iff $\varphi = \mathbf{o}$.

Now let $\lambda \in \sigma(F)$. We have $\lambda \ne 1$ and $F[\varphi] = \lambda\varphi$ iff $\varphi = F^{-1}[\lambda\varphi]$ iff $\varphi = \lambda F^{-1}[\varphi]$ iff $\varphi(x) = \lambda\varphi(x) - \alpha\lambda\varphi(\beta x)$ for every $x \in \mathbb{R}$ iff

$$\varphi(x) = (\lambda\alpha/(\lambda-1))\varphi(\beta x) \qquad \text{for every} \quad x \in \mathbb{R}. \tag{3}$$

For $\gamma := \alpha\lambda/(\lambda-1) = 1$, i.e., $\lambda = 1/(1-\alpha) = \lambda_1$, (3) becomes

$$\varphi(x) = \varphi(\beta x) \qquad \text{for every} \quad x \in \mathbb{R}. \tag{4}$$

The general bounded solution of (4) is described in the subsequent Remark 2. Iterative functional equations like (2), (4), (4') are treated in Kuczma's monograph [8] 1968. Some of our results regarding those equations are given here for the convenience of the reader, they could as well be deduced from [8]. For $\gamma = -1$, i.e., $\lambda = 1/(1+\alpha) = \lambda_2$, (3) becomes

$$\varphi(x) = -\varphi(\beta x) \qquad \text{for every} \quad x \in \mathbb{R}. \tag{4'}$$

In the remaining cases (3) becomes

$$\varphi(x) = \gamma\varphi(\beta x) \qquad \text{for every} \quad x \in \mathbb{R} \tag{5}$$

with $\gamma \ne 0$, $|\gamma| \ne 1$.

The only bounded solution of (5) is the zero function $\mathbf{o}$. This is proved by contradiction:

Let $\varphi(\xi) \ne 0$ for some $\xi \in \mathbb{R}$ and, say $0 < |\gamma| < 1$. Iteration of (5) gives $\varphi(\xi) = \gamma^m\varphi(\beta^m\xi)$ for every $m \in \mathbb{N}$. It follows that the sequence $(|\varphi(\beta^m\xi)|)_{m\in\mathbb{N}} = (|\gamma^{-m}\varphi(\xi)|)_{m\in\mathbb{N}}$ is unbounded, which contradicts $\varphi \in B$. A similar reasoning applies to the case $|\gamma| > 1$. Collecting these results we see that λ_1 and λ_2 are the only eigenvalues of F and that the elements of the eigenspaces $E(F, \lambda_1)$ respectively $E(F, \lambda_2)$ are exactly the solutions of (4) respectively (4').

(b) $F^{-1} : B \to B$ exists by (a) and is linear.
Let $\varphi \in B$ and $\psi := F[\varphi]$. By Proposition 1, ψ is a bounded solution of (2), i.e.,

$$\varphi(x) = \psi(x) - \alpha\psi(\beta x) = F^{-1}[\psi](x) \qquad \text{for every} \quad x \in \mathbb{R}.$$

So the inversion of the rather complicated operator F, given by (1), is performed by the simple de Rham type functional equation (2).

For $\psi \in B$ with $\|\psi\|_B \le 1$ we have

$$\|F^{-1}[\psi]\|_B = \sup\{|\psi(x) - \alpha\psi(\beta x)|;\ x \in \mathbb{R}\} \le 1 + \alpha,$$

hence

$$\|F^{-1}\| \le 1 + \alpha.$$

On the other hand, let $\psi_0 \in B$ with $\|\psi_0\|_B \le 1$, $\psi_0(1) = 1$, $\psi_0(\beta) = -1$. Clearly such a function ψ_0 exists (recall that $\beta \ne 1$). It follows

$$\|F^{-1}[\psi_0]\|_B = \sup\{|\psi_0(x) - \alpha\psi_0(\beta x)|;\ x \in \mathbb{R}\}$$
$$\ge \psi_0(1) - \alpha\psi_0(\beta) = 1 + \alpha,$$

hence

$$\|F^{-1}\| \ge 1 + \alpha.$$

Finally, because of $F[\varphi] = \lambda\varphi$ iff $\varphi = F^{-1}[\lambda\varphi]$ iff $F^{-1}[\varphi] = \lambda^{-1}\varphi$ ($\lambda \ne 0$) we have

$$\sigma(F^{-1}) = \{\lambda_1^{-1},\ \lambda_2^{-1}\} = \{1 - \alpha,\ 1 + \alpha\}$$

and

$$E(F^{-1}, \lambda_\rho^{-1}) = \{\varphi \in B;\ F^{-1}[\varphi] = \lambda_\rho^{-1}\varphi\} = E(F, \lambda_\rho)$$

for $\rho \in \{1, 2\}$. □

Remark 2. (a) The general bounded solution $\widetilde{\varphi}$ of (4) is obtained as follows: Take any bounded $\widetilde{\varphi}_0$ on $\{0\} \cup (1, \beta] \cup [-\beta, -1)$ in case $\beta > 1$ respectively on $\{0\} \cup (\beta, 1] \cup [-1, -\beta)$ in case $\beta < 1$ and extend it uniquely by (4) to $\widetilde{\varphi} : \mathbb{R} \to \mathbb{R}$. So $\dim E(F, \lambda_1) = \infty$, but $E(F, \lambda_1) \ne B$ in contrast to the situation for $\beta = 1$ (see Remark 1). Similarly, $\dim E(F, \lambda_2) = \infty$, but $E(F, \lambda_2) \ne B$.

(b) For fixed $\alpha \in (0, 1)$, consider $\|F\|$ as a function of $\beta \in \mathbb{R}_+$. Then $\|F\|(\beta) = 1/(1 - \alpha)$ for every $\beta \in \mathbb{R}_+$ by Remark 1 and Theorem 1, (a).

(c) For fixed $\alpha \in (0, 1)$, consider $\|F^{-1}\|$ as a function of $\beta \in \mathbb{R}_+$. Then

$$\|F^{-1}\|(1) = 1 - \alpha$$

by Remark 1,

$$\|F^{-1}\|(\beta) = 1 + \alpha$$

for $\beta \ne 1$ by Theorem 1, (b). Consequently $\|F^{-1}\|$ does not depend continuously on β, in severe contrast to $\|F\|$.

(d) The operator $F : B \to B$ is not compact, as $\dim B = \infty$ and F^{-1} is continuous (F compact and F^{-1} continuous would imply $\dim B < \infty$).

4. PROPERTIES OF $F_\nu : B_\nu \to B_\nu$

Motivated by the historical background and the examples in Section 1 we define

$B_1 := \{\varphi \in B;\ \varphi\ \text{1-periodic}\}$,
$B_2 := \{\varphi \in B;\ \varphi\ \text{1-periodic and even}\}$,
$B_3 := \{\varphi \in B;\ \varphi\ \text{continuous}\}$,
$B_4 := \{\varphi \in B_3;\ \varphi\ \text{1-periodic}\}$,
$B_5 := \{\varphi \in B_3;\ \varphi\ \text{1-periodic and even}\}$.

To unify notation we write from now on B_0 instead of B.

Remark 3. (a) Equip all the spaces B_ν, $1 \le \nu \le 5$ with the norm $\|\ldots\|_B$. Then B_ν is a closed subspace of B_0, hence a Banach space, for $1 \le \nu \le 5$.

(b) Let $\nu \in \{0,3\}$. Then the operator F given by (1) clearly maps B_ν to B_ν for every choice of $\alpha \in (0,1)$, $\beta \in (0,\infty)$. This is no longer true for $\nu \in \{1,2,4,5\}$. Take for example $\alpha = \frac{1}{2}$, $\beta = \frac{1}{4}$ and let $\varphi_0 : \mathbb{R} \to \mathbb{R}$ be polygonal with vertices $(0,0), (1/4,0), (1/2,1), (3/4,0), (1,0)$ and 1-periodic. It is easily checked that for $x \in [0,1)$, we have $F[\varphi_0](x) = \varphi_0(x)$ and for $x \in [1,2)$ we have $F[\varphi_0](x) = \varphi_0(x) + \frac{1}{2}\varphi_0(\frac{1}{4}x) \ne \varphi_0(x)$. So $F[\varphi_0]$ is not 1-periodic, i.e., $F[\varphi_0] \notin B_\nu$ for $\nu \in \{1,2,4,5\}$ though $\varphi_0 \in B_\nu$ for $\nu \in \{1,2,4,5\}$.

(c) However the situation changes if we allow β to assume only values from $\mathbb{N}$. Then it is obvious that $F : B_\nu \to B_\nu$ also in the cases $\nu \in \{1,2,4,5\}$. We denote the restriction of F onto B_ν by F_ν.

Theorem 2. *Let* $\alpha \in (0,1)$, $\beta \in \mathbb{N}\backslash\{1\}$, $\lambda_1 = 1/(1-\alpha)$, $\lambda_2 = 1/(1+\alpha)$. *Then*

(a) $F_\nu : B_\nu \to B_\nu$ *is a continuous Banach space automorphism for* $1 \le \nu \le 5$ *with* $\|F_\nu\| = 1/(1-\alpha)$. *We have* $\sigma(F_\nu) = \{\lambda_1, \lambda_2\}$ *for* $1 \le \nu \le 2$ *with*

$E(F_1, \lambda_\rho) = \{\varphi \in B_0;\ \varphi(x) = (-1)^{\rho-1}\varphi(\beta x),\ \varphi(x) = \varphi(x+1),\ x \in \mathbb{R}\}$,

$E(F_2, \lambda_\rho) = \{\varphi \in B_0;\ \varphi(x) = (-1)^{\rho-1}\varphi(\beta x),\ \varphi(x) = \varphi(x+1), \varphi(x) = \varphi(-x),\ x \in \mathbb{R}\}$

and $\sigma(F_\nu) = \{\lambda_1\}$ *for* $3 \le \nu \le 5$ *with*

$E(F_3, \lambda_1) = \{\varphi \in B_3;\ \varphi(x) = \varphi(\beta x),\ x \in \mathbb{R}\}$,
$E(F_4, \lambda_1) = \{\varphi \in B_3;\ \varphi(x) = \varphi(\beta x),\ \varphi(x) = \varphi(x+1),\ x \in \mathbb{R}\}$,
$E(F_5, \lambda_1) = \{\varphi \in B_3;\ \varphi(x) = \varphi(\beta x),\ \varphi(x) = \varphi(x+1),\ \varphi(x) = \varphi(-x),\ x \in \mathbb{R}\}$.

(b) $F_\nu^{-1} : B_\nu \to B_\nu$ *is given by* $F_\nu^{-1}[\psi](x) = \psi(x) - \alpha\psi(\beta x)$. F_ν^{-1} *is continuous as well with* $\|F_\nu^{-1}\| = 1 + \alpha$. *We have* $\sigma(F_\nu^{-1}) = \{\lambda_1^{-1}, \lambda_2^{-1}\}$ *for* $1 \le \nu \le 2$ *and*
$\sigma(F_\nu^{-1}) = \{\lambda_1^{-1}\}$ *for* $3 \le \nu \le 5$ *with* $E(F_\nu^{-1}, \lambda_\rho^{-1}) = E(F_\nu, \lambda_\rho)$.

Proof. We treat the case $\nu = 2$, the situation for $\nu \in \{1, 3, 4, 5\}$ is very similar and the proof is essentially the same as for $\nu = 0$ (Theorem 1). We have $B_2 = \{\varphi : \mathbb{R} \to \mathbb{R}; \ \varphi \text{ bounded, 1-periodic and even}\}$. For $\varphi \in B_2$

$$F_2[\varphi](x+1) = \sum_{k=0}^{\infty} \alpha^k \varphi(\beta^k(x+1)) = F_2[\varphi](x),$$

$$F_2[\varphi](-x) = \sum_{k=0}^{\infty} \alpha^k \varphi(\beta^k(-x)) = F_2[\varphi](x),$$

hence $F_2 : B_2 \to B_2$. F_2 is clearly linear, the injectivity is proved exactly as in Theorem 1, the surjectivity with a slight modification:

Let $\psi \in B_2$. Then ψ is bounded. Define $\varphi(x) := \psi(x) - \alpha\psi(\beta x)$ for $x \in \mathbb{R}$. Clearly φ is bounded, 1-periodic and even, i.e., $\varphi \in B_2$. By Proposition 1 $\psi = F[\varphi] = F_2[\varphi]$ with $\varphi \in B_2$. $\|F_2\| = 1/(1-\alpha)$ is shown exactly as in Theorem 1, note that the test function $\mathbf{1} \in B_0$ also belongs to B_2.

Again as in the proof of Theorem 1, F_2 has exactly two eigenvalues λ_1, λ_2 and the eigenfunctions are exactly the elements of B_2 which satisfy the functional equation (4), respectively (4') i.e., $E(F_2, \lambda_\rho) = \{\varphi : \mathbb{R} \in \mathbb{R}; \ \varphi \text{ a bounded, 1-periodic, even solution of } \varphi(x) = (-1)^{\rho-1}\varphi(\beta x)\}$. λ_2 is no eigenvalue of F_ν for $3 \leq \nu \leq 5$, because the only continuous solution of

$$\varphi(x) = -\varphi(\beta x) \qquad (x \in \mathbb{R}),$$

is the zero function. $F_2^{-1} : B_2 \to B_2$ is linear and given by $F_2^{-1}[\psi](x) = \psi(x) - \alpha\psi(\beta x)$ (Proposition 1). We have $\|F_2^{-1}\| \leq 1 + \alpha$ as in Theorem 1. The reverse inequality is obtained as follows.

Let ψ_1 the polygon function in $[0, 1]$ with vertices $(0, -1), (1/\beta, 1)$, $(1 - 1/\beta, 1), (1, -1)$ and extend by $\psi_1(x+1) = \psi_1(x)$. (For $\beta = 2$ the trapezium becomes a triangle.) Clearly $\psi_1 \in B_2$, $\|\psi_1\|_B = 1$ and

$$\|F_2^{-1}[\psi_1]\|_B = \sup\{|\psi_1(x) - \alpha\psi_1(\beta x)|; x \in \mathbb{R}\}$$
$$\geq \psi_1(1/\beta) - \alpha\psi_1(1) = 1 + \alpha.$$

The remaining properties of F_2^{-1} follow as in Theorem 1. $\qquad\square$

Remark 4. (a) For $\beta \in \mathbb{N} \setminus \{1\}$ the only continuous solutions $\varphi : \mathbb{R} \to \mathbb{R}$ of (4) are the constant ones. Consequently $E(F_3, \lambda_1) = E(F_4, \lambda_1) = E(F_5, \lambda_1) = \text{span } \{\mathbf{1}\}$.

(b) In severe contrast, $E(F_1, \lambda_1)$ and $E(F_2, \lambda_1)$ have the same dimension as $E(F_0, \lambda_1)$, namely ∞. To see this, let $\{v_\lambda; \ \lambda \in \Lambda\} \cup \{1\}$ be a Hamel basis of $\mathbb{R}$ over $\mathbb{Q}$ and $U_\lambda := \text{span }_\mathbb{Q} \{1, v_\lambda\}$.

Define $\varphi_\lambda : \mathbb{R} \to \mathbb{R}$ by $\varphi_\lambda(x) := 1$ for $x \in U_\lambda$, $\varphi_\lambda(x) := 0$ elsewhere. Clearly $x \in U_\lambda$ implies $\beta x \in U_\lambda (\beta \in \mathbb{N} \setminus \{1\})$, $x + 1 \in U_\lambda$, $-x \in U_\lambda$ and $x \notin U_\lambda$ implies $\beta x \notin U_\lambda$, $x + 1 \notin U_\lambda$, $-x \notin U_\lambda$. Consequently every φ_λ, $\lambda \in \Lambda$, is an element of $E(F_\nu, \lambda_1)$, $0 \le \nu \le 2$, and obviously $\{\varphi_\lambda; \lambda \in \Lambda\}$ is linearly independent over $\mathbb{R}$ and has cardinality **c** (continuum). Let us note that $U := \text{span}_\mathbb{R} \{\varphi_\lambda; \lambda \in \Lambda\} \ne E(F_\nu, \lambda_1)$ for $0 \le \nu \le 2$ (it is readily checked that $\mathbf{1} \notin U$). An extension of $\{\varphi_\lambda; \lambda \in \Lambda\}$ to a basis of $E(F_\nu, \lambda_1)$ is not known in either case.

(c) We now show that also $\dim E(F_1, \lambda_2) = \dim E(F_2, \lambda_2) = \infty$. To this end assume that $\xi \in \mathbb{R}$ is irrational. Let $P := \{\beta^{2m}\xi + \beta^{2n}(-\xi) + r; \ m, n \in \mathbb{Z}, \ r \in \mathbb{Q}_\beta\}$, where $\mathbb{Q}_\beta$ is the set of β-adic rationals (recall that $\beta \in \mathbb{N} \setminus \{1\}$).

It is readily checked that $x \in P$ implies $x + 1 \in P, -x \in P$ and $\beta^{2j} x \in P$ for any $j \in \mathbb{Z}$. Moreover, we have $P \cap \beta P = \emptyset$ and similarly $x \in \beta P$ implies $x + 1 \in \beta P, -x \in \beta P$ and $\beta^{2j} x \in \beta P$ for any $j \in \mathbb{Z}$.

Define $\varphi_\xi : \mathbb{R} \to \mathbb{R}$ by $\varphi_\xi(x) := 1$ for $x \in P$, $\varphi_\xi(x) := -1$ for $x \in \beta P$ and $\varphi_\xi(x) := 0$ elsewhere. Then φ_ξ satisfies all the equations $\varphi(x) = -\varphi(\beta x)$, $\varphi(x + 1) = \varphi(x)$ and $\varphi(-x) = \varphi(x)$. Hence $\varphi_\xi \in E(F_\nu, \lambda_2)$ for $1 \le \nu \le 2$. Note that $\{\varphi_\xi; \xi \in A\}$ is linearly independent, if A is algebraically independent over $\mathbb{Q}$.

5. IMAGES WITH RESPECT TO F AND F^{-1}

Let S be some prescribed set of functions from $\mathbb{R}$ to $\mathbb{R}$ and define

$$S_\nu := B_\nu \cap S \tag{6}$$

with $0 \le \nu \le 5$. We have for instance
$S_0 = \{\varphi : \mathbb{R} \to \mathbb{R}; \ \varphi \in S, \ \varphi \text{ bounded}\}$,
$S_1 = \{\varphi : \mathbb{R} \to \mathbb{R}; \ \varphi \in S, \ \varphi \text{ bounded and 1-periodic}\}$,
$S_2 = \{\varphi : \mathbb{R} \to \mathbb{R}; \ \varphi \in S, \ \varphi \text{ bounded, 1-periodic and even}\}$.
In this note we restrict S to be one of the following sets:
$A := \{\varphi : \mathbb{R} \to \mathbb{R}; \ \varphi \text{ analytic}\}$,
$C^n := \{\varphi : \mathbb{R} \to \mathbb{R}; \ \varphi \text{ } n \text{ times continuously differentiable }\}$, $n \in \mathbb{N}$,
$H := \{\varphi : \mathbb{R} \to \mathbb{R}; \ \varphi \text{ absolutely continuous}\}$,
$N := \{\varphi : \mathbb{R} \to \mathbb{R}; \ \varphi \text{ cnd (continuous and nowhere differentiable)}\}$.
In all four cases we have $S_0 = S_3, S_1 = S_4, S_2 = S_5$ since all the members in these cases are continuous. Therefore it is sufficient to consider S_ν for $\nu \in \{0, 1, 2\}$.

We have $S_2 \subset S_1 \subset S_0 \subset B_0$ in all cases. The inclusions are proper, all the A_ν, C^n_ν and H_ν are subspaces, the N_ν just subsets. The subsequent remarks are concerned with images with respect to F respectively F^{-1}. They contain some facts but focus much more on open problems.

Remark 5. Let $\alpha \in (0,1)$, $\beta \in \mathbb{N} \setminus \{1\}$ and F given by (1).

(a) Clearly $F[A_\nu]$, $F[C_\nu^n]$, $F[H_\nu]$, $0 \leq \nu \leq 2$, are subspaces of B_0.

Problem 1. *Given $\alpha \in (0,1), \beta \in \mathbb{N} \setminus \{1\}$, find an intrinsic characterization or a basis of $F[A_\nu]$, $F[C_\nu^n]$, $F[H_\nu]$ for $0 \leq \nu \leq 2$.*

Note that $F[c] = w_{\alpha,\beta}$ (see Section 1), hence $F[S_\nu]$ contains cnd functions in all the above cases provided that $\alpha\beta \geq 1$. This example also shows that none of the spaces A_ν, C_ν^n, H_ν, $0 \leq \nu \leq 2$, is F-invariant if $\alpha\beta \geq 1$.

A complete solution of the problem would answer the question: *For which $\alpha \in (0,1)$, $\beta \in \mathbb{N} \setminus \{1\}$ is $F[A_\nu] \subset A_\nu$ respectively $F[C_\nu^n] \subset C_\nu^n$ respectively $F[H_\nu] \subset H_\nu$?*

Results of Behrend [1] and Mikolás [9] indicate that there is an abundance of $\varphi \in C_\nu^n$ or $\varphi \in H_\nu$ such that $F[\varphi]$ is cnd.

(b) $F[N_\nu]$ is not a subspace of B_0 for $0 \leq \nu \leq 2$.

Problem 2. *Given $\alpha \in (0,1)$, $\beta \in \mathbb{N} \setminus \{1\}$, find an intrinsic characterization of $F[N_\nu]$ for $0 \leq \nu \leq 2$.*

$F[N_\nu]$ cannot contain an element of "high" regularity: $F[\varphi] = \psi$ and ψ differentiable at two points ξ and $\beta\xi$ implies $\varphi \notin N_\nu$. In fact, φ then is differentiable at ξ because of $\varphi(x) = F^{-1}[\psi](x) = \psi(x) - \alpha\psi(\beta x)$.

Remark 6. (a) Recall that F preserves boundedness, boundedness and continuity but not mere continuity or differentiability or any of the $A-, C^n-, H-$ properties. This is seen by the counterexample $\varphi(x) = x$ in case $\alpha\beta \geq 1$.

In severe contrast, F^{-1} preserves all these types of regularity. This follows from the representation $F^{-1}[\psi](x) = \psi(x) - \alpha\psi(\beta x)$ and holds for any choice of $\alpha \in (0,1)$, $\beta \in (0,\infty)$.

(b) $F^{-1}[A_\nu]$, $F^{-1}[C_\nu^n]$ and $F^{-1}[H_\nu]$ are subspaces of B_0, $0 \leq \nu \leq 2$. The A_ν, C_ν^n, H_ν are F^{-1} - invariant but all of the inclusions $F^{-1}[A_\nu] \subset A_\nu$, $F^{-1}[C_\nu^n] \subset C_\nu^n$, $F^{-1}[H_\nu] \subset H_\nu$ are proper in the case $\alpha\beta \geq 1$. This follows from $F^{-1}[w_{\alpha,\beta}] = c$.

Problem 3. *Given $\alpha \in (0,1)$, $\beta \in \mathbb{N} \setminus \{1\}$, find an intrinsic characterization or a basis of $F^{-1}[A_\nu]$, $F^{-1}[C_\nu^n]$, $F^{-1}[H_\nu]$ for $0 \leq \nu \leq 2$.*

(c) The images $F^{-1}[N_\nu]$ (which are no subspaces) are especially interesting. For instance, let $\alpha = \frac{1}{2}, \beta = 2$. Then $F^{-1}[N_2]$ contains elements which are in our context maximal regular ($c \in F^{-1}[N_2]$) and maximal irregular ($w_{\frac{1}{2},2} \in F^{-1}[N_2]$).

The nonobvious fact that $F[w_{\frac{1}{2},2}]$ is cnd will be proved in a forthcoming paper. The above examples show also that the sets N_ν are in general not F^{-1} - invariant. Clearly the members of $F^{-1}[N_\nu]$ must be continuous: $\varphi \in F^{-1}[N_\nu]$ implies the continuity of $\psi = F[\varphi]$ and because of $\varphi(x) = \psi(x) - \alpha\psi(\beta x)$, φ must be continuous as well.

Problem 4. *Given* $\alpha \in (0,1)$, $\beta \in \mathbb{N} \setminus \{1\}$, *find an intrinsic characterization of* $F^{-1}[N_\nu]$ *for* $0 \leq \nu \leq 2$.

Remark 7. (a) The operator F given by (1) makes sense also for functions $\varphi : \mathbb{C} \to \mathbb{C}$ and complex parameters α, β.

(b) There are other subspaces of B_0 on which F may act, generating interesting spectral properties. Moreover we could discuss F as well as a mapping on $L^p(\mathbb{R})$ and various of its subspaces.

(c) The list of suitable sets S is by no means exhausted by our examples A, C^n, H, N.

References

[1] F. A. Behrend, *Some remarks on the construction of continuous nondifferentiable functions*, Proc. London Math. Soc. **50** (1949), 463–481.

[2] Z. Gajda, *On stability of additive mappings*, Internat. J. Math. & Math. Sci. **14** (1991), 431–434.

[3] R. Girgensohn, *Functional equations and nowhere differentiable functions*, Aequationes Math. **46** (1993), 243–256.

[4] R. Girgensohn, *Nowhere differentiable solutions of a system of functional equations*, Aequationes Math. **47** (1994), 89–99.

[5] G. H. Hardy, *Weierstrass's nondifferentiable function*, Trans. Amer. Math. Soc. **17** (1916), 301–325.

[6] H.-H. Kairies, *On a Banach space automorphism and its connections to functional equations and cnd functions*, Wyz. Szkola Ped. Krakow. Rocznik Nauk.-Dydakt. Prace Mat. To appear.

[7] K. Knopp, *Ein einfaches Verfahren zur Bildung stetiger nirgends differenzierbarer Funktionen*, Math. Z. **2** (1918), 1–26.

[8] M. Kuczma, *Functional equations in a single variable*, Monografie Mat. 46. Polish Scientific Publishers, Warsaw 1968.

[9] M. Mikolás, *Construction des familles de fonctions partout continues non dérivables*, Acta Sci. Math. (Szeged) **17** (1956), 49–62.

[10] G. de Rham, *Sur un example de fonction continue sans dérivée*, Enseign. Math. **3** (1957), 71–72.

[11] T. Takagi, *A simple example of the continuous function without derivative*, Proc. Phys. Math. Soc. Japan **1** (1903), 176–177.

[12] B. L. v. d. Waerden, *Ein einfaches Beispiel einer nichtdifferenzier-baren stetigen Funktion*, Math. Z. **32** (1930), 474–475.

[13] K. Weierstrass, *Über continuierliche Funktionen eines reellen Ar-guments, die für keinen Werth des letzteren einen bestimmten Dif-ferentialquotienten besitzen*, Adademievortrag 1872. Mathematische Werke II, 71–74. Mayer und Müller, Berlin 1895.

IV
COMPOSITE FUNCTIONAL EQUATIONS AND THEORY OF MEANS

A MATKOWSKI-SUTÔ TYPE PROBLEM FOR QUASI-ARITHMETIC MEANS OF ORDER α

Zoltán Daróczy and Zsolt Páles

Institute of Mathematics and Informatics, University of Debrecen,
H-4010 Debrecen, Pf. 12, Hungary

daroczy@math.klte.hu, pales@math.klte.hu

Dedicated to the 70th birthday of Professor Mátyás Arató

Abstract   In this paper the Matkowski–Sutô problem for quasi-arithmetic means of order α, that is, the functional equation

$$A_\varphi^{(\alpha)}(x,y) + A_\psi^{(\alpha)}(x,y) = x + y \qquad (x,y \in I) \qquad (*)$$

is considered, where φ and ψ are continuous stricly monotonic functions defined on an open real interval I and the mean $A_\varphi^{(\alpha)}$ is defined by

$$A_\varphi^{(\alpha)}(x,y) := \varphi^{-1}\left(\frac{\varphi(x) + \varphi(y) + \alpha\varphi\left(\frac{x+y}{2}\right)}{2+\alpha} \right) \qquad (x,y \in I).$$

Assuming continuous differentiability of one of the functions φ and ψ, all solutions of the equation (*) are determined.

Keywords: quasi-arithmetic mean, quasi-arithmetic mean of order α, Matkowski-Sutô problem

Mathematics Subject Classification (2000): 39B22, 39B12, 26A18, 26A51

[0]This research has been supported by the Hungarian National Research Science Foundation (OTKA) Grant T-030082 and by the Higher Education, Research, and Development Fund (FKFP) Grant 0310/1997.

Z. Daróczy and Z. Páles (eds.),
Functional Equations - Results and Advances, 189–200.
© 2002 *Kluwer Academic Publishers.*

1.   INTRODUCTION

Let $I \subset \mathbb{R}$ be an open interval. We say that a function $M : I^2 \to I$ is a *mean* on I if it satisfies the following conditions

(i) If $x \neq y$ and $x, y \in I$ then $\min\{x, y\} < M(x, y) < \max\{x, y\}$;

(ii) $M(x, y) = M(y, x)$ if $x, y \in I$;

(iii) M is continuous on I^2.

The best-known mean is the *arithmetic mean*

$$A(x, y) := \frac{x + y}{2} \qquad (x, y \in I), \tag{1}$$

which is defined on any interval I.

Let $\mathcal{CM}(I)$ denote the set of all *continuous* and *strictly monotonic* real functions on I.

Definition 1. Let $\alpha \geq -1$. A function $M : I^2 \to I$ is called a *quasi-arithmetic mean of order α* if there exists $\varphi \in \mathcal{CM}(I)$ such that

$$M(x, y) = \varphi^{-1} \left(\frac{\varphi(x) + \varphi(y) + \alpha\varphi\left(\frac{x+y}{2}\right)}{2 + \alpha} \right) =: A_\varphi^{(\alpha)}(x, y) \tag{2}$$

for all $x, y \in I$. Then the function $\varphi \in \mathcal{CM}(I)$ is called the *generating function* of the quasi-arithmetic mean of order α (2).

Remark 1. (1) It is easy to see that $A_\varphi^{(\alpha)} : I^2 \to I$ is a *mean* on I if $\alpha \geq -1$ and $\varphi \in \mathcal{CM}(I)$.
(2) If $\alpha = 0$, then

$$A_\varphi^{(0)}(x, y) = \varphi^{-1} \left(\frac{\varphi(x) + \varphi(y)}{2} \right)$$

is the well-known quasi-arithmetic mean with the generating function $\varphi \in \mathcal{CM}(I)$ (see [1], [2], [8], [12], [13]).
(3) If $\alpha = -1$, then

$$A_\varphi^{(-1)}(x, y) = \varphi^{-1} \left(\varphi(x) + \varphi(y) - \varphi\left(\frac{x+y}{2}\right) \right)$$

is the so-called *conjugate arithmetic mean* with the generating function $\varphi \in \mathcal{CM}(I)$ (see [3], [4], [7]).
(4) If $\alpha = 1$, then

$$A_\varphi^{(1)}(x, y) = \varphi^{-1} \left(\frac{\varphi(x) + \varphi(y) + \varphi\left(\frac{x+y}{2}\right)}{3} \right)$$

is the so-called *mixed arithmetic mean* with the generating function $\varphi \in \mathcal{CM}(I)$.

Definition 2. Let $\varphi, \psi \in \mathcal{CM}(I)$. If there exist real constants $a \neq 0$ and b such that

$$\varphi(x) = a\psi(x) + b \qquad \text{for all} \quad x \in I, \tag{3}$$

then we say that φ is *equivalent* to ψ on I; and, in this case, we write: $\varphi \sim \psi$ on I or $\varphi(x) \sim \psi(x)$ if $x \in I$.

Proposition 1. *Let $\varphi, \psi \in \mathcal{CM}(I)$ and $\alpha \geq -1$. If φ is equivalent to ψ on I, then the means $A_\varphi^{(\alpha)}$ and $A_\psi^{(\alpha)}$ are identical on I^2.*

Matkowski [13] (see earlier Sutô [14], [15]) raised the following problem: For which quasi-arithmetic means M and N does the equality

$$M + N = 2A, \tag{4}$$

hold on I^2? This problem has been partially answered by Sutô [14], [15], Matkowski [13], Daróczy-Páles [7], Daróczy–Maksa [5], and Daróczy–Maksa–Páles [6]). Daróczy [4], investigated the analogous problem for conjugate arithmetic means M and N.

In this paper we examine the Matkowski-Sutô type problem in the class of quasi-arithmetic means of order α (where $\alpha \geq -1$). The results extend those obtained for the cases $\alpha = 0$ and $\alpha = -1$ by Daróczy [3], Daróczy-Páles [7], and Daróczy [4], respectively.

2.   THE FUNCTIONAL EQUATION OF THE PROBLEM, LEMMAS

The subject of this paper is the investigation of the following question. Let $\alpha \geq -1$. Characterize the quasi-arithmetic means M and N of order α satisfying the equation

$$M + N = 2A \tag{5}$$

on I^2. By taking generating functions $\varphi, \psi \in \mathcal{CM}(I)$ such that $M = A_\varphi^{(\alpha)}$ and $N = A_\psi^{(\alpha)}$, (5) is equivalent to the functional equation

$$A_\varphi^{(\alpha)}(x, y) + A_\psi^{(\alpha)}(x, y) = x + y \qquad (x, y \in I). \tag{6}$$

Clearly, it is enough to solve (6) up to equivalence of the generating functions φ and ψ.

Now we prove some lemmas. In the first such result we show that continuous differentiablity of one of the generating functions yields similar regularity property of the other generating function. The proof is based on Lebesgue's differentiablity theorem for monotonic functions and Járai's regularity theorem on measurable solutions of functional equations.

Lemma 1. *If $\varphi, \psi \in \mathcal{CM}(I)$ satisfy the functional equation (6) and ψ is continuously differentiable on I with nonvanishing first derivative, then φ is also continuously differentiable on I.*

Proof. By (6), the function $\varphi \in \mathcal{CM}(I)$ satisfies the functional equation

$$\varphi(x) = (2 + \alpha)\varphi(g_2(x, y)) - \alpha\varphi\left(\frac{x + y}{2}\right) - \varphi(y) \qquad (x, y \in I), \quad (7)$$

where

$$g_2(x, y) := x + y - \psi^{-1}\left(\frac{\psi(x) + \psi(y) + \alpha\psi\left(\frac{x+y}{2}\right)}{2 + \alpha}\right) \qquad (8)$$

if $x, y \in I$. By the assumptions of the lemma, $g_2 : I^2 \to I$ is continuously differentiable and

$$\frac{\partial g_2(x, y)}{\partial x} = 1 - \frac{\psi'(x) + \frac{\alpha}{2}\psi'\left(\frac{x+y}{2}\right)}{(2 + \alpha)\psi'(A_\psi^{(\alpha)}(x, y))} \qquad (x, y \in I) \qquad (9)$$

and a similar expression is valid for $\partial g_2(x, y)/\partial y$.

Now we intend to show that φ is differentiable at each point of I. For, let $x_0 \in I$ be arbitrarily fixed. Then, from (7), we have

$$\varphi(x_0) = (2 + \alpha)\varphi(g_2(x_0, y)) - \alpha\varphi\left(\frac{x_0 + y}{2}\right) - \varphi(y) \qquad (y \in I). \quad (10)$$

Define the function $h : I \to \mathbb{R}$ by

$$h(y) := g_2(x_0, y).$$

Then h is continuously differentiable and $h'(x_0) = 1/2$. Therefore h is invertible in a neighbourhood U of x_0 and its inverse is also continuously differentiable on $h(U) = V$, which is a neighbourhood of $h(x_0) = x_0$. Hence h^{-1} is locally Lipschitz in V, whence it follows that $h^{-1}(H)$ is of measure zero whenever $H \subset V$ is of measure zero.

By Lebesgue's theorem, the function φ is differentiable almost everywhere, that is, the complement of the set D where φ is differentiable is of measure zero. Hence $h^{-1}(V \setminus D)$ is also of measure zero. Thus the set $[(2D - x_0) \cap U] \setminus h^{-1}(V \setminus D)$ cannot be empty. Let y_0 be an arbitrary element of this set. Then $y_0 \in U$, $\frac{x_0+y_0}{2} \in D$, and $y_0 \notin h^{-1}(V \setminus D)$, i.e., $h(y_0) \in D$. Thus φ is differentiable at the points $\frac{x_0+y_0}{2}$ and $g_2(x_0, y_0)$. The functions $x \mapsto \frac{x+y_0}{2}$ and $x \mapsto g_2(x, y_0)$ are differentiable at x_0 and, by (7),

$$\varphi(x) = (2 + \alpha)\varphi(g_2(x, y_0)) - \alpha\varphi\left(\frac{x + y_0}{2}\right) - \varphi(y_0) \qquad (x \in I),$$

from which, by the chain rule, we obtain that φ is differentiable at x_0. Thus φ is differentiable on I everywhere.

Differentiating the equation (7) with respect to x, and using (9), we get

$$\varphi'(x) = (2 + \alpha)\varphi'(g_2(x, y)) \left(1 - \frac{\psi'(x) + \frac{\alpha}{2}\psi'\left(\frac{x+y}{2}\right)}{\psi'(A_\psi^{(\alpha)}(x, y))}\right) - \frac{\alpha}{2}\varphi'\left(\frac{x + y}{2}\right)$$

$$(11)$$

for all $x, y \in I$. Equation (11) can be written as

$$\varphi'(x) = f(x, y; \varphi'(g_1(x, y)), \varphi'(g_2(x, y))),$$

where $g_1(x, y) := \frac{x+y}{2}$, g_2 is the function defined in (8), and

$$f(x, y; u, v) = -\frac{\alpha}{2}u + (2 + \alpha)v\frac{\partial g_2(x, y)}{\partial x}$$

for all $x, y \in I$ and $u, v \in \mathbb{R}$. Since $\varphi' : I \to \mathbb{R}$ is a *measurable* function and $\frac{\partial g_1}{\partial y}(x, x) = \frac{1}{2} \neq 0$ and $\frac{\partial g_2}{\partial y}(x, x) = \frac{1}{2} \neq 0$, moreover f is continuous, thus, it follows from the theorem of Járai [10, Theorem 2] that φ' is continuous on I. (See also the papers [9] and [11].) $\square$

Lemma 2. *Let $f, g : I \to \mathbb{R}$ be continuous functions with $g(x) \neq 0$ if $x \in I$. If $M(x, y) := x \circ y$ and $N(x, y) := x * y$ are means on I and the functional equation*

$$\frac{f(x * y)}{g(x \circ y)}(g(y) - g(x)) + f(y) - f(x) = 0 \qquad (12)$$

holds for all $x, y \in I$ then there exists a constant $c \in \mathbb{R}$ such that

$$f(x)g(x) = c \qquad if \quad x \in I. \qquad (13)$$

Proof. If $g(y) = g(x)$ then by (12) $f(y) = f(x)$ thus there exists a function $F : g(I) \to \mathbb{R}$ such that

$$f(x) = F(g(x)) \qquad \text{if} \quad x \in I. \tag{14}$$

If g is constant on I then f is constant on I, too and then (13) obviously holds. Therefore we can assume that g is not constant on I. Then, by the continuity of g, $J := g(I)$ is a nonvoid open interval and $0 \notin J$. We shall show that $F : J \to \mathbb{R}$ is differentiable on J.

Let $u \in J$ and let $(u_n) \subset J$ $(n \in \mathbb{N})$ be any sequence converging to u from the *left* $(u_n < u)$ (or from the right $(u_n > u)$). It is enough to show that, for any sequence with this property,

$$\frac{F(u_n) - F(u)}{u_n - u}$$

tends to the same number (depending on u). Let

$$u_0 := \inf\{u_n \mid n \in \mathbb{N}\} = \min\{u_n\}.$$

Then there exist $x_0 \in I$ for which $g(x_0) = u_0$ and $x^* \in I$ for which $g(x^*) = u$. We can assume that $x_0 < x^*$ (the other case can be handled similarly). Let

$$H := \{t \in I \mid x_0 \le t \le x^* \text{ and } g(t) = u\}.$$

Then H is a closed set and, since $x^* \in H$, we have $H \ne \emptyset$. Therefore there exists $x := \inf H$ and $x > x_0$. The continuity of g yields that $g(x) = u$ and if $x_0 \le t < x$ then $g(t) \ne u$. The function g takes every value between u and u_0 in the closed interval $[x_0, x]$ thus there exist $x_n \in [x_0, x[$ for which $g(x_n) = u_n$ $(n \in \mathbb{N})$. We show that $x_n \to x$ $(n \to \infty)$. If this were not the case, then there would exist a convergent subsequence (x_{n_k}) $(n_1 < n_2 < n_3 \ldots)$ tending to $\bar{x} \ne x$ and for which necessarily $\bar{x} < x$ would hold. Then, by the continuity of g, $g(x_{n_k}) \to g(\bar{x})$ $(k \to \infty)$ and $g(x_{n_k}) = u_{n_k} \to u = g(\bar{x})$ $(k \to \infty)$. Thus $g(\bar{x}) = g(x)$, which contradicts the definition of x. Hence $x_n \to x$ as $n \to \infty$.

It follows from equation (12) that

$$\frac{f(x_n * x)}{g(x_n \circ x)} = -\frac{f(x_n) - f(x)}{g(x_n) - g(x)} = -\frac{F(u_n) - F(u)}{u_n - u}. \tag{15}$$

If $n \to \infty$ in (15) then, by the mean property,

$$\lim_{n \to \infty} \frac{f(x_n * x)}{g(x_n \circ x)} = \frac{f(x)}{g(x)} = \frac{F(g(x))}{g(x)} = \frac{F(u)}{u}.$$

Therefore, the limit of the right hand side of (15) exists. Thus

$$\frac{F(u)}{u} = -\lim_{n\to\infty} \frac{F(u_n) - F(u)}{u_n - u}$$

follows. Hence, F is differentiable (we obtain the same for the right derivative) and

$$\frac{F(u)}{u} = -F'(u) \qquad (u \in J = g(I)).$$

This implies $(uF(u))' = 0$, that is, $F(u) = \dfrac{c}{u}$ ($u \in J, 0 \notin J$, and c is a constant). Therefore, from (14), we get $f(x) = \dfrac{c}{g(x)}$ ($x \in I$), i.e., $f(x)g(x) = c$ holds for all $x \in I$. □

Lemma 3. *Let* $l : I \to \mathbb{R}$ *be a continuous function and* $M(x,y) := x \circ y$ *be a mean on* I. *Then the functional equation*

$$(l(x + y - x \circ y) + l(x \circ y) - l(x) - l(y))(l(x) - l(y)) = 0 \qquad (16)$$

holds for all $x, y \in I$ *if and only if there exist* $a, b \in \mathbb{R}$ *such that*

$$l(x) = ax + b \qquad (x \in I). \qquad (17)$$

Proof. It is clear that the function (17) is a continuous solution of (16).

We suppose that the continuous function $l : I \to \mathbb{R}$ satisfies (16) and — contrary to the assertion of the Lemma — l is nonaffine on I (i.e., l is not of the form (17)). Then there exist $\alpha < \beta$, $(\alpha, \beta \in I)$ such that $l(\alpha) \neq l(\beta)$ and

$$\text{Graph}\, l \not\equiv \text{Graph}\, L,$$

where

$$L(x) := \frac{l(\beta) - l(\alpha)}{\beta - \alpha} x + \frac{l(\beta)\alpha - l(\alpha)\beta}{\beta - \alpha} \qquad (x \in I)$$

and

$$\text{Graph}\, l := \{(x, l(x)) \mid x \in I\}, \qquad \text{Graph}\, L := \{(x, L(x)) \mid x \in I\}.$$

It is clear that

$$(\alpha, l(\alpha)), (\beta, l(\beta)) \in \text{Graph}\, l \cap \text{Graph}\, L,$$

therefore — by the continuity of l — there exist $\alpha^* < \beta^*$, $(\alpha^*, \beta^* \in I)$ such that

$$(\alpha^*, l(\alpha^*)), (\beta^*, l(\beta^*)) \in \text{Graph}\, l \cap \text{Graph}\, L$$

but, for all $t \in]\alpha^*, \beta^*[$, we have that $(t, l(t)) \notin \operatorname{Graph} L$. For example, we may assume that $l(t) > L(t)$ for all $t \in]\alpha^*, \beta^*[$ (the opposite case is similar). Putting $x = \alpha^*, y = \beta^*$ in (16), we have

$$
\begin{aligned}
L(\alpha^*) + L(\beta^*) &= l(\alpha^*) + l(\beta^*) \\
&= l(\alpha^* + \beta^* - \alpha^* \circ \beta^*) + l(\alpha^* \circ \beta^*) \\
&> L(\alpha^* + \beta^* - \alpha^* \circ \beta^*) + L(\alpha^* \circ \beta^*) \\
&= L(\alpha^*) + L(\beta^*),
\end{aligned}
$$

which is an obvious contradiction. $\qquad\qquad\qquad\qquad\qquad\qquad\square$

3.   THE MAIN RESULTS

Theorem 1. *If $\varphi, \psi \in \mathcal{CM}(I)$ satisfy the functional equation (6) and ψ is continuously differentiable on I and $\psi'(x) \neq 0$ if $x \in I$ then there exists $p \in \mathbb{R}$ such that*

$$
\varphi(x) \sim \chi_p(x) \qquad \text{and} \qquad \psi(x) \sim \chi_{-p}(x) \qquad \text{if } x \in I, \qquad (18)
$$

where

$$
\chi_p(x) := \begin{cases} x & \text{if } p = 0 \\ e^{px} & \text{if } p \neq 0 \end{cases} \qquad (x \in I). \qquad (19)
$$

Proof. Let

$$
x \circ y := A_\psi^{(\alpha)}(x, y) = \psi^{-1} \left(\frac{\psi(x) + \psi(y) + \alpha\psi\left(\frac{x+y}{2}\right)}{2 + \alpha} \right)
$$

for all $x, y \in I$, where $\alpha \geq -1$.

By Lemma 1, φ is continuously differentiable on I, and thus both sides of the equation (6)

$$
\varphi(x) + \varphi(y) + \alpha\varphi\left(\frac{x+y}{2}\right) = (2 + \alpha)\varphi(x + y - x \circ y) \qquad (x, y \in I)
$$

are differentiable on I with respect to x, that is,

$$
\varphi'(x) + \frac{\alpha}{2}\varphi'\left(\frac{x+y}{2}\right)
$$

$$
= (2 + \alpha)\varphi'(x + y - x \circ y) \left(1 - \frac{\psi'(x) + \frac{\alpha}{2}\psi'\left(\frac{x+y}{2}\right)}{(2 + \alpha)\psi'(x \circ y)} \right) \qquad (20)
$$

for all $x, y \in I$. From this, we have

$$\varphi'(x) + \frac{\varphi'(x + y - x \circ y)}{\psi'(x \circ y)} \psi'(x) = : x \Box y = y \Box x$$

$$= \varphi'(y) + \frac{\varphi'(x + y - x \circ y)}{\psi'(x \circ y)} \psi'(y),$$

which implies

$$\frac{\varphi'(x + y - x \circ y)}{\psi'(x \circ y)} (\psi'(y) - \psi'(x)) + \varphi'(y) - \varphi'(x) = 0 \qquad (21)$$

for all $x, y \in I$. By Lemma 1 and Lemma 2, with $f := \varphi', g := \psi'$ and $x * y := x + y - x \circ y$ we have $\varphi'(x)\psi'(x) = c$ if $x \in I$, where $c \neq 0$, since φ and ψ are strictly monotonic.

It is trivial that $\varphi'(x) \neq 0$ if $x \in I$ and, by $\psi' = \dfrac{c}{\varphi'}$, from (21) we obtain

$$\left(\frac{\varphi'(x + y - x \circ y)\varphi'(x \circ y)}{\varphi'(x)\varphi'(y)} - 1 \right) (\varphi'(x) - \varphi'(y)) = 0 \qquad (22)$$

for all $x, y \in I$. Let $\varepsilon_\varphi := 1$ if $\varphi'(x) > 0$ for all $x \in I$ and $\varepsilon_\varphi := -1$ if $\varphi'(x) > 0$ for all $x \in I$. Then the function l defined by

$$l(x) := \log \varepsilon_\varphi \varphi'(x) \qquad (x \in I) \qquad (23)$$

is continuous and satisfies the functional equation

$$\left(e^{l(x+y-x\circ y)+l(x\circ y)-l(x)-l(y)} - 1 \right) \left(e^{l(x)-l(y)} \right) e^{l(y)} = 0$$

for all $x, y \in I$, i.e., the functional equation

$$(l(x + y - x \circ y) + l(x \circ y) - l(x) - l(y))(l(x) - l(y)) = 0 \qquad (24)$$

holds for all $x, y \in I$. By Lemma 3, there exist $p, q \in \mathbb{R}$ such that

$$l(x) = px + q \qquad \text{if} \quad x \in I. \qquad (25)$$

From (23) and (25) we have

$$\varphi'(x) = \frac{1}{\varepsilon_\varphi} e^{px+q} \qquad \text{if} \quad x \in I,$$

from which we obtain

$$\varphi(x) = \begin{cases} \frac{1}{\varepsilon_\varphi} e^q x + c & \text{if } p = 0 \\ \frac{1}{\varepsilon_\varphi} \frac{1}{p} e^q e^{px} + c & \text{if } p \neq 0, \end{cases}$$

i.e., $\varphi(x) \sim \chi_p(x)$ if $x \in I$. From (6) we have $\psi(x) \sim \chi_{-p}(x)$ if $x \in I$. $\square$

Theorem 2. *If $\varphi, \psi \in \mathcal{CM}(I)$ satisfy the functional equation (6) and φ or ψ is continuously differentiable on I, then there exists $p \in \mathbb{R}$ such that (18) holds.*

Proof. By the symmetry of (6), we can suppose that ψ is continuously differentiable on I. Let

$$N_\psi := \{x \mid x \in I, \ \psi'(x) = 0\}.$$

Then N_ψ is closed (since ψ' is continuous on I). Thus $I \setminus N_\psi$ is open. There are two possible cases: either $N_\psi = \emptyset$ or $N_\psi \neq \emptyset$. We show that the latter is impossible.

Suppose that $N_\psi \neq \emptyset$. Then there exists a maximal component $K \subset I \setminus N_\psi$, that is, an open interval $K =]a, b[$ such that at least one of its endpoints — let it be b — belongs to I, and so $b \in N_\psi$.

Then $\varphi, \psi \in CM(K)$ and they satisfy the functional equation (6) for all $x, y \in K$ and $\psi'(x) \neq 0$ if $x \in K$. By Theorem 1, there exists $p \in \mathbb{R}$ such that $\psi(x) \sim \chi_{-p}(x)$ if $x \in K$. From this assertion, we have

$$\psi(x) = \begin{cases} Ax + B & \text{if } p = 0 \\ Ae^{-px} + B & \text{if } p \neq 0 \end{cases} \quad (x \in K), \tag{26}$$

where $A \neq 0$. By the continuity of ψ' on I ($K \subset I$), we have

$$\lim_{x \to b, x \in K} \psi'(x) = \psi'(b) = 0.$$

However, this is a contradiction, because by (26), we obtain

$$\psi'(b) = \lim_{x \to b, x \in K} \psi'(x)$$
$$= \lim_{x \to b, x \in K} \begin{cases} A & \text{if } p = 0 \\ -pAe^{-px} & \text{if } p \neq 0 \end{cases}$$
$$= \begin{cases} A & \text{if } p = 0 \\ -pAe^{-pb} & \text{if } p \neq 0 \end{cases} \neq 0.$$

Thus $N_\psi = \emptyset$, that is, $\psi'(x) \neq 0$ for all $x \in I$. Theorem 2 follows from Theorem 1 immediately. $\square$

Now we can state the main result of our investigations.

Theorem 3. *If the quasi-arithmetic means M and N of order α ($\alpha \geq -1$) satisfy the functional equation (5) on I, and either of them has a* *continuously differentiable generating function, then there exists $p \in \mathbb{R}$ such that*

$$M(x,y) = D_p^{(\alpha)}(x,y) \quad \text{and} \quad N(x,y) = D_{-p}^{(\alpha)}(x,y) \qquad (x,y \in I), \quad (27)$$

where

$$D_p^{(\alpha)}(x,y) := \begin{cases} \dfrac{x+y}{2} & \text{if } p = 0 \\ \dfrac{1}{p}\log\left(\dfrac{e^{px} + e^{py} + \alpha e^{p\frac{x+y}{2}}}{2+\alpha}\right) & \text{if } p \neq 0 \end{cases} \qquad (x,y \in I).$$

$$(28)$$

Proof. There exist generating functions $\psi, \varphi \in \mathcal{CM}(I)$ such that $M = A_\varphi^{(\alpha)}$ and $N = A_\psi^{(\alpha)}$, furthermore, the functional equation (6) is true. By Theorem 2, there exists $p \in \mathbb{R}$ such that $\varphi \sim \chi_p$ and $\psi \sim \chi_{-p}$ on I. Obviously $M = A_\varphi^{(\alpha)} = A_{\chi_p}^{(\alpha)} = D_p^{(\alpha)}$ and $N = A_\psi^{(\alpha)} = A_{\chi_{-p}}^{(\alpha)} = D_{-p}^{(\alpha)}$, where $D_p^{(\alpha)}$ is the mean defined in (28), that is, (27) holds. $\qquad \square$

References

[1] J. Aczél, *Lectures on Functional Equations and Their Applications*, Academic Press, New York–London, 1966.

[2] J. Aczél and J. Dhombres, *Functional Equations in Several Variables (With Applications to Mathematics, Information Theory and to the Natural and Social Sciences)*, Cambridge University Press, Cambridge, 1989.

[3] Z. Daróczy, *On a class of means of two variables*, Publ. Math. Debrecen **55** (1999), 177–197.

[4] Z. Daróczy, *Matkowski-Sutô type problem for conjugate arithmetic means*, Rocznik Nauk.-Dydakt. Prace Mat. **17** (2000), 89–100, Dedicated to Professor Zenon Moszner on his 70th birthday.

[5] Z. Daróczy and Gy. Maksa, *On a problem of Matkowski*, Colloq. Math. **82** (1999), 117–123.

[6] Z. Daróczy, Gy. Maksa, and Zs. Páles, *Extension theorems for the Matkowski-Sutô problem*, Demonstr. Math. **33** (2000), 547–556.

[7] Z. Daróczy and Zs. Páles, *On means that are both quasi-arithmetic and conjugate arithmetic*, Acta Math. Hungar. **90** (2001), 271–282.

[8] G. H. Hardy, J. E. Littlewood, and G. Pólya, *Inequalities*, Cambridge University Press, Cambridge, 1934.

[9] A. Járai, *On measurable solutions of functional equations*, Publ. Math. Debrecen **26** (1979), 17–35.

[10] A. Járai, *Regularity properties of functional equations*, Aequationes Math. **25** (1982), 52–66.

[11] A. Járai, *On regular solutions of functional equations*, Aequationes Math. **30** (1986), 21–54.

[12] M. Kuczma, *An Introduction to the Theory of Functional Equations and Inequalities*, Państwowe Wydawnictwo Naukowe & Uniwersytet Śląski, Warszawa–Kraków–Katowice, 1985.

[13] J. Matkowski, *Invariant and complementary quasi-arithmetic means*, Aequationes Math. **57** (1999), 87–107.

[14] O. Sutô, *Studies on some functional equations I*, Tôhoku Math. J. **6** (1914), 1–15.

[15] O. Sutô, *Studies on some functional equations II*, Tôhoku Math. J. **6** (1914), 82–101.

AN EXTENSION THEOREM FOR THE MATKOWSKI–SUTÔ PROBLEM FOR CONJUGATE ARITHMETIC MEANS

Gabriella Hajdu

Institute of Mathematics and Informatics, University of Debrecen,
H-4010 Debrecen, Pf. 12, Hungary

hajdug@math.klte.hu

Abstract   Let $I \subset \mathbb{R}$ be an open interval, and $\mathcal{CM}(I)$ denote the set of all continuous and strictly monotonic real functions.

We consider the Matkowski–Sutô type functional equation for conjugate arithmetic means, that is,

$$\varphi^{-1}\left(\varphi(x) + \varphi(y) - \varphi\left(\frac{x+y}{2}\right) \right)$$
$$+ \psi^{-1}\left(\psi(x) + \psi(y) - \psi\left(\frac{x+y}{2}\right) \right) = x + y$$

for all $x, y \in I$, where $\varphi, \psi \in \mathcal{CM}(I)$.

In this paper we prove that if $K =]a, b[\subset I$ then the solutions of the equation can be uniquely extended from K to I.

Keywords: quasi–arithmetic mean, conjugate arithmetic mean, functional equation

Mathematics Subject Classification (2000): 39B22, 39B12, 26A18

1.   INTRODUCTION

Let $I \subset \mathbb{R}$ be an open interval. We say that a function $M : I^2 \to I$ is a *mean* on I if it satisfies the following conditions

(i) If $x \neq y$ and $x, y \in I$ then $\min\{x, y\} < M(x, y) < \max\{x, y\}$;

---

[0]This research has been supported by the Hungarian National Research Science Foundation (OTKA) Grant T-0300082 and by the Higher Education, Research, and Development Fund (FKFP) Grant 0310/1997.

Z. Daróczy and Z. Páles (eds.),
Functional Equations - Results and Advances, 201–208.
© 2002 Kluwer Academic Publishers.

(ii) $M(x, y) = M(y, x)$ if $x, y \in I$;

(iii) M is continuous on I^2.

The best-known mean is the arithmetic mean

$$A(x, y) := \frac{x + y}{2} \qquad (x, y \in I),$$

which is defined on any interval I.

Let $\mathcal{CM}(I)$ denote the set of all continuous and strictly monotonic real functions on I.

Definition 1. ([3]) A function $M : I^2 \to I$ is called a *conjugate arithmetic mean* on I if there exists $\varphi \in \mathcal{CM}(I)$ such that

$$M(x, y) = \varphi^{-1}\left(\varphi(x) + \varphi(y) - \varphi\left(\frac{x + y}{2}\right)\right) =: A_\varphi^*(x, y) \qquad (x, y \in I).$$

Then the function $\varphi \in \mathcal{CM}(I)$ is called the *generating function* of the conjugate arithmetic mean on I.

Definition 2. Let $\varphi, \psi \in \mathcal{CM}(I)$. If there exist real constants $\alpha \neq 0$ and β such that

$$\varphi(x) = \alpha\psi(x) + \beta \qquad \text{if} \quad x \in I,$$

then we say that φ is *equivalent* to ψ on I; and, in this case, we write: $\varphi \sim \psi$ on I or $\varphi(x) \sim \psi(x)$ $(x \in I)$.

The following result is known concerning the equivalence of generating functions (cf. [3]).

Theorem 1. *If $\varphi, \psi \in \mathcal{CM}(I)$ then the equation $A_\varphi^* \equiv A_\psi^*$ holds on I^2 if and only if φ is equivalent to ψ on I.*

In a paper Daróczy [2] raised the following problem.
Which conjugate arithmetic means M and N satisfy the equation

$$M(x, y) + N(x, y) = x + y?$$

It means that the generating function φ of M and the generating function ψ of N satisfy

$$\varphi^{-1}\left(\varphi(x) + \varphi(y) - \varphi\left(\frac{x + y}{2}\right)\right) +$$

$$\psi^{-1}\left(\psi(x) + \psi(y) - \psi\left(\frac{x + y}{2}\right)\right) = x + y \tag{1}$$

for all $x, y \in I$.

Daróczy proved the following

Theorem 2. *If $\varphi, \psi \in \mathcal{CM}(I)$ satisfy* (1) *and either φ or ψ is twice continuously differentiable on I then there exists $p \in \mathbb{R}$ for which*

$$\varphi(x) \sim \chi_{-p}(x), \qquad \psi(x) \sim \chi_p(x) \qquad (x \in I), \tag{2}$$

where

$$\chi_p(x) := \begin{cases} x & \text{if } p = 0 \\ e^{px} & \text{if } p \neq 0 \end{cases} \qquad (x \in I).$$

Earlier Sutô [9], [10] later Matkowski [8], Daróczy, Maksa, and Páles [5], [7] examined the same question for quasi–arithmetic means (see [1]). A mean M is called quasi–arithmetic if there exists $\varphi \in \mathcal{CM}(I)$ such that

$$M(x, y) = \varphi^{-1}\left(\frac{\varphi(x) + \varphi(y)}{2}\right) =: A_\varphi(x, y) \qquad (x, y \in I).$$

The question in this case is which quasi–arithmetic means A_φ and A_ψ satisfy the equation

$$A_\varphi(x, y) + A_\psi(x, y) = x + y \tag{3}$$

for all $x, y \in I$?

They proved that under some regularity or comparison conditions if functions $\varphi, \psi \in \mathcal{CM}(I)$ satisfy (3) then there exists $p \in \mathbb{R}$ such that (2) holds.

Daróczy, Maksa, and Páles also proved an extension theorem [4].

Theorem 3. *If $\varphi, \psi \in \mathcal{CM}(I)$ satisfy* (3) *on I and there exists a nonvoid open interval $K \subset I$ such that $\varphi \sim \chi_p$ and $\psi \sim \chi_{-p}$ on K for some $p \in \mathbb{R}$ then $\varphi \sim \chi_p$ and $\psi \sim \chi_{-p}$ on I.*

In this paper we prove an extension theorem for conjugate arithmetic means. In more detail, we examine the following problem.

Suppose that $\varphi, \psi \in \mathcal{CM}(I)$ satisfy (1) on I and there exists a nonvoid open interval $K \subset I$ such that $\varphi \sim \chi_p$ and $\psi \sim \chi_{-p}$ on K for some $p \in \mathbb{R}$. Is it true then that $\varphi \sim \chi_p$ and $\psi \sim \chi_{-p}$ on I? In other words, can the solutions be extended from K to I?

2.   PRELIMINARY EXAMINATIONS

Let $\varphi, \psi \in \mathcal{CM}(I)$ satisfy (1) on I and let $K =]a, b[\subset I$ on which $\varphi \sim \chi_p$ and $\psi \sim \chi_{-p}$ for some $p \in \mathbb{R}$. Then without loss of generality we can suppose that

$$\varphi(x) = \chi_p(x), \qquad \psi(x) = \chi_{-p}(x) \qquad (x \in K).$$

Our aim is to show that if K is a proper subset of I then there exists an open interval $K_1 \subset I$ such that K is a proper subset of K_1 and

$$\varphi(x) = \chi_p(x), \qquad \psi(x) = \chi_{-p}(x) \qquad (x \in K_1).$$

This implies $\varphi = \chi_p$ and $\psi = \chi_{-p}$ on I for the solutions can be extended from K to K_1, from K_1 to K_2, etc., and the union $J := \bigcup_{i \in \mathbb{N}} K_i$ cannot be a proper subinterval of I because $J \subset I$ is an open interval therefore the solutions could be extended from J.

We assume that $K =]a, b[$ is a proper subinterval of I and $a \in I$ (the other case when $b \in I$ is similar). We can also assume that $b \in I$, for if it were not so there would be $b' \in I$, $a < b' < b$ and it is enough to prove our assertion for $]a, b'[\subset I$.

First we prove a lemma.

Lemma. *If $\varphi, \psi \in \mathcal{CM}(I)$, $K =]a, b[\subset I$, $a, b \in I$ then there exists $\delta > 0$ such that for all $x \in [a - \delta, a] \subset I$ and $y \in]b - \delta, b[\subset]a, b[$ the following hold*

$$\frac{x + y}{2} \in \,]a, b[,$$

$$\varphi(x) + \varphi(y) - \varphi\left(\frac{x + y}{2}\right) \in \varphi(]a, b[), \tag{4}$$

$$\psi(x) + \psi(y) - \psi\left(\frac{x + y}{2}\right) \in \psi(]a, b[).$$

Proof. If $\delta_0 < \dfrac{b - a}{2}$ then for all $x \in [a - \delta_0, a]$ and $y \in]b - \delta_0, b[$, $\dfrac{x + y}{2} \in \,]a, b[$.

Now suppose that φ is *increasing*. Then $\varphi(x) + \varphi(y) - \varphi\left(\dfrac{x + y}{2}\right) \in$ $\varphi(]a, b[)$ if and only if $\varphi(a) < \varphi(x) + \varphi(y) - \varphi\left(\dfrac{x + y}{2}\right) < \varphi(b)$. For any $\delta > 0$ if $x \in [a - \delta, a]$, $y \in]b - \delta, b[$

$$\varphi(x) + \varphi(y) - \varphi\left(\frac{x + y}{2}\right) > \varphi(a - \delta) + \varphi(b - \delta) - \varphi\left(\frac{a + b}{2}\right)$$

and $\varphi(a - \delta) + \varphi(b - \delta) - \varphi\left(\dfrac{a + b}{2}\right) > \varphi(a)$ if and only if

$$\varphi(b - \delta) - \varphi\left(\frac{a + b}{2}\right) > \varphi(a) - \varphi(a - \delta).$$

Since the right hand side tends to 0 as $\delta \to 0$ (φ is continuous) and the left hand side is positive if δ is small enough, there exists $\delta_1 > 0$ such that for all $x \in [a - \delta_1, a]$ and $y \in]b - \delta_1, b[$ $\varphi(a) < \varphi(x) + \varphi(y) - \varphi\left(\dfrac{x+y}{2}\right)$. On the other hand, $\varphi(x) + \varphi(y) - \varphi\left(\dfrac{x+y}{2}\right) < \varphi(a) + \varphi(b) - \varphi\left(\dfrac{a - \delta_1 + b - \delta_1}{2}\right) < \varphi(a) + \varphi(b) - \varphi(a) = \varphi(b)$ if $\delta_1 < \dfrac{b-a}{2}$.

We can use the same argument if ψ is increasing and a similar one if ψ is decreasing. In this case $\psi(x) + \psi(y) - \psi\left(\dfrac{x+y}{2}\right) \in \psi(]a, b[)$ if and only if $\psi(b) < \psi(x) + \psi(y) - \varphi\left(\dfrac{x+y}{2}\right) < \psi(a)$. For any $\delta > 0$ if $x \in [a - \delta, a]$, $y \in]b - \delta, b[$

$$\psi(x) + \psi(y) - \psi\left(\frac{x+y}{2}\right) < \psi(a - \delta) + \psi(b - \delta) - \psi\left(\frac{a+b}{2}\right)$$

and $\psi(a - \delta) + \psi(b - \delta) - \psi\left(\dfrac{a+b}{2}\right) < \psi(a)$ if and only if

$$\psi(a - \delta) - \psi(a) < \psi\left(\frac{a+b}{2}\right) - \psi(b - \delta).$$

Since the left hand side tends to 0 as $\delta \to 0$ (ψ is continuous) and the right hand side is positive if δ is small enough, there exists $\delta_2 > 0$ such that for all $x \in [a - \delta_2, a]$ and $y \in]b - \delta_2, b[$ $\psi(x) + \psi(y) - \psi\left(\dfrac{x+y}{2}\right) < \psi(a)$. On the other hand, $\psi(x) + \psi(y) - \psi\left(\dfrac{x+y}{2}\right) > \psi(a) + \psi(b) - \psi\left(\dfrac{a - \delta_2 + b - \delta_2}{2}\right) > \psi(a) + \psi(b) - \psi(a) = \psi(b)$ if $\delta_2 < \dfrac{b-a}{2}$.

$\delta := \min\{\delta_0, \delta_1, \delta_2\}$ satisfies the conditions. $\square$

3.   THE MAIN RESULT

Theorem. Let $\varphi, \psi \in \mathcal{CM}(I)$ satisfy (1) on I and let $K =]a, b[\subset I$ such that $\varphi \sim \chi_p$ and $\psi \sim \chi_{-p}$ on K for some $p \in \mathbb{R}$. Then $\varphi \sim \chi_p$ and $\psi \sim \chi_{-p}$ on I.

Proof. By Section 2, we can suppose that $\varphi(x) = \chi_p(x)$ and $\psi(x) = \chi_{-p}(x)$ ($x \in K$), and it is enough to show that the solutions φ, ψ can be extended from $]a, b[$ to $]a - \delta, b[\subset I$, where $\delta > 0$. Therefore we choose δ by the Lemma.

Now there are two possible cases. Either *(i)* $p \neq 0$ or *(ii)* $p = 0$. First we examine case *(i)*.

Then $\varphi(x) = e^{px}$ and $\psi(x) = e^{-px}$ for all $x \in]a, b[$. Let $x \in [a - \delta, a], y \in]b - \delta, b[$. Then, by (4), we can rewrite (1) as

$$\frac{1}{p} \log \left(\varphi(x) + e^{py} - e^{p\frac{x+y}{2}} \right) - \frac{1}{p} \log \left(\psi(x) + e^{-py} - e^{-p\frac{x+y}{2}} \right) = x + y,$$

from which we have

$$\frac{\varphi(x) + e^{py} - e^{p\frac{x+y}{2}}}{\psi(x) + e^{-py} - e^{-p\frac{x+y}{2}}} = e^{px+py},$$

that is,

$$\varphi(x) + e^{py} - e^{p\frac{x+y}{2}} = e^{px} e^{py} \psi(x) + e^{px} - e^{p\frac{x+y}{2}},$$

so

$$\varphi(x) - e^{px} = e^{py} (e^{px} \psi(x) - 1).$$

Since y takes its values from an interval, this yields

$$e^{px} \psi(x) - 1 = 0$$

and from this we have

$$\psi(x) = e^{-px}, \qquad \varphi(x) = e^{px} \qquad (x \in [a - \delta, a]).$$

Thus $\psi(x) = e^{-px}$, $\varphi(x) = e^{px}$ on the whole of $]a - \delta, b[$, which was to be proved.

In case *(ii)*, when $p = 0$, $\varphi(x) = \psi(x) = x$ $(x \in]a, b[)$. Again, let $x \in [a - \delta, a], y \in]b - \delta, b[$ then by (4), (1) can be rewritten as

$$\varphi(x) + y - \frac{x + y}{2} + \psi(x) + y - \frac{x + y}{2} = x + y,$$

from which

$$\varphi(x) + \psi(x) = 2x,$$

that is,

$$\psi(x) = 2x - \varphi(x) \qquad (x \in [a - \delta, a]). \tag{5}$$

Now let $x, y \in [a - \delta, a]$. Then, by (5), (1) implies

$$\varphi^{-1} \left(\varphi(x) + \varphi(y) - \varphi \left(\frac{x + y}{2} \right) \right)$$

$$+ \psi^{-1} \left(2x - \varphi(x) + 2y - \varphi(y) + \varphi \left(\frac{x + y}{2} \right) - x - y \right) = x + y,$$

which implies

$$\psi^{-1}\left(x - \varphi(x) + y - \varphi(y) + \varphi\left(\frac{x+y}{2}\right)\right) = x + y - A_\varphi^*(x,y),$$

from which, applying ψ to both sides then (5) again,

$$\varphi(x + y - A_\varphi^*(x,y)) - (x + y - A_\varphi^*(x,y)) = \varphi(A_\varphi^*(x,y)) - A_\varphi^*(x,y)$$

follows for all $x, y \in [a - \delta, a]$. Introducing the notation $f(t) := \varphi(t) - t$, we have

$$f(x + y - A_\varphi^*(x,y)) = f(A_\varphi^*(x,y)) \qquad (x, y \in [a - \delta, a]).$$

By a result of Daróczy and Ng [6], there exists $\alpha \neq 0$ and β such that $\varphi(x) = \alpha x + \beta$ for all $x \in [a - \delta, a]$. We shall show that $\alpha = 1$ and $\beta = 0$. By (5), then also $\psi(x) = x$ ($x \in [a - \delta, a]$).

From (5) we have that

$$\psi(x) = 2x - \varphi(x) = (2 - \alpha)x - \beta \qquad (x \in [a - \delta, a]).$$

Now let $x \in [a - \delta, a]$ and let $y \in [a, b]$ such that $\frac{x+y}{2} \in [a - \delta, a]$, $\varphi(x) + \varphi(y) - \varphi\left(\frac{x+y}{2}\right) \in \varphi([a - \delta, a])$, and $\psi(x) + \psi(y) - \psi\left(\frac{x+y}{2}\right) \in \psi([a - \delta, a])$. Then $\varphi^{-1}(t) = \frac{t - \beta}{\alpha}$ and $\psi^{-1}(t) = \frac{t + \beta}{2 - \alpha}$, and (1) implies

$$\frac{1}{\alpha}\left(\alpha x + \beta + y - \alpha\frac{x+y}{2} - \beta - \beta\right)$$
$$+ \frac{1}{2-\alpha}\left((2-\alpha)x - \beta + y - (2-\alpha)\frac{x+y}{2} + \beta + \beta\right) = x + y,$$

from which

$$\left(\frac{1}{\alpha} + \frac{1}{2-\alpha} - 2\right)y = \frac{\beta}{\alpha} - \frac{\beta}{2-\alpha}$$

follows. Since this equation for y holds on an interval, this yields

$$\frac{1}{\alpha} + \frac{1}{2-\alpha} - 2 = \frac{\beta}{\alpha} - \frac{\beta}{2-\alpha} = 0,$$

and necessarily $\alpha = 1$, $\beta = 0$.

And this completes the proof of the theorem. $\qquad\qquad\square$

References

[1] J. Aczél, *Lectures on Functional Equations and Their Applications*, Academic Press, New York–London, 1966.

[2] Z. Daróczy, *Matkowski-Sutô type problem for conjugate arithmetic means*, Rocznik Nauk.-Dydakt. Prace Mat. **17** (2000), 89–100, Dedicated to Professor Zenon Moszner on his 70th birthday.

[3] Z. Daróczy, *On a class of means of two variables*, Publ. Math. Debrecen **55** (1999), 177–197.

[4] Z. Daróczy, Gy. Maksa, and Zs. Páles, *Extension theorems for the Matkowski–Sutô problem*, Demonstr. Math. **33** (2000), 547–556.

[5] Z. Daróczy and Gy. Maksa, *On a problem of Matkowski*, Colloq. Math. **82** (1999), 117–123.

[6] Z. Daróczy and C. T. Ng, *A functional equation on complementary means*, Acta Sci. Math. (Szeged) **66** (2000), 603–611.

[7] Z. Daróczy and Zs. Páles, *On means that are both quasi–arithmetic and conjugate arithmetic*, Acta Math. Hungar. **90** (2001), 271–282.

[8] J. Matkowski, *Invariant and complementary quasi–arithmetic means*, Aequationes Math. **57** (1999), 87–107.

[9] O. Sutô, *Studies on some functional equations I*, Tôhoku Math. J. **6** (1914), 1–15.

[10] O. Sutô, *Studies on some functional equations II*, Tôhoku Math. J. **6** (1914), 82–101.

HOMOGENEOUS CAUCHY
MEAN VALUES

László Losonczi

Department of Mathematics and Computer Science, Kuwait University,
P.O.Box 5969 Safat, 13060 Kuwait

losonczi@mcs.sci.kuniv.edu.kw

Abstract   Suppose that $x_1 \leq \cdots \leq x_n$ and $f^{(n-1)}, g^{(n-1)}$ exist, $g^{(n-1)} \neq 0$, and $\dfrac{f^{(n-1)}}{g^{(n-1)}}$ is invertible on $[x_1, x_n]$. Denote by $[x_1, \ldots, x_n]_f$ the divided difference of f on the points $x_1, \ldots, x_n$ then

$$t = \left(\frac{f^{(n-1)}}{g^{(n-1)}} \right)^{-1} \left(\frac{[x_1, \ldots, x_n]_f}{[x_1, \ldots, x_n]_g} \right)$$

is a mean value of $x_1, \ldots, x_n$. It is called the *Cauchy mean or difference mean of the numbers* $x_1, \ldots, x_n$ and will be denoted by $D_{f,g}(x_1, \ldots, x_n)$ (see Leach-Scholander [4], Losonczi [5]).

Here we find all homogeneous Cauchy means, i.e. we determine all Cauchy means satisfying the functional equation

$$D_{f,g}(tx_1, \ldots, tx_n) = t D_{f,g}(x_1, \ldots, x_n)$$

for all $x_1, \ldots, x_n, t \in \mathbb{R}_+ = (0, \infty)$ if $n \geq 3$ is fixed.

Keywords: Cauchy mean value, divided difference, homogeneous function

Mathematics Subject Classification (2000): 39B22

1.   INTRODUCTION

For a function $f : I \to \mathbb{R}$, on a (proper) real interval I, the divided differences of f on *distinct* points $x_i \in I$ are usually defined inductively

---

[0]This research has been supported by Kuwait University Grant SM04/99

Z. Daróczy and Z. Páles (eds.),
Functional Equations - Results and Advances, 209–218.
© 2002 *Kluwer Academic Publishers.*

by

$$[x_1]_f := f(x_1),$$

$$[x_1, \ldots, x_n]_f := \frac{[x_1, \ldots, x_{n-1}]_f - [x_2, \ldots, x_n]_f}{x_1 - x_n} \qquad (n = 2, 3, \ldots).$$

This definition must be modified if two or more points of $\{x_1, \ldots, x_n\}$ coincide: if at most r points coincide, then the definition is framed on the assumption that f is $(r-1)$-times differentiable on I. For example in the case $n = 2$ we obtain

$$[x_1, x_2]_f := \begin{cases} \dfrac{f(x_1) - f(x_2)}{x_1 - x_2} & (x_1 \neq x_2), \\ f'(x_1) & (x_1 = x_2). \end{cases}$$

A full definition, as the ratio of two determinants, can be found in Schumaker [10].

$[x_1, \ldots, x_n]_f$ is independent of the order of its arguments and the second line of the above inductive definition remains valid provided only that $x_1 \neq x_n$.

The Cauchy mean value theorem for divided differences is due to Leach and Sholander [4] (see also Rätz and Russell [9], Páles [7], [8]).

Theorem LS. Let $x_1 \leq \cdots \leq x_n$ and let $f^{(n-1)}$, $g^{(n-1)}$ exist, with $g^{(n-1)}(u) \neq 0$, on $[x_1, x_n]$. Then there exists a $t \in [x_1, x_n]$ (moreover $t \in (x_1, x_n)$ if $x_1 < x_n$) such that

$$\frac{[x_1, \ldots, x_n]_f}{[x_1, \ldots, x_n]_g} = \frac{f^{(n-1)}(t)}{g^{(n-1)}(t)}.$$

Supposing that the function $u \to \dfrac{f^{(n-1)}(u)}{g^{(n-1)}(u)}$ is invertible we get that

$$t = \left(\frac{f^{(n-1)}}{g^{(n-1)}}\right)^{-1} \left(\frac{[x_1, \ldots, x_n]_f}{[x_1, \ldots, x_n]_g}\right)$$

is a mean value of $x_1, \ldots, x_n$ which is symmetric in its variables. It is called the *Cauchy or difference mean of* $\mathbf{x} = (x_1, \ldots, x_n)$ and will be denoted by $D_{f,g}(x_1, \ldots, x_n)$ or by $D_{f,g}(\mathbf{x})$. This mean value was first defined and examined by Leach and Sholander [4] (they called it *extended* (f, g) *mean of* $x_1, \ldots, x_n$).

The aim of this paper is to find all Cauchy means satisfying the *homogeneity equation* on $\mathbb{R}_+ = (0, \infty)$:

$$D_{f,g}(t\mathbf{x}) = tD_{f,g}(\mathbf{x}) \qquad (\mathbf{x} \in \mathbb{R}_+^n, \ t \in \mathbb{R}_+)$$

for a fixed $n \geq 3$.

2.   REDUCTION TO A VECTOR-MATRIX EQUATION

First we show that the homogeneity equation can be reduced to a vector matrix equation.

We need the following result, proved in [5].

Theorem L. *Suppose that I is a real interval, $n \geq 3$ is a fixed natural number,*

(i) *$f, g, F, G : I \to \mathbb{R}$ are $n + 2$ times continuously differentiable on I,*
(ii) *$g^{(n-1)}(u) \neq 0, G^{(n-1)}(u) \neq 0$ for $u \in I$ and*
(iii) *the functions $\frac{f^{(n-1)}}{g^{(n-1)}}, \frac{F^{(n-1)}}{G^{(n-1)}}$ have non-vanishing first derivative on the interval I.*

The functional equation

$$D_{f,g}(\mathbf{x}) = D_{F,G}(\mathbf{x}) \qquad (\mathbf{x} \in I^n) \tag{1}$$

holds if and only if there exist constants $\alpha, \beta, \gamma, \delta$ with $\alpha\delta - \beta\gamma \neq 0$ such that for all $x \in I$

$$\begin{cases} f^{(n-1)}(x) = \alpha F^{(n-1)}(x) + \beta G^{(n-1)}(x) \\ g^{(n-1)}(x) = \gamma F^{(n-1)}(x) + \delta G^{(n-1)}(x) \end{cases} \tag{2}$$

is satisfied.

Theorem 1. *Suppose that $f, g : \mathbb{R}_+ \to \mathbb{R}$ satisfy conditions (i)-(iii) of Theorem L on the interval $I = \mathbb{R}_+$. Then the homogeneity equation*

$$D_{f,g}(t\mathbf{x}) = t D_{f,g}(\mathbf{x}) \qquad (\mathbf{x} \in \mathbb{R}_+^n, \ t \in \mathbb{R}_+) \tag{3}$$

holds for a fixed $n \geq 3$ if and only if there are functions $\alpha, \beta, \gamma, \delta : \mathbb{R}_+ \to \mathbb{R}$ with $\alpha(t)\delta(t) - \beta(t)\gamma(t) \neq 0$ $(t \in \mathbb{R}_+)$ for which the matrix equation

$$\begin{pmatrix} f^{(n-1)}(tx) \\ g^{(n-1)}(tx) \end{pmatrix} = \begin{pmatrix} \alpha(t)t^{1-n} & \beta(t)t^{1-n} \\ \gamma(t)t^{1-n} & \delta(t)t^{1-n} \end{pmatrix} \begin{pmatrix} f^{(n-1)}(x) \\ g^{(n-1)}(x) \end{pmatrix} \tag{4}$$

$$(x, t \in \mathbb{R}_+)$$

is satisfied.

Proof. First we show that (3) is equivalent to

$$D_{f,g}(\mathbf{x}) - D_{f_t,g_t}(\mathbf{x}) \qquad (\mathbf{x} \subset \mathbb{R}_+^n, \ t \in \mathbb{R}_+), \tag{5}$$

where

$$f_t(x) := f(tx), \qquad g_t(x) := g(tx) \qquad (x, t \in \mathbb{R}_+).$$

From

$$[x_1, \ldots, x_n]_{f_t} = [tx_1, \ldots, tx_n]_f$$

it follows that

$$
\begin{aligned}
D_{f_t, g_t}(\mathbf{x}) &= \left(\frac{f_t^{(n-1)}}{g_t^{(n-1)}} \right)^{-1} \left(\frac{[x_1, \ldots, x_n]_{f_t}}{[x_1, \ldots, x_n]_{g_t}} \right) \\
&= \left(\frac{f_t^{(n-1)}}{g_t^{(n-1)}} \right)^{-1} \left(\frac{[tx_1, \ldots, tx_n]_f}{[tx_1, \ldots, tx_n]_g} \right).
\end{aligned}
\tag{6}
$$

Since $t > 0$ we get from the equation

$$\frac{f_t^{(n-1)}(x)}{g_t^{(n-1)}(x)} = \frac{t^{n-1} f^{(n-1)}(tx)}{t^{n-1} g^{(n-1)}(tx)} = \frac{f^{(n-1)}(tx)}{g^{(n-1)}(tx)} = u$$

that

$$x = \left(\frac{f_t^{(n-1)}}{g_t^{(n-1)}} \right)^{-1} (u) = \frac{tx}{t} = \frac{1}{t} \left(\frac{f^{(n-1)}}{g^{(n-1)}} \right)^{-1} (u)$$

therefore we can continue (6) as

$$D_{f_t, g_t}(\mathbf{x}) = \frac{1}{t} \left(\frac{f^{(n-1)}}{g^{(n-1)}} \right)^{-1} \left(\frac{[tx_1, \ldots, tx_n]_f}{[tx_1, \ldots, tx_n]_g} \right), = \frac{1}{t} D_{f,g}(t\mathbf{x})$$

which proves the equivalence we claimed.

Let us fix a value $t \in \mathbb{R}_+$ and apply Theorem L for the equation (5). For each t there exists constants $\alpha, \beta, \gamma, \delta \in \mathbb{R}$ with $\alpha\delta - \beta\gamma \neq 0$ for which

$$f_t^{(n-1)}(x) = t^{n-1} f^{(n-1)}(tx) = \alpha f^{(n-1)}(x) + \beta g^{(n-1)}(x)$$
$$g_t^{(n-1)}(x) = t^{n-1} g^{(n-1)}(tx) = \gamma f^{(n-1)}(x) + \delta g^{(n-1)}(x)$$

is satisfied for all $x \in \mathbb{R}_+$. If we let t vary in $\mathbb{R}_+$ then the same holds, but the constants have to be replaced by functions $\alpha, \beta, \gamma, \delta : \mathbb{R}_+ \to \mathbb{R}$ with $\alpha(t)\delta(t) - \beta(t)\gamma(t) \neq 0$ $(t \in \mathbb{R}_+)$ and our system goes over into (4). $\square$

3.   SOLUTION OF A VECTOR-MATRIX FUNCTIONAL EQUATION

Equation (4) is a special case of the vector-matrix equation

$$\mathbf{f}(tx) = \mathbf{A}(t)\mathbf{f}(x) \qquad (x, t \in \mathbb{R}_+) \tag{7}$$

where

$$\mathbf{f}(x) = \begin{pmatrix} f_1(x) \\ f_2(x) \end{pmatrix} \quad (x \in \mathbb{R}_+), \quad \mathbf{A}(t) = \begin{pmatrix} a_{11}(t) & a_{12}(t) \\ a_{21}(t) & a_{22}(t) \end{pmatrix} \quad (t \in \mathbb{R}_+)$$

and we may suppose that the components of $\mathbf{f}$ are continuously differentiable (in fact comparing (7) with (4) these components in (4) are 3 times continuously differentiable) and $\mathbf{A}$ is a regular matrix. The continuous solutions of (7) were found by Aczél and Daróczy [1] (under slightly different conditions that we have here) by applying results of Balogh [3] where again the assumption on the range of the variables were different from our case. Due to these deviations in the assumptions and also for the sake of completeness we are going to present the solutions of (7) here (and unlike in [1] *we give the exact form of the constants* in the solutions).

Let $\mathcal{F}$ denote the set of all vector-valued functions $\mathbf{f}$ on the interval $\mathbb{R}_+$ whose components f_1, f_2 are

(j) continuously differentiable on $\mathbb{R}_+$

(jj) $f_2(x) \neq 0$ $(x \in \mathbb{R}_+)$ and f_1/f_2 is invertible on $\mathbb{R}_+$.

A consequence of (jj) is that if $x_1 \neq x_2$ then

$$\frac{f_1(x_1)}{f_2(x_1)} \neq \frac{f_1(x_2)}{f_2(x_2)}.$$

Lemma 1. *Suppose that* $\mathbf{f} \in \mathcal{F}$ *and* (7) *is satisfied. Then* $\mathbf{A}$ *is continuously differentiable on* $\mathbb{R}_+$.

Proof. Taking any fixed point $t_0 \in \mathbb{R}_+$, also a neighborhood $U(t_0, \epsilon) = (t_0 - \epsilon, t_0 + \epsilon)$ $(\epsilon > 0)$ is in $\mathbb{R}_+$. For any two values $x_1, x_2 \in \mathbb{R}_+$, $x_1 \neq x_2$ we have $x_i \cdot U(t_0, \epsilon) \subseteq \mathbb{R}_+$ $(i = 1, 2)$. Therefore

$$\begin{aligned} \mathbf{f}(tx_1) &= \mathbf{A}(t)\mathbf{f}(x_1) \\ \mathbf{f}(tx_2) &= \mathbf{A}(t)\mathbf{f}(x_2) \end{aligned} \qquad (t \in U(t_0, \epsilon))$$

hold. This is a linear inhomogeneous system of equations for the entries of $\mathbf{A}$. The determinant of the system is

$$D = -\frac{1}{f_2(x_1)^2 f_2(x_2)^2} \left(\frac{f_1(x_1)}{f_2(x_1)} - \frac{f_1(x_2)}{f_2(x_2)} \right) \neq 0$$

by $x_1 \neq x_2$ and condition (jj). Thus, on each interval $U(t_0, \epsilon)$ the entries of $\mathbf{A}$ can be expressed as the linear combination of $f_i(tx_j)$ $(i, j = 1, 2)$, which shows that the regularity properties of the functions f_i are inherited by the entries of $\mathbf{A}$, proving our statement. $\qquad\square$

Lemma 2. *Suppose that* $\mathbf{f} \in \mathcal{F}$ *and* (7) *is satisfied. Then for all* $x \in \mathbb{R}_+$ *both components* f_1, f_2 *of* $\mathbf{f}$ *are twice differentiable and satisfy the same Euler's differential equation*

$$x^2 y'' + (1 - T)xy' + Dy = 0, \tag{8}$$

where $T = a'_{11}(1) + a'_{22}(1)$ *is the trace and* D *is the determinant of* $\mathbf{A}'(1)$.

Proof. Differentiating (7) with respect to t and substituting $t = 1$ we get $x\,\mathbf{f}'(x) = \mathbf{A}'(1)\mathbf{f}(x)$, or

$$
\begin{aligned}
x f'_1(x) &= a'_{11}(1)f_1(x) + a'_{12}(1)f_2(x) \\
x f'_2(x) &= a'_{21}(1)f_1(x) + a'_{22}(1)f_2(x)
\end{aligned}
\qquad (x \in \mathbb{R}_+). \tag{9}
$$

The right hand sides of these equations are differentiable hence so are the left hand sides. Thus for $x > 0$ the functions f'_1, f'_2 are differentiable too. Differentiating the first equation of (9), multiplying by x substituting $x f'_2(x)$ from the second equation and adding the first equation multiplied by $-a'_{22}(1)$ we obtain that f_1 satisfies the equation

$$x^2 y'' + (1 - a'_{11}(1) - a'_{22}(1))xy' + (a'_{11}(1)a'_{22}(1) - a'_{12}(1)a'_{21}(1))y = 0,$$

as we claimed. The coefficient of y is the determinant D of the matrix $\mathbf{A}'(1)$. The proof for f_2 is similar. $\qquad\square$

Theorem 2. *Suppose that* $\mathbf{f} \in \mathcal{F}$. *The functional equation* (7) *holds if and only if* $\mathbf{f}, \mathbf{A}$ *have one of the forms, for* $t \in \mathbb{R}_+$,

$$\mathbf{f}(t) = \mathbf{C} \begin{pmatrix} t^\alpha \\ t^\beta \end{pmatrix}, \qquad \mathbf{A}(t) = \mathbf{C} \begin{pmatrix} t^\alpha & 0 \\ 0 & t^\beta \end{pmatrix} \mathbf{C}^{-1}, \tag{10}$$

$$\mathbf{f}(t) = \mathbf{C} \begin{pmatrix} t^\beta \ln t \\ t^\beta \end{pmatrix}, \qquad \mathbf{A}(t) = \mathbf{C} \begin{pmatrix} t^\beta & t^\beta \ln t \\ 0 & t^\beta \end{pmatrix} \mathbf{C}^{-1}, \tag{11}$$

or

$$\mathbf{f}(t) = \mathbf{C} \begin{pmatrix} t^\beta \sin(\ln t^\gamma) \\ t^\beta \cos(\ln t^\gamma) \end{pmatrix},$$

$$\mathbf{A}(t) = \mathbf{C} \begin{pmatrix} t^\beta \cos(\ln t^\gamma) & t^\beta \sin(\ln t^\gamma) \\ -t^\beta \sin(\ln t^\gamma) & t^\beta \cos(\ln t^\gamma) \end{pmatrix} \mathbf{C}^{-1}, \tag{12}$$

where

$$\mathbf{C} = \begin{pmatrix} c_{11} & c_{12} \\ c_{21} & c_{22} \end{pmatrix}$$

is an arbitrary regular constant matrix, $\alpha, \beta, \gamma \in \mathbb{R}$ are arbitrary constants apart from the restrictions $\alpha \neq \beta, \gamma \neq 0$.

Proof. By Lemma 3 f_1, f_2 satisfy the Euler's equation (8) on $\mathbb{R}_+$ hence we have

$$\begin{pmatrix} f_1 \\ f_2 \end{pmatrix} = \begin{pmatrix} c_{11}y_1 + c_{12}y_2 \\ c_{21}y_1 + c_{22}y_2 \end{pmatrix} = \mathbf{C} \begin{pmatrix} y_1 \\ y_2 \end{pmatrix},$$

where y_1, y_2 are linearly independent solutions of (8) and $\mathbf{C}$ is a regular matrix (otherwise (jj) would not be satisfied). The indicial equation of (8) is

$$m^2 - Tm + D = 0$$

thus, corresponding to the cases $T^2 - 4D > 0$, $T^2 - 4D = 0$ and $T^2 - 4D < 0$ we have for $t \in \mathbb{R}_+$

$$\mathbf{y}(t) = \begin{pmatrix} y_1(t) \\ y_2(t) \end{pmatrix} = \begin{pmatrix} t^\alpha \\ t^\beta \end{pmatrix}, \qquad \mathbf{y}(t) = \begin{pmatrix} t^\beta \ln t \\ t^\beta \end{pmatrix}, \qquad \text{and}$$

$$\mathbf{y}(t) = \begin{pmatrix} t^\beta \sin(\ln t^\gamma) \\ t^\beta \cos(\ln t^\gamma) \end{pmatrix}$$

respectively, where $\alpha, \beta, \gamma \in \mathbb{R}, \alpha \neq \beta, \gamma \neq 0$. Write (7) in the form

$$\mathbf{y}(tx) = \mathbf{B}(t)\mathbf{y}(x), \tag{13}$$

where

$$\mathbf{y}(t) = \mathbf{C}^{-1}\mathbf{f}(t), \qquad \mathbf{B}(t) = \mathbf{C}^{-1}\mathbf{A}(t)\mathbf{C}.$$

A direct calculation shows that in the above 3 cases (13) is satisfied by the matrices

$$\mathbf{B}(t) = \begin{pmatrix} t^\alpha & 0 \\ 0 & t^\beta \end{pmatrix}, \qquad \mathbf{B}(t) = \begin{pmatrix} t^\beta & t^\beta \ln t \\ 0 & t^\beta \end{pmatrix}, \qquad \text{and}$$

$$\mathbf{B}(t) = \begin{pmatrix} t^\beta \cos(\ln t^\gamma) & t^\beta \sin(\ln t^\gamma) \\ -t^\beta \sin(\ln t^\gamma) & t^\beta \cos(\ln t^\gamma) \end{pmatrix}$$

respectively. The end of the proof of Lemma 1 shows that $\mathbf{B}$ is uniquely determined by $\mathbf{y}$. Thus the above matrices are the only ones, and by $\mathbf{A}(t) = \mathbf{C}\mathbf{B}(t)\mathbf{C}^{-1}$ we obtain solutions (10), (11), (12). $\qquad \square$

4.   THE HOMOGENEOUS CAUCHY MEAN VALUES

Theorem 3. *Suppose that $n \geq 3$ is a fixed natural number,*
(k) $f, g : \mathbb{R}_+ \to \mathbb{R}$ *are $n+2$ times continuously differentiable on $\mathbb{R}_+$,*
(kk) $g^{(n-1)}(u) \neq 0$ *for $u \in \mathbb{R}_+$ and*
(kkk) *the function $\frac{f^{(n-1)}}{g^{(n-1)}}$ has non-vanishing first derivative on $\mathbb{R}_+$.*
Then all Cauchy mean values $D_{f,g}$ satisfying the homogeneity equation

$$D_{f,g}(t\mathbf{x}) = tD_{f,g}(\mathbf{x}) \qquad (\mathbf{x} \in \mathbb{R}_+^n, t \in \mathbb{R}_+)$$

are generated by the functions f, g for which

$$f^{(n-1)}(t) = t^\alpha, \qquad g^{(n-1)} = t^\beta, \qquad (t \in \mathbb{R}_+), \qquad (14)$$

$$f^{(n-1)}(t) = t^\beta \ln t, \qquad g^{(n-1)} = t^\beta, \qquad (t \in \mathbb{R}_+), \qquad (15)$$

where $\alpha, \beta \in \mathbb{R}$ are arbitrary constants apart from the restrictions $\alpha \neq \beta$.

Proof. We have seen that the vector

$$\mathbf{f} = \begin{pmatrix} f^{(n-1)} \\ g^{(n-1)} \end{pmatrix}$$

satisfies (7) and conditions (k)-(kkk) guarantee that this vector is in $\mathcal{F}$. By Theorem L the vectors

$$\begin{pmatrix} f^{(n-1)} \\ g^{(n-1)} \end{pmatrix} \qquad \text{and} \qquad \mathbf{C} \begin{pmatrix} f^{(n-1)} \\ g^{(n-1)} \end{pmatrix}$$

give the same mean value. Thus, we may take the solutions (10), (11) with $\mathbf{C}$ replaced by the unit matrix and obtain the solutions (14), (15). One can check that the conditions (k)-(kkk) are satisfied by these solutions.

From the solution (12) we obtain

$$f^{(n-1)}(t) = t^\beta \sin(\ln t^\gamma), \qquad g^{(n-1)} = t^\beta \cos(\ln t^\gamma), \qquad (t \in \mathbb{R}_+). \quad (16)$$

Here the condition (kkk) (the function $t \to \tan(\ln t^\gamma)$ must have non-vanishing derivative on $\mathbb{R}_+$) is not satisfied. $\square$

Remark 1. From the solution (12) we do not get a homogeneous mean on $\mathbb{R}_+$. But (16) satisfies the condition (kkk) on the interval

$$I_\gamma = (e^{-\frac{\pi}{2|\gamma|}}, e^{\frac{\pi}{2|\gamma|}}),$$

and this is the maximal subinterval of $\mathbb{R}_+$ containing the point 1 on which (kkk) holds.

Remark 2. The condition $n \geq 3$ was needed in Theorem 1 only to prove the implication (3)$\Rightarrow$(4) (via Theorem L), the reverse implication holds for $n = 2$ too. Therefore (4) is sufficient for the homogeneity of $D_{f,g}$ in the two variable case and thus (14), (15) generate homogeneous Cauchy means in the two variable case too (however *there may exist other two variable* homogeneous Cauchy means). The two variable Cauchy means have a very simple form:

$$D_{f,g}(x,y) = \left(\frac{f'}{g'}\right)^{-1}\left(\frac{f(x) - f(y)}{g(x) - g(y)}\right) \qquad (x \neq y)$$

and it is worth to calculate the homogeneous ones from (14), (15). They are

$$D_{f,g}(x,y) = \left(\frac{(x^a - y^a)b}{(x^b - y^b)a}\right)^{1/(a-b)} \qquad (ab(a - b) \neq 0)$$

$$D_{f,g}(x,y) = \left(\frac{x^a - y^a}{(\ln x - \ln y)a}\right)^{1/(a)} \qquad (a \neq 0)$$

$$D_{f,g}(x,y) = \left(\frac{(\ln x - \ln y)b}{x^b - y^b}\right)^{1/(-b)} \qquad (b \neq 0)$$

$$D_{f,g}(x,y) = e^{\left(\frac{x^a \ln(x^a/e) - y^a \ln(y^a/e)}{(x^a - y^a)a}\right)} \qquad (a \neq 0)$$

$$D_{f,g}(x,y) = \sqrt{xy},$$

where $a, b \in \mathbb{R}$ are constants with the restrictions indicated, $x \neq y$ and $x, y \in \mathbb{R}_+$.

From (16) we get the Cauchy mean

$$D_{f,g}(x,y) = e^{\frac{1}{c}\arctan\frac{p(x) - p(y)}{q(x) - q(y)}} \qquad (c \neq 0)$$

for

$$x, y \in I_c = (e^{-\frac{\pi}{2|c|}}, e^{\frac{\pi}{2|c|}}),$$

where

$$p(x) = x^a \left(-c\cos(\ln x^c) + a\sin(\ln x^c)\right) \qquad (a, c \in \mathbb{R}, c \neq 0)$$
$$q(x) = x^a \left(a\cos(\ln x^c) + c\sin(\ln x^c)\right) \qquad (a, c \in \mathbb{R}, c \neq 0).$$

This mean value satisfies a *generalized homogeneity* equation:

$$D_{f,g}(tx, ty) = tD_{f,g}(x, y)$$

for all $x, y \in I_c$ and for those t's for which $tx, ty \in I_c$ holds.

To obtain the above means first we find f, g by integration from (14), (15), (16) distinguishing the cases $\alpha, \beta = -1$ and $\alpha, \beta \neq -1$ and introducing new constants a, b, c. The first five of these means have been known and investigated by many authors (see e.g. [6] and the references there), the last one seems to be a new mean.

References

[1] J. Aczél and Z. Daróczy, *Über verallgemeinerte quasilineare Mittelwerte, die mit Gewichtsfuntionen gebildet sind*, Publ. Math. Debrecen **10** (1963), 171–190.

[2] G. Aumann and O. Haupt, *Einführung in die reelle Analysis*, vol. II, W. de Gruyter, Berlin-New York, 1979.

[3] A. Balogh, *On the determination of geometric objects with special transformation formulae*, Mathematica Cluj **24** (1959), 199–219.

[4] E. Leach and M. Sholander, *Multi-variable extended mean values*, J. Math. Anal. Appl. **104** (1984), 390–407.

[5] L. Losonczi, *Equality of Cauchy mean values*, Publ. Math. Debrecen **57** (2000), 217–230.

[6] L. Losonczi and Zs. Páles, *Minkowski's inequality for two variable difference means*, Proc. Amer. Math. Soc. **126** (1998), 779–789.

[7] Zs. Páles, *Notes on mean value theorems*, (manuscript).

[8] Zs. Páles, *A unified form of the classical mean value theorems*, in: *Inequalities and Applications* (ed. by R. P. Agarwal), World Scientific Publ., Singapore–New Jersey–London–Hong Kong, 1994, 493–500.

[9] J. Rätz and D. Russell, *An extremal problem related to probability*, Aequationes Math. **34** (1987), 316–324.

[10] L. Schumaker, *Spline functions*, Wiley, New York-Toronto, 1981.

[11] J. F. Steffenson, *Interpolation*, 2nd ed., Chelsea, New York, 1950.

ON INVARIANT GENERALIZED
BECKENBACH-GINI MEANS

Janusz Matkowski

Institute of Mathematics, Technical University,
Podgórna 50, PL-65-246 Zielona Góra, Poland
and
Institute of Mathematics, Silesian University,
Bankowa 14, PL-40-007 Katowice, Poland
jmatk@lord.wsp.zgora.pl

Dedicated to Professor Peter Kahlig on the occasion of his sixtieth
birthday

Abstract   A functional equation that characterizes generalized Beckenbach-Gini means which are invariant with respect to Beckenbach-Gini mean-type mappings is considered. In the case when an invariant mean is either arithmetic or geometric or harmonic, without any regularity conditions, all solutions are found. In the general case, under some regularity assumptions, a necceasary condition is given. For positively homogeneous Beckenbach-Gini means a complete list of solutions is established. Translative Beckenbach-Gini means are also examined.

Keywords: mean-type mapping, iteration, invariant mean, Beckenbach-Gini mean, homogeneous mean, translative mean

Mathematics Subject Classification (2000): 39B22; 26A18

Let $I \subseteq \mathbb{R}$ be an interval. By a *mean* we mean a function $M : I^2 \to I$ such that

$$\min(x, y) \leq M(x, y) \leq \max(x, y) \qquad (x, y \in I).$$

If for all $x, y \in I, x \neq y$, these inequalities are sharp, we call M to be a *strict mean*. It is proved in [6] that if $M, N : I^2 \to I$ are strict continuous

Z. Daróczy and Z. Páles (eds.),
Functional Equations - Results and Advances, 219–230.
© 2002 *Kluwer Academic Publishers.*

means then the sequence of iterates of the mean-type mapping (M, N) : $I^2 \to I^2$ converges to a mean-type mapping $(K, K) : I^2 \to I^2$ where K is a unique (M, N)-invariant mean, i.e.

$$K\left(M(x, y), N(x, y)\right) = K(x, y) \qquad (x, y \in I).$$

There are some important special classes of means, for instance, quasi-arithmetic means, Gini means, Stolarsky means (cf. [3]). In this connection the following general question arises. Given a class of means, determine all pairs (M, N) from this class such that their (unique) (M, N)-invariant mean K is also a member of this class. This problem, under some regularity assumption, has been solved for the class of quasi-arithmetic means in [7]. In the present paper we examine this problem for the class of generalized Beckenbach-Gini means $M_f : I^2 \to I$ which are of the form

$$M_f(x, y) := \frac{xf(x) + yf(y)}{f(x) + f(y)} \qquad (x, y \in I),$$

where $f : I \to (0, \infty)$ is a function, called a *generator* of the mean. This mean is a special quasi-arithmetic weighted mean. In the case when f is a power function, the mean M_f was considered by Gini [5] and Beckenbach [2]. Thus the above invariant mean relation reduces to the functional equation

$$M_h\left(M_f(x, y), M_g(x, y)\right) = M_h(x, y) \qquad (x, y \in I), \tag{1}$$

with three unknown functions $f, g, h : I \to (0, \infty)$.

As the arithmetic mean A is a Beckenbach-Gini mean, in section 2 we determine all pairs of functions (f, g) such that A is (M_f, M_g)-invariant. In this context the translative Beckenbach-Gini means are considered. Since the geometric mean G and harmonic mean H are also of Beckenbach-Gini type, in section 3 we establish all the pair of functions (f, g) such that G and H are (M_f, M_g)-invariant.

In section 4 we show that, under some regularity assumption, if M_h is (M_f, M_g)-invariant, i.e. eq. (1) is satisfied, then, necessarily, for some $c > 0$,

$$h = c\sqrt{fg}.$$

Using this result, in section 5, we find all positively homogeneous means M_f, M_g and M_h such that M_h is (M_f, M_g)-invariant.

1.   AUXILIARY RESULTS

Some properties of Beckenbach-Gini means can be found in [3]. We begin with recalling the following easy to verify

Remark 1. Let $I \subseteq \mathbb{R}$ be an interval and $f, g : I \to (0, \infty)$. Then $M_f = M_g$ if, and only if, $g = cf$ for some $c > 0$.

An important role is played by

Lemma 1. *Let $I \subseteq \mathbb{R}$ be an interval. If $f : I \to (0, \infty)$ is continuously differentiable then*

$$\lim_{y \to x} \frac{\frac{\partial M_f}{\partial x}(x, y) - \frac{\partial M_f}{\partial y}(x, y)}{x - y} = \frac{f'(x)}{f(x)} \qquad (x \in I). \tag{2}$$

Proof. From the definition of M_f we have

$$\frac{\partial M_f}{\partial x}(x, y) = \frac{f(x)^2 + f(x)f(y) + xf'(x)f(y) - yf'(x)f(y)}{(f(x) + f(y))^2} \qquad (x, y \in I),$$

and, by the symmetry of M_f,

$$\frac{\partial M_f}{\partial y}(x, y) = \frac{\partial M_f}{\partial x}(y, x) \qquad (x, y \in I).$$

Hence, by simple calculations, we get

$$\frac{\frac{\partial M_f}{\partial x}(x, y) - \frac{\partial M_f}{\partial y}(x, y)}{x - y}$$

$$= \frac{\frac{f(x) - f(y)}{x - y}(f(x) + f(y)) + f'(x)f(y) + f(x)f'(y)}{(f(x) + f(y))^2},$$

and, letting y tend to x, we obtain formula (2). $\qquad \square$

In section 4 we need the following.

Lemma 2. *Let $f : (0, \infty) \to (0, \infty)$ be an arbitrary function. The Beckenbach-Gini mean $M_f : (0, \infty)^2 \to (0, \infty)$ is positively homogeneous, i.e.*

$$M_f(tx, ty) = tM_f(x, y) \qquad (x, y, t > 0),$$

if, and only if, the function $\frac{f}{f(1)}$ is multiplicative. If moreover f is measurable or the graph of f is not dense in $(0, \infty)^2$ then $f(x) = f(1)x^p$, $x > 0$, for some $p \in \mathbb{R}$, and

$$M_f(x, y) = \frac{x^{p+1} + y^{p+1}}{x^p + y^p} \qquad (x, y > 0).$$

Proof. For every $t > 0$ define $f_t : (0,\infty) \to (0,\infty)$ by $f_t(x) := f(tx)$, $x > 0$. Then, by Remark 1, M_f is homogeneous iff $M_f = M_{f_t}$ for all $t > 0$, that is, there exists a function $c : (0,\infty) \to (0,\infty)$ such that

$$f(tx) = c(t)f(x) \qquad (x,t > 0).$$

Setting here $x = 1$ we get $f(t) = f(1)c(t)$ for all $t > 0$, and, consequently,

$$f(1)f(tx) = f(t)f(x) \qquad (x,t > 0),$$

which means that the function $\frac{f}{f(1)}$ is multiplicative. The converse implication is obvious. The second part follows from the first one (cf. [1], Theorem 3, p. 14). $\square$

2.   WHEN THE ARITHMETIC MEAN IS INVARIANT; TRANSLATIVE BECKENBACH-GINI MEANS

Taking a constant generator in the definition of Beckenbach-Gini mean we get the arithmetic mean $A(x,y) = \frac{x+y}{2}$, $x,y \in \mathbb{R}$.

Theorem 1. *Let $I \subseteq \mathbb{R}$ be an interval and $f,g : I \to (0,\infty)$. The following conditions are equivalent:*

(1) the arithmetic mean A is (M_f, M_g)-invariant, i.e.

$$M_f(x,y) + M_g(x,y) = x + y \qquad (x,y \in I); \tag{3}$$

(2) there is a $c > 0$ such that

$$f(x)g(x) = c \qquad (x \in I);$$

(3) $M_g = M_{1/f}$, i.e.

$$M_g(x,y) = \frac{yf(x) + xf(y)}{f(x) + f(y)} \qquad (x,y \in I).$$

Moreover, for every continuous function $f : I \to (0,\infty)$, the sequence of iterates of the mapping $(M_f, M_{1/f})$ converges to the arithmetic mean.

Proof. Setting $h(x) = 1$, $x \in I$, in (1) gives the equation $A(M_f, M_g) = A$, i.e. (3), which is equivalent to

$$\frac{xf(x) + yf(y)}{f(x) + f(y)} + \frac{xg(x) + yg(y)}{g(x) + g(y)} = x + y \qquad (x,y \in I).$$

This functional equation reduces to the relation

$$f(x)g(x) = f(y)g(y) \qquad (x,y \in I),$$

and the proofs of equivalences of the conditions 1-3 are obvious. Since the means $M_f, M_{1/f}$ are strict, the remaining part of the theorem is a consequence of Theorem 1 in [6]. $\qquad\qquad\qquad\qquad\qquad\qquad\qquad\square$

Remark 2. A counterpart of equation (3) for quasi-arithmetic means (which is much more difficult and yet not completely solved) leads to an interesting one-parameter family of translative quasi-arithmetic means (cf. Z. Daróczy [4]).

In this connection recall that a mean $M : \mathbb{R}^2 \to \mathbb{R}$ is called *translative* if

$$M(x+t, x+t) = M(x,y) + t \qquad (x,y,t \in \mathbb{R}).$$

Lemma 3. *Let $f : R \to (0,\infty)$. The Beckenbach-Gini mean M_f is translative if, and only if, the function $\frac{f}{f(0)}$ is exponential, i.e.*

$$f(0)f(x+y) = f(x)f(y) \qquad (x,y \in \mathbb{R}).$$

Proof. For every $t \in \mathbb{R}$ define $f_t : \mathbb{R} \to (0,\infty)$ by $f_t(x) := f(x+t)$, $x \in \mathbb{R}$. Then, by Remark 1, M_f is translative iff $M_f = M_{f_t}$ for all $t \in \mathbb{R}$, that is, when there is a function $c : \mathbb{R} \to (0,\infty)$ such that

$$f(x+t) = c(t)f(x) \qquad (x,t \in \mathbb{R}).$$

Setting here $x = 0$ gives $f(t) = f(0)g(t)$, $t \in \mathbb{R}$. Hence we get

$$f(0)f(x+t) = f(x)f(t) \qquad (x,y,t \in \mathbb{R}),$$

which means that the function $\frac{f}{f(0)}$ is exponential.

Since the converse implication is easy to verify, the proof is completed. $\qquad\qquad\qquad\qquad\qquad\qquad\qquad\square$

Hence we obtain (cf. [1], Theorem 3, p. 14)

Corollary 1. *Let $f : \mathbb{R} \to (0,\infty)$ be measurable or its graph be not dense in $(0,\infty) \times \mathbb{R}$. The Beckenbach-Gini mean M_f is translative if, and only if, there exists a constant $p > 0$ such that*

$$M_f(x,y) = \frac{xp^x + yp^y}{p^x + p^y} \qquad (x,y \in \mathbb{R}).$$

For arbitrary $p > 0$, we define $T_{[p]} : \mathbb{R} \times \mathbb{R} \to \mathbb{R}$ by

$$T_{[p]}(x,y) := \frac{xp^x + yp^y}{p^x + p^y} \qquad (x,y \in \mathbb{R}).$$

Thus $\{T_{[p]} : p > 0)\}$ is a one-parameter family of translative Beckenbach Gini means. Applying Theorem 1 we obtain

Corollary 2. *For all $p,r > 0$ the arithmetic mean A is $(T_{[p]}, T_{[r]})$ invariant, if and only if, $pr = 1$.*

3.   THE INVARIANCE OF THE GEOMETRIC AND HARMONIC MEANS

Taking the generator $x \to x^{-1/2}$ in the definition of Beckenbach-Gini mean we get the geometric mean $G(x,y) := \sqrt{xy}$, $x, y > 0$.

Theorem 2. *Let $I \subseteq (0,\infty)$ be an interval and $f, g : I \to (0,\infty)$. The geometric mean is (M_f, M_g)-invariant if, and only if, there is a $c > 0$ such that*

$$f(x)g(x) = \frac{c}{x} \qquad (x \in I).$$

Moreover, for all continuous functions $f : I \to (0,\infty)$, $c > 0$, and $g(x) := \frac{c}{xf(x)}$, $x \in I$, the sequence of iterates of the mapping (M_f, M_g) converges to the geometric mean.

Proof. Writing in the explicit form the equation $G(M_f, M_g) = G$ we get

$$\frac{xf(x) + yf(y)}{f(x) + f(y)} \frac{xg(x) + yg(y)}{g(x) + g(y)} = xy \qquad (x, y \in I),$$

which, after simple calculations reduces to the equivalent condition

$$f(x)g(x)x = f(y)g(y)y \qquad (x, y \in I),$$

and the result follows. The second statement is an immediate consequence of Theorem 1 in [6]. $\qquad\square$

Taking the generator $x \to x^{-1}$ in the definition of Beckenbach-Gini mean we get the harmonic mean H.

Theorem 3. *Let $I \subseteq (0,\infty)$ be an interval and $f, g : I \to (0,\infty)$. The harmonic mean is (M_f, M_g)-invariant if, and only if, there is a $c > 0$ such that*

$$f(x)g(x) = \frac{c}{x^2} \qquad (x \in I).$$

Moreover, for all continuous functions $f : I \to (0,\infty)$, $c > 0$, and $g(x) := \frac{c}{f(x)}$, $x \in I$, the sequence of iterates of the mapping (M_f, M_g) converges to the geometric mean.

Proof. Writing the equation $H(M_f, M_g) = H$ in the explicit form we get

$$\frac{\frac{xf(x)+yf(y)}{f(x)+f(y)} \cdot \frac{xg(x)+yg(y)}{g(x)+g(y)}}{\frac{xf(x)+yf(y)}{f(x)+f(y)} + \frac{xg(x)+yg(y)}{g(x)+g(y)}} = \frac{xy}{x+y} \qquad (x, y \in I).$$

Simple calculation shows that this functional equation is equivalent to the relation

$$f(x)g(x)x^2 = f(y)g(y)y^2 \qquad (x, y \in I),$$

which completes the proof. □

Remark 3. Taking $h(x) = x, x \in I$, gives a Beckenbach-Gini mean $M_h(x,y) = \frac{x^2+y^2}{x+y}$ (which is the contra-harmonic one). It is not difficult to show that this mean is (M_f, M_g)-invariant for some functions $f, g : I \to (0, \infty)$, i.e.

$$\frac{\left(\frac{xf(x)+yf(y)}{f(x)+f(y)}\right)^2 + \left(\frac{xg(x)+yg(y)}{g(x)+g(y)}\right)^2}{\frac{xf(x)+yf(y)}{f(x)+f(y)} + \frac{xg(x)+yg(y)}{g(x)+g(y)}} = \frac{x^2+y^2}{x+y} \qquad (x,y \in I),$$

if, and only if, $f = ah$ and $g = bh$, for some positive $a, b \in \mathbb{R}$, i.e., if, and only if, $M_f = M_h = M_g$ (cf. Remark 1).

This remark shows that the classical means A, G and H play a special role in the theory of invariant Beckenbach-Gini means.

4.   A NECESSARY CONDITION

In this section we prove the following

Theorem 4. *Let* $I \subseteq \mathbb{R}$ *be an interval. Suppose that* $f, g : I \to (0, \infty)$ *are differentiable and* $h : I \to (0, \infty)$ *is twice differentiable. If the mean* M_h *is* (M_f, M_g)-*invariant then there is a constant* $c > 0$ *such that*

$$h = c\sqrt{fg}.$$

Proof. Suppose that a mean M_h is (M_f, M_g)-invariant. Then the functions f, g and h satisfy equation (1) which, by the definition of Beckenbach-Gini mean, can be written in the form

$$[M_f(x,y)h\left(M_f(x,y)\right) + M_g(x,y)h\left(M_g(x,y)\right)]\left(h(x) + h(y)\right)$$
$$= (xh(x) + yh(y))\left[h\left(M_f(x,y)\right) + h\left(M_g(x,y)\right)\right] \qquad (x,y \in I).$$

Denote, for convenience, the expressions of the left and right hand sides of this equation by $L(x,y)$ and $R(x,y)$, respectively. Since $L = R$ we have

$$\frac{\partial L}{\partial x} = \frac{\partial R}{\partial x}, \qquad \frac{\partial L}{\partial y} = \frac{\partial R}{\partial y},$$

and, consequently,

$$\frac{\frac{\partial L}{\partial x}(x,y) - \frac{\partial L}{\partial y}(x,y)}{x-y} = \frac{\frac{\partial R}{\partial x}(x,y) - \frac{\partial R}{\partial y}(x,y)}{x-y} \qquad (x,y \in I).$$

By simple calculations we have

$$\frac{\frac{\partial L}{\partial x}(x,y) - \frac{\partial L}{\partial y}(x,y)}{x - y} = \left[h\left(M_f\right) + M_f h'\left(M_f\right)\right] \frac{\frac{\partial M_f}{\partial x} - \frac{\partial M_f}{\partial y}}{x - y}[h(x) + h(y)]$$

$$+ \left[h(M_g) + M_g h'(M_g)\right] \frac{\frac{\partial M_g}{\partial x} - \frac{\partial M_g}{\partial y}}{x - y}[h(x) + h(y)]$$

$$+ \left[M_f h\left(M_f\right) + M_g h(M_g)\right] \frac{h'(x) - h'(y)}{x - y},$$

and

$$\frac{\frac{\partial R}{\partial x}(x,y) - \frac{\partial R}{\partial y}(x,y)}{x - y} = \left[h\left(M_f\right) + h\left(M_g\right)\right] \frac{h(x) + xh'(x) - h(y) - yh'(y)}{x - y}$$

$$+ \left[xh(x) + yh(y)\right]\left[h'(M_f)\frac{\frac{\partial M_f}{\partial x} - \frac{\partial M_f}{\partial y}}{x - y} + h'(M_g)\frac{\frac{\partial M_g}{\partial x} - \frac{\partial M_g}{\partial y}}{x - y}\right],$$

where, for short, $M_f = M_f(x,y)$ and $M_g = M_g(x,y)$. Letting here $y \to x$ and applying Lemma 1, the continuity of the means M_f, M_g, and the relation $M_f(x,x) = x = M_g(x,x)$, $x \in I$, we get

$$h(x) + xh'(x)\left[\frac{f'(x)}{f(x)} + \frac{g'(x)}{g(x)}\right] h(x) + xh(x)h''(x)$$

$$= h(x)[2h'(x) + xh''(x)] + xh(x)h'(x)\left[\frac{f'(x)}{f(x)} + \frac{g'(x)}{g(x)}\right],$$

which, after reduction, can be written in the form

$$2\frac{h'(x)}{h(x)} = \frac{f'(x)}{f(x)} + \frac{g'(x)}{g(x)} \qquad (x \in I).$$

It follows that there is a $c > 0$ such that

$$h(x)^2 = cf(x)g(x) \qquad (x \in I),$$

which completes the proof. □

Remark 4. The condition given in this theorem is not sufficient. To show that the converse implication is not true take $f, g, h : (0, \infty) \to (0, \infty)$ defined by

$$f(x) := \frac{x}{x+1}, \qquad g(x) := x(x+1), \qquad h(x) := x \qquad (x > 0).$$

Then $h = \sqrt{fg}$ and, it is easy to verify the mean M_h is not (M_f, M_g)-invariant (cf. also Remark 3.

Thus the problem to determine some neccesary and sufficient conditions is open.

5.   HOMOGENEOUS BECKENBACH-GINI MEANS

In this section we present the following

Theorem 5. *Let each of the functions* $f, g, h : (0, \infty) \to (0, \infty)$ *be measurable or its graph be not dense in* $(0, \infty)^2$. *Suppose that the Beckenbach-Gini means* M_f, M_g, M_h *are positively homogeneous. Then* M_h *is* (M_f, M_g)-*invariant if, and only if, one of the following cases occurs:*

(1) there is a $p \in R$ *such that*

$$\frac{f(x)}{f(1)} = \frac{g(x)}{g(1)} = \frac{h(x)}{h(1)} = x^p \qquad (x > 0),$$

and, consequently,

$$M_f(x, y) = M_g(x, y) = M_h(x, y) = \frac{x^{p+1} + y^{p+1}}{x^p + y^p} \qquad (x, y > 0);$$

(2) there exists a $p \in \mathbb{R}$ *such that*

$$f(x) = f(1)x^p, \qquad g(x) = g(1)x^{-p}, \qquad h(x) = h(1) \qquad (x > 0),$$

and, consequently,

$$M_f(x, y) = \frac{x^{p+1} + y^{p+1}}{x^p + y^p}, \qquad M_g(x, y) = \frac{x^{1-p} + y^{1-p}}{x^{-p} + y^{-p}}, \qquad M_h = A$$

(3) there is a $p \in \mathbb{R}$ *such that*

$$f(x) = f(1)x^p, \qquad g(x) = g(1)x^{-p-1}, \qquad h(x) = h(1)x^{-1/2} \quad (x > 0),$$

and, consequently,

$$M_f(x, y) = \frac{x^{p+1} + y^{p+1}}{x^p + y^p}, \qquad M_g(x, y) = \frac{x^{-p} + y^{-p}}{x^{-p-1} + y^{-p-1}}, \qquad M_h = G;$$

(4) there is a $p \in \mathbb{R}$ *such that*

$$f(x) = f(1)x^p, \qquad g(x) = g(1)x^{-p-2}, \qquad h(x) = h(1)x^{-1} \qquad (x > 0),$$

and, consequently,

$$M_f(x,y) = \frac{x^{p+1}+y^{p+1}}{x^p+y^p}, \qquad M_g(x,y) = \frac{x^{-p-1}+y^{-p-1}}{x^{-p-2}+y^{-p-2}}, \qquad M_h = H.$$

Proof. Without any loss of generality we may assume that at least two of the means M_f, M_g, M_h are not the same. The homogeneity of M_f, M_g, M_h, in view of Lemma 2, implies that there exist $p, q, r \in \mathbb{R}$ such that

$$f(x) = f(1)x^p, \qquad g(x) = g(1)x^q, \qquad h(x) = h(1)x^r \qquad (x > 0).$$

Since f, g, h satisfy the assumptions of Theorem 5, we infer that

$$h(x) = c\sqrt{f(x)g(x)} \qquad (x > 0),$$

for a positive $c > 0$, and consequently,

$$r = \frac{p+q}{2}.$$

Now the (M_f, M_g)-invariance of the mean M_h, i.e. equation (1), can be written in the form

$$\left(\frac{x^{p+1}+y^{p+1}}{x^p+y^p}\right)^{\frac{p+q+2}{2}} + \left(\frac{x^{q+1}+y^{q+1}}{x^q+y^q}\right)^{\frac{p+q+2}{2}}$$

$$= \frac{x^{\frac{p+q+2}{2}}+y^{\frac{p+q+2}{2}}}{x^{\frac{p+q}{2}}+y^{\frac{p+q}{2}}}\left[\left(\frac{x^{p+1}+y^{p+1}}{x^p+y^p}\right)^{\frac{p+q}{2}} + \left(\frac{x^{q+1}+y^{q+1}}{x^q+y^q}\right)^{\frac{p+q}{2}}\right],$$

for all $x, y > 0$.

Suppose that some real numbers p, q satisfy this equation for all $x, y > 0$. Setting here $y = 1$ gives

$$\left(\frac{x^{p+1}+1}{x^p+1}\right)^{\frac{p+q+2}{2}} + \left(\frac{x^{q+1}+1}{x^q+1}\right)^{\frac{p+q+2}{2}}$$

$$- \frac{x^{\frac{p+q+2}{2}}+1}{x^{\frac{p+q}{2}}+1}\left[\left(\frac{x^{p+1}+1}{x^p+1}\right)^{\frac{p+q}{2}} + \left(\frac{x^{q+1}+1}{x^q+1}\right)^{\frac{p+q}{2}}\right] = 0$$

for all $x > 0$. Denote by $F_{p,q}(x)$ the left hand side of this identity. Then, of course,

$$\frac{d^k}{dx^k}F_{p,q}(1) = 0$$

for all nonnegative integers k. Careful calculations show that

$$\frac{d^k}{dx^k}F_{p,q}(1) = 0 \qquad (k = 0, 1, ..., 5),$$

for all $p, q \in \mathbb{R}$. Only the condition

$$\frac{d^6}{dx^6} F_{p,q}(1) = 0$$

reduces to the condition

$$\frac{15}{32}(p - q)^2(p + q)(p + q + 1)(p + q + 2) = 0$$

(we omit long and tedious calculations). Consequently, either $q = p$ or $q = -p$ or $q = -p - 1$ or $q = -p - 2$. Since the converse implication is easy to verify, the proof is completed. $\square$

Recall that for every $p \in \mathbb{R}$, a power mean $M^{[p]} : (0, \infty)^2 \to (0, \infty)$ is defined by

$$M^{[p]}(x, y) := \left(\frac{x^p + y^p}{2}\right)^{1/p} \quad (p \neq 0); \qquad M^{[0]}(x, y) = \sqrt{xy}.$$

Theorem 6. *Let* $p, q \in R$. *Then*

$$M^{[p]}(x, y) = \frac{x^{q+1} + y^{q+1}}{x^q + y^q} \qquad (x, y > 0), \tag{4}$$

if, and only if, either $q = 0, p = 1$ *or* $q = -1 = p$, *or* $q = -\frac{1}{2}, p = 0$.

Proof. Setting $y = 1$ in (4) gives

$$\left(\frac{x^p + 1}{2}\right)^{1/p} = \frac{x^{q+1} + 1}{x^q + 1} \qquad (x > 0).$$

Calculating the second and the forth derivatives of both sides and then substituting $x = 1$ gives the system of equations

$$p = 2q + 1, \qquad 8q(4 - q^2) = (1 - p)(2p^2 - p - 15).$$

Hence

$$q(q + 1)(2q + 1) = 0,$$

and consequently, either $q = 0$ and $p = 1$, or $q = -1$ and $p = -1$, or $q = -\frac{1}{2}$ and $p = 0$. The converse implication is easy to verify. $\square$

As an immediate consequence we get the following

Corollary 3. *The Beckonbach-Gini and power means coincide if, and only if, they are equal either to A, or to G or to H.*

Applying Theorem 6 we obtain

Remark 5. A Beckenbach-Gini mean is invariant with respect to a homogeneous Beckenbach-Gini mean-type mapping if, and only if, it is a power mean i.e., if, and only if, it is either A, or G or H.

Acknowledgement. The author is indebted to the referees for their valuable remarks.

References

[1] J. Aczél and J. Dhombres, *Functional equations in several variables*, Encyclopedia of Mathematics and its Applications **31**, Cambridge University Press, Cambridge–New York–New Rochelle–Melbourne–Sydney, 1989.

[2] E. F. Beckenbach, *A class of mean value functions*, Amer. Math. Monthly **57** (1950), 1–6.

[3] P. S. Bullen, D. S. Mitrinović, and P. M. Vasić, *Means and their inequalities*, Mathematics and its Applications, D. Reidel Publishing Company, Dodrecht-Boston - Lancaster-Tokyo 1988.

[4] Z. Daróczy, *On a class of means of two variables*, Publ. Math. Debrecen **55** (1999), 177–197.

[5] C. Gini, *Di una formula comprensiva delle medie*, Metron **13** (1938), 3–22.

[6] J. Matkowski, *Iterations of mean-type mappings and invariant means*, Ann. Math. Sil. **13** (1999), 211–226.

[7] J. Matkowski, *Invariant and complementary quasi-arithmetic means*, Aequationes Math. **57** (1999), 87–107.

FINAL PART OF THE ANSWER
TO A HILBERT'S QUESTION

Maciej Sablik

Institute of Mathematics, Silesian University,

Bankowa 14, 40 007, Katowice, Poland

mssablik@us.edu.pl

Abstract   We present new results concerning the following functional equation of Abel

$$\psi\left(xf(y) + yf(x)\right) = \phi(x) + \phi(y).$$

D. Hilbert in the second part of his fifth problem asked whether it can be solved without differentiability assumption on the unknown functions ψ, f and φ. We gave earlier (cf. [9] and [10]) a positive answer assuming however that 0 is either in the domain or the range of f. Now we solve the equation in the remaining case and thus complete the answer to Hilbert's question.

Keywords: functional equation of Abel, continuous solution, Hilbert's fifth problem

Mathematics Subject Classification (2000): 39B22

1.   INTRODUCTION

In the second part of his fifth problem D. Hilbert (cf. [7]) dealt with functional equations, *usually investigated only under the assumption of the differentiability of functions involved,* and asked the following: *In how far are the assertions which we can make in the case of differentiable functions true under proper modifications without this assumption?* Among the equations explicitly mentioned by Hilbert one finds the following one

$$\phi(x) + \phi(y) = \psi\left(xf(y) + yf(x)\right) \tag{1}$$

with which we will be concerned in this paper. The equation (1) was considered by N. Abel (cf. [1]). Hilbert's question was recalled by J. Aczél

231

Z. Daróczy and Z. Páles (eds.),

Functional Equations - Results and Advances, 231–242.

© 2002 *Kluwer Academic Publishers.*

during the Twenty-fifth International Symposium on Functional Equations in 1987 (see [3]). In 1989 Aczél published the article [2], where he reported on the state of the second part of Hilbert's fifth problem, and the reader is referred to that survey for more information on other equations solved in the spirit of Hilbert's question.

Following Abel, we assume that the variables x and y are in a real interval, and the unknown functions ϕ, ψ and f are real-valued. We will assume continuity of the functions which can partially be relaxed, as we point out in concluding remarks.

In our papers [9] and [10] we solved (1) for continuous functions, assumming however that 0 belongs either to the domain or to the range of the unknown function f. Let us note that in Abel's paper, although it is not precised, 0 is at least an accumulation point of the domain. Abel gave his solution in an implicit form, quoting however few examples of explicit solutions. All of them but one were included in the list we gave in [9]. The exceptional one, where the functions cannot be defined at 0, was among the solutions enumerated in [10]. In the present paper we deal with the remaining case.

Let us adopt the following notation. By I we denote the non-degenerate interval which is the domain of f and ϕ. Observe that I is also to be determined throughout our investigation. For any function $p : I \to \mathbb{R}$ let us define the operation $A_p : I \times I \to \mathbb{R}$ by

$$A_p(x, y) = xp(y) + yp(x).$$

Thus the unknown function ψ is defined in $A_f(I \times I)$. Let us fix a $t \in I$, put $\hat{I} := t^{-1}I$, and define $\hat{f}, \hat{\phi} : \hat{I} \to \mathbb{R}$ by

$$\hat{f}(u) = \frac{f(tu)}{2f(t)}, \qquad \hat{\phi}(u) = \phi(tu) - \phi(t),$$

and $\hat{\psi} : A_{\hat{f}}(\hat{I} \times \hat{I}) \to \mathbb{R}$ by

$$\hat{\psi}(z) = \psi(2tf(t)z) - \psi(2tf(t)).$$

It is easy to observe that if the triple (ψ, ϕ, f) solves (1) for all $x, y \in I$ then so does the triple $(\hat{\psi}, \hat{\phi}, \hat{f})$ for all $x, y \in \hat{I}$. Conversely, let $(\hat{\psi}, \hat{\phi}, \hat{f})$ solve (1) on an interval $\hat{I}$ containing 1 and assume that $\hat{f}(1) = \frac{1}{2}$, $\hat{\phi}(1) = \hat{\psi}(1) = 0$. Then for every interval $I \subset \mathbb{R} \setminus \{0\}$, every $t \in I$ and every real constants $a \neq 0$ and b, the triple (ψ, ϕ, f) defined by

$$\psi(z) = \hat{\psi}\left(\frac{z}{at}\right) + 2b, \qquad \phi(x) = \hat{\phi}\left(\frac{x}{t}\right) + b, \qquad f(x) = a\hat{f}\left(\frac{x}{t}\right)$$

solves (1) for all $x, y \in I$. Therefore, in order to simplify the reasoning, in the sequel we will admit the following hypothesis

(H) I is a nondegenerate interval, $1 \in I$, $0 \notin I$, $f : I \to \mathbb{R}$, $\phi : I \to \mathbb{R}$ and $\psi : A_f(I \times I) \to \mathbb{R}$ are continuous. Moreover $0 \notin f(I)$, $f(1) = \frac{1}{2}$, $\phi(1) = \psi(1) = 0$.

Let us also note that (ψ, ϕ, f) is a solution of (1) with invertible f : $I \to \mathbb{R}$ if and only if $(\psi, \phi \circ f^{-1}, f^{-1})$ solves (1) for $x, y \in f(I)$. Of course, if $0 \notin I \cup f(I)$ then $0 \notin f(I) \cup f^{-1}(f(I))$. We will use this observation to shorten the statement of our main result in section 4.

2. INVERTIBLE SOLUTIONS

In this section we consider the case where ϕ is invertible. Since the function is also continuous, we see that ϕ is strictly monotonic.

Invertibility of ϕ implies invertibility of A_f with respect to both variables. Indeed, fix $x \in I$ and let $y, y' \in I$ be such that $A_f(x, y) = A_f(x, y')$. Then (1) implies $\phi(y) = \phi(y')$, whence $y = y'$. Symmetry of A_f yields its invertibility with respect to the first variable. Again, continuity of f implies strict monotonicity of A_f with respect to each variable. We can admit w.l.o.g. that A_f is strictly increasing in each variable, and so $B_f : I \to \mathbb{R}$ given by

$$B_f(x) = A_f(x, x)$$

is strictly increasing. Moreover, for every $x, y \in I$ we have

$$B_f(\min\{x, y\}) \le A_f(x, y) \le B_f(\max\{x, y\}),$$

whence it follows that

$$B_f(I) = A_f(I \times I). \tag{2}$$

Let us prove now that also ψ has to be invertible. Indeed, let $z, z' \in A_f(I \times I)$ be such that $\psi(z) = \psi(z')$. In view of (2) we may write $z = B_f(x)$ and $z' = B_f(x')$ for some $x, x' \in I$. Applying (1), we get

$$\phi(x) = \frac{1}{2}\psi(z) = \frac{1}{2}\psi(z') = \phi(x').$$

Hence in view of invertibility of ϕ we get $x = x'$, which obviously implies $z = z'$.

Now, in view of invertibility of ϕ and ψ we may rewrite (1) in the following form

$$\phi^{-1}(u)f(\phi^{-1}(v)) + \phi^{-1}(v)f(\phi^{-1}(u)) = \psi^{-1}(u + v),$$

or, denoting $p = \phi^{-1}$, $q = f \circ \phi^{-1}$ and $g = \psi^{-1}$,

$$g(u + v) = p(u)q(v) + p(v)q(u), \tag{3}$$

where u, v are in the interval $J := \phi(I)$. Of course, in view of our assumptions functions p, q and g are continuous and thus solving (3) reduces to make use of results proved by J. Aczél and J. K. Chung in [4]. As a matter of fact, in view of A. Járai's remark [5, Section 8.9], we should consider two cases: (α) p and q are linearly dependent, and (β) p and q are linearly independent.

The case (α) leads to the equality $f(x) = cx$ for all $x \in I$, where the constant $c = \frac{1}{2}$ in view of (H). Then (1) becomes

$$\psi(xy) = \phi(x) + \phi(y),$$

which is a local pexiderization of the logarithm equation. It follows from (H) and well known results on local additivity that there exists a constant $d \neq 0$ such that $\psi(z) = d \ln z$ for all $z \in A_f(I \times I) = I \cdot I$, and $\phi(x) = d \ln x$ for all $x \in I$. There are no further restrictions on the interval I than those enumerated in (H), i.e. $1 \in I \subset (0, \infty)$.

In the case (β) the equation (3) is a particular case of the situation considered in [4, Section 4], where the authors give a list of Lebesgue integrable solutions of the equation

$$g(u + v) + h(u - v) = p_1(u)q_1(v) + p_2(u)q_2(v).$$

Inserting these solutions into (3), and going back to ψ, ϕ and f, we obtain the list of invertible solutions of (1) which is given in the Theorem 4.1 below. The procedure of selecting solutions of (1) among the list given in [4] is not difficult although rather tedious since many cases have to be considered. We omit the detailed proof, let us only remark that all the invertible solutions are essentially restrictions of those known already from [9] and [10].

3.   NON-INVERTIBLE SOLUTIONS

Let us now assume that ϕ is non-invertible. We shall start by showing that ϕ is *non-philandering*[1], i.e. ϕ has an interval of constancy. Suppose to the contrary that ϕ is philandering. There are $x, y \in I$ such that $x < y$, $\phi(x) = \phi(y)$ and $\phi(u) \neq \phi(x)$ for every $u \in (x, y)$. Put $I_1 = [x, y]$. To fix our attention we may admit that

$$\phi(u) > \phi(x) \tag{4}$$

---

[1]The notion of philandering function has been introduced by A. Lundberg, see e.g. [6]

for every $u \in \mathrm{Int} I_1 =: I_1^o$. From (1) and (4) we get for every $u \in I_1^o$ and every $z \in I$

$$A_f(x, z) \neq A_f(u, z) \neq A_f(y, z). \tag{5}$$

Let us consider two cases. At first, we assume that

$$A_f(u, z) > \max\{A_f(x, z), A_f(y, z)\} \tag{6}$$

for every $u \in I_1^o$ and $z \in I$, noting that the subcase

$$A_f(u, z) < \min\{A_f(x, z), A_f(y, z)\}$$

may be treated analogously. From (6) and the definition of A_f we infer that for every $u \in I_1^o$ and $z \in I$ we have

$$uf(z) + zf(u) > \max\{xf(z) + zf(x), yf(z) + zf(y)\}$$

whence, taking into account **(H)** we get

$$\frac{f(x) - f(u)}{u - x} < \frac{f(z)}{z} < \frac{f(y) - f(v)}{v - y}$$

for every $u, v \in I_1^o$ and $z \in I$. Hence, letting $u \to y$ and $v \to x$ in the above inequalities we obtain

$$\frac{f(z)}{z} = \frac{f(x) - f(y)}{y - x}$$

for every $z \in I$, which means that f is linear, and in view of **(H)**, $f(z) = \frac{1}{2}z$, $z \in I$. This however, as we have seen in the previous section implies that ϕ equals $d \ln |I$ which means that either ϕ is invertible or it vanishes identically. Thus (6) is impossible.

In view of (5), for every $u \in I_1^o$ and $z \in I$, the value $A_f(u, z)$ lies strictly between $A_f(x, z)$ and $A_f(y, z)$. In view of continuity of A_f this implies that either

$$A_f(x, z) < A_f(u, z) < A_f(y, z) \tag{7}$$

for every $u \in I_1^o$ and $z \in I$, or the reverse inequalities hold, the situation then may be treated analogously as below. In particular, we get from (7) for every $u, v \in I_1^o$:

$$A_f(x, x) < A_f(v, x) = A_f(x, v) < A_f(u, v)$$
$$< A_f(y, v) = A_f(v, y) < A_f(y, y).$$

Hence $A_f(I_1 \times I_1) = [A_f(x,x), A_f(y,y)]$. Now choose a $u \in I_1^o$ so that $A_f(u,u) = A_f(x,y)$. Then (1) and (4) imply

$$\psi(A_f(x,y)) = \phi(x) + \phi(y) < 2\phi(u) = \psi(A_f(u,u)) = \psi(A_f(x,y)).$$

This contradiction shows indeed, that if ϕ is non-invertible then it is non-philandering, i.e. it has an interval of constancy.

Let us denote by I_2 a maximal interval of constancy of ϕ, and let $K_x := A_f(\{x\} \times I_2)$, $x \in I$. By the definition of I_2 and (1) the function ψ is constant on any of the interval K_x, $x \in I$. By continuity of ϕ, the interval I_2 is of the form $I_2 = [m, M] \cap I$ for some numbers m and M, with $0 \leq m < M \leq \infty$. In the next step we will make sure that either $m = \inf I$ or $M = \sup I$. Indeed, if $M < \sup I$ then for every $\delta > 0$ such that $[M, M+\delta] \subset I$ the interval $\phi([M, M+\delta)) = \psi(A_f([M, M+\delta) \times I_2))$ is non-degenerated, whence it follows that K_{x_δ} is degenerated for at least one $x_\delta \in (M, M+\delta)$, for otherwise ψ could admit only denumerably many values in $A_f([M, M+\delta) \times I_2)$. Thus there exists a sequence $(x_n)_{n \in \mathbb{N}}$ decreasing to M and such that K_{x_n} reduces to one point for any $n \in \mathbb{N}$. In particular, we get

$$x_n f(y) + y f(x_n) = x_n f(M) + M f(x_n)$$

for every $y \in I_2$ and $n \in \mathbb{N}$ which implies, after letting $n \to \infty$, that

$$f(y) = -\frac{f(M)}{M} y + 2f(M) \tag{8}$$

for every $y \in I_2$. An analogous argument shows that if $m > \inf I$ then

$$f(y) = -\frac{f(m)}{m} y + 2f(m)$$

for every $y \in I_2$. This would imply however that $m = M$, a contradiction.

Thus only the following three cases are possible for a maximal interval of constancy of ϕ.

(i) $I \subset [m, M]$;
(ii) $\inf I = m < M < \sup I$;
(iii) $\inf I < m < M = \sup I$.

In the case (i) we easily see that both ϕ and ψ are constant (and hence vanish identically in view of (**H**)). Suppose now that (ii) holds. Then arguing as above we see that (8) holds for all $y \in I_2$. Hence for every $x \in I$ and $y \in I_2$ we get

$$A_f(x,y) = \left(f(x) - \frac{f(M)}{M} x \right) y + 2f(M)x. \tag{9}$$

In view of (1), ψ is constant on any interval $K_x = A_f(\{x\} \times I_2)$ while by definition of M the interval $\psi(A_f(((M,\infty) \cap I) \times I_2)) = \phi((M,\infty) \cap I) + \phi(M)$ is non-degenerate. This implies that there is an $x \in (M,\infty) \cap I$ such that the interval K_x is degenerate, which by (9) is possible only if

$$f(x) = \frac{f(M)}{M} x.$$

In particular, the set $Z := \left\{ x \in (M,\infty) \cap I : f(x) = \frac{f(M)}{M} x \right\}$ is a non-empty closed set, and therefore $((M,\infty) \cap I) \setminus Z$ is an at most denumerable union of pairwise disjoint open intervals, say J_n, $n \in \mathbb{N}$. Now, if $x \in J_n$ for some $n \in \mathbb{N}$ then K_x is a non-degenerate interval. On the other hand, if $y \in J_n$ and $\phi(x) \neq \phi(y)$ then $K_x \cap K_y = \emptyset$ which follows from (1) because, as we have noticed, ψ is constant on both K_x and K_y. Thus $\phi|J_n$ can take on at most denumerably many values, whence by continuity we infer that ϕ is constant on J_n. But taking into account that a constancy intervals of ϕ can only have one of three possible forms we see that in the present situation $((M,\infty) \cap I) \setminus Z$ is either empty set or reduces to just one interval of the form $(d,\infty) \cap I$ where $M < d < \sup I$. In the latter case we have $[M,d] \subset Z$, or

$$f(x) = \frac{f(M)}{M} x$$

for $x \in [M,d]$, and similarly as above we get also the formula

$$f(x) = -\frac{f(d)}{d} x + 2f(d),$$

which holds for $x \in (d,\infty) \cap I$. Taking into account continuity of f we obtain

$$f(x) = \begin{cases} -\alpha x + 2\alpha M & \text{if } x \leq M, \\ \alpha x & \text{if } M < x < d, \\ -\alpha x + 2\alpha d & \text{if } x \in [d,\infty) \cap I, \end{cases} \qquad (10)$$

where $\alpha = \frac{f(M)}{M} = \frac{f(d)}{d}$. Moreover $\phi|(0,M] \cap I = \text{const} = c_1$, $\phi|[d,\infty) \cap I = \text{const} = c_2$ and ϕ is philandering in (M,d). This, however, leads to a contradiction. Indeed, in view of (10) there exists an $\epsilon \in (0, 2\alpha d^2)$ such that

$$(\epsilon, 2\alpha d^2) \subset B_f((M,d)) \cap B_f((d,\infty) \cap I)).$$

Let $\delta > 0$ be such that $B_f((d-\delta,d)) \subset (\epsilon, 2\alpha d^2)$ and let $x \in (d-\delta,d)$ be arbitrary. Choose $y > d$ so that $B_f(x) = B_f(y)$. Then we get from

(1)
$$\phi(x) = \frac{1}{2}\psi(B_f(x)) = \frac{1}{2}\psi(B_f(y)) = \phi(y) = c_2,$$

which means that ϕ is constant in $(d - \delta, \infty) \cap I$ and thus contradicts the definition of d. This contradiction shows that $Z = (M, \infty) \cap I$ and consequently

$$f(x) = \begin{cases} -\frac{f(M)}{M}x + 2f(M) & \text{if } x \in (0, M] \cap I, \\ \frac{f(M)}{M}x & \text{if } x \in (M, \infty) \cap I. \end{cases} \tag{11}$$

Moreover $\phi|(0, M] \cap I = \text{const} = c_1$. Taking into account (11) we see that (1) supposed to hold for $x, y \in (M, \infty) \cap I$ becomes a local pexiderized logarithmic equation which is easy to solve in the class of continuous functions.

Summarizing, in the case (ii) solutions of (1) are given by (11) and

$$\phi(x) = \begin{cases} c_1 & \text{if } x \in (0, M] \cap I, \\ c\ln\frac{x}{M} + c_1 & \text{if } x \in (M, \infty) \cap I, \end{cases} \tag{12}$$

$$\psi(z) = \begin{cases} 2c_1 & \text{if } z \in A_f(((0, M] \cap I) \times ((0, M] \cap I)), \\ c\ln\frac{z}{2Mf(M)} + 2c_1 & \text{if } x \in (M, \infty) \cap I, \end{cases} \tag{13}$$

where $c \neq 0$ and c_1 are some constants.

Finally, we will show that the case (iii) cannot occur. Indeed, if $m > \inf I$ then using an analogous argument as in the case (ii) we would get the following form of f

$$f(x) = \begin{cases} -\frac{f(m)}{m}x + 2f(m) & \text{if } x \in (m, \infty) \cap I, \\ \frac{f(m)}{m}x & \text{if } x \in (0, m] \cap I. \end{cases}$$

Thus there exists an $\epsilon \in (0, 2mf(m))$ such that

$$(\epsilon, 2mf(m)) \subset B_f((0, m) \cap I) \cap B_f((m, \infty) \cap I).$$

Let $\delta \in (\inf I, m)$ be such that $B_f((m - \delta, m)) \subset (\epsilon, 2mf(m))$ and let $x \in (m - \delta, m)$ be arbitrary. Choose a $y > m$ so that $B_f(x) = B_f(y)$. Then

$$\phi(x) = \frac{1}{2}\psi(B_f(x)) = \frac{1}{2}\psi(B_f(y)) = \phi(y),$$

whence it follows that ϕ is constant in $(m - \delta, \infty) \cap I$, contrary to the definition of m. This contradiction shows that the case (iii) cannot occur indeed.

4. MAIN RESULT

We are now in a position to state the main theorem of the paper. Let us introduce some notation first. Namely we define the following functions:

- $\gamma : (0, \infty) \to \mathbb{R}$ given by $\gamma(u) = u \ln u$.

- $g_\alpha : (0, \infty) \to \mathbb{R}$ given by $g_\alpha(u) = \frac{1}{2}(u^\alpha - u)$, where $\alpha \in \mathbb{R} \setminus \{0, 1\}$. Note that g_α is invertible, if $\alpha < 0$ as well as $g_{\alpha,l} := g_\alpha | (0, q_\alpha]$ and $g_{\alpha,r} := g_\alpha | [q_\alpha, \infty)$, if $\alpha > 0$, where $q_\alpha = \alpha^{1/(1-\alpha)}$.

- $h_\alpha : (0, +\infty) \to \mathbb{R}$, given by $h_\alpha(u) = \frac{1}{2}(u^\alpha + u)$, where $\alpha \in \mathbb{R} \setminus \{0, 1\}$. If $1 \neq \alpha > 0$ then h_α is invertible, and if $\alpha < 0$ then $h_{\alpha,l} := h_\alpha | (0, p_\alpha]$ and $h_{\alpha,r} := h_\alpha | [p_\alpha, +\infty)$ are invertible, where $p_\alpha = (-\alpha)^{1/(1-\alpha)}$.

- $\lambda_D : \mathbb{R} \to \mathbb{R}$ given by $\lambda_D(u) = \frac{1}{\sqrt{1+u^2}} \exp(D \arctan u)$, where $D \in \mathbb{R}$. Note that $\lambda_{D,0} = \lambda_D | (-\infty, D]$ and $\lambda_{D,1} = \lambda_D | [D, +\infty)$ are invertible functions.

- $\sigma_D : \mathbb{R} \to \mathbb{R}$ given by $\sigma_D(u) = u \lambda_D(u)$. Functions $\sigma_{D,1} = \sigma_D | \frac{1}{D}[-1, +\infty)$ and $\sigma_{D,0} = \sigma_D | \frac{1}{D}(-\infty, -1]$ are invertible, as well as $\sigma_{0,0} = \sigma_{0,1} = \sigma_0$. Let us also adopt the convention $\sigma_{D,1}^{-1}\left(\exp\left(\pm\frac{D\pi}{2}\right)\right) = \pm\infty$ and $\arctan(\pm\infty) = \pm\frac{\pi}{2}$.

Using the above notation let us present the list of, say, basic solutions of (1) under the assumption **(H)**. The remaining solutions will be expressed in terms of the following ones.

(S.0) $1 \in I_0 \subset (0, +\infty)$ is an arbitrary interval, f_0 is an arbitrary continuous function with $f_0(1) = \frac{1}{2}, \psi_0 = 0 = \phi_0$.

(S.1) $1 \in I_1 \subset (0, \infty)$ is an arbitrary interval, $f_1 = \frac{1}{2}, \psi_1(z) = 2(z-1), \phi_1(x) = x - 1$.

(S.2) $1 \in I_2 \subset I_\alpha$ is an arbitrary interval, $f_2(x) = \alpha(x-1) + \frac{1}{2}, \psi_2(z) = \ln\left(\frac{8\alpha(z-1)}{(1+2\alpha)^2} + 1\right), \phi_2(x) = \ln\left(\frac{4\alpha(x-1)}{1+2\alpha} + 1\right)$, where $\alpha \notin \{0, -\frac{1}{2}\}$ is a constant, and $I_\alpha = (0, \infty) \cap (1 + \frac{1}{2\alpha}(-1, \infty))$.

(S.3) $1 \in I_3 \subset I_\beta$ is an arbitrary interval, $f_3(x) = \frac{\gamma(\beta x)}{2\gamma(\beta)}$,

$$\psi_3(z) = \ln \beta^{-2} \gamma^{-1}\left(\gamma(\beta^2)z\right), \qquad \phi_3(x) = \ln x,$$

where $0 < \beta \neq 1$ is a constant, and

$$
I_\beta = \begin{cases}
(\beta^{-1}, \infty) & \text{if } \beta > 1, \\
\beta^{-1}\left[\frac{1}{\sqrt{e}}, 1\right) & \text{if } \frac{1}{\sqrt{e}} \leq \beta < 1, \\
\beta^{-1}\left(0, \frac{1}{\sqrt{e}}\right) & \text{if } 0 < \beta < \frac{1}{\sqrt{e}}.
\end{cases}
$$

(S.4) $1 \in I_4 \subset \beta^{-1} h_\alpha((0, \infty))$, $f_4(x) = \frac{1}{2g_\alpha(h_{\alpha,r}^{-1}(\beta))} g_\alpha\big(h_{\alpha,r}^{-1}(\beta x)\big)$,

$$\psi_4(z) = \ln g_\alpha^{-1}\big(2\beta g_\alpha(h_{\alpha,r}^{-1}(\beta)z\big), \qquad \phi_4(x) = \ln h_{\alpha,r}^{-1}(\beta x),$$

where $\beta \in h_\alpha((0, \infty) \setminus \{1\})$, and $\alpha < 0$ are constants.

(S.5) $1 \in I_5 \subset \beta^{-1} h_\alpha((0, \infty))$, $f_5(x) = \frac{1}{2g_\alpha(h_{\alpha,l}^{-1}(\beta))} g_\alpha\big(h_{\alpha,l}^{-1}(\beta x)\big)$,

$$\psi_5(z) = \ln g_\alpha^{-1}\big(2\beta g_\alpha(h_{\alpha,l}^{-1}(\beta)z\big), \qquad \phi_5(x) = \ln h_{\alpha,l}^{-1}(\beta x),$$

where $\beta \in h_\alpha((0, \infty) \setminus \{1\}$, and $\alpha < 0$ are constants.

(S.6) $1 \in I_6 \subset \beta^{-1} h_\alpha\left(0, \sqrt{q_\alpha}\right]$, $f_6(x) = \frac{1}{2g_{\alpha,l}(h_\alpha^{-1}(\beta))} g_{\alpha,l}\big(h_\alpha^{-1}(\beta x)\big)$,

$$\psi_6(z) = \ln g_{\alpha,l}^{-1}\big(2\beta g_{\alpha,l}(h_\alpha^{-1}(\beta)z\big), \qquad \phi_6(x) = \ln h_\alpha^{-1}(\beta x),$$

where $\beta \in h_\alpha((0, \sqrt{q_\alpha}]$, and $1 \neq \alpha > 0$ are constants.

(S.7) $1 \in I_7 \subset \beta^{-1} h_\alpha\left[\sqrt{q_\alpha}, 1\right)$, $f_7(x) = \frac{1}{2g_{\alpha,r}(h_\alpha^{-1}(\beta))} g_{\alpha,r}\big(h_\alpha^{-1}(\beta x)\big)$,

$$\psi_7(z) = \ln g_{\alpha,r}^{-1}\big(2\beta g_{\alpha,r}(h_\alpha^{-1}(\beta)z\big), \qquad \phi_7(x) = \ln h_\alpha^{-1}(\beta x),$$

where $\beta \in h_\alpha([\sqrt{q_\alpha}, 1))$, and $1 \neq \alpha > 0$ are constants.

(S.8) $1 \in I_8 \subset \beta^{-1} h_\alpha((1, \infty))$, $f_8(x) = \frac{1}{2g_{\alpha,r}(h_\alpha^{-1}(\beta))} g_{\alpha,r}\big(h_\alpha^{-1}(\beta x)\big)$,

$$\psi_8(z) = \ln g_{\alpha,r}^{-1}\big(2\beta g_{\alpha,r}(h_\alpha^{-1}(\beta)z\big), \qquad \phi_8(x) = \ln h_\alpha^{-1}(\beta x),$$

where $\beta \in h_\alpha((1, \infty))$, and $1 \neq \alpha > 0$ are constants.

(S.9) $1 \in I_9 \subset \beta^{-1}\lambda_D I_D$, $f(x) = \alpha^{-1} \frac{1}{2\sigma_D(\lambda_{D,k}^{-1}(\beta))} \sigma_{D,p}\big(\lambda_{D,k}^{-1}(\beta x)\big)$;

$$
\psi(z) = \begin{cases}
\arctan \sigma_{D,p}^{-1}(\alpha\beta z) & \text{if } z = B_f(x) \text{ and } |\sigma_{D,p}^{-1}(\beta x)| \leq 1, \\
(\mathrm{sgn}D)\pi + \arctan \sigma_{D,1-p}^{-1}\left(\frac{-\alpha\beta z}{\exp|D|\pi}\right) \\
\qquad \text{if } z = B_f(x) \text{ and } |\sigma_{D,p}^{-1}(\beta x)| > 1;
\end{cases}
$$

$\phi(x) = \frac{1}{2}\psi(2xf(x))$, where D is a real constant, I_D is any interval not containing 0 and such that $\sigma_D|I_D$ is invertible, $\beta \in I_D$ is a constant, $\alpha = 2\sigma_D\left(\lambda_{D,k}^{-1}(\beta)\right)$, $p, k \in \{0, 1\}$ are suitably chosen according to which invertible branch of σ_D or λ_D is used.

(S.10) $1 \in I \subset (0, \infty)$,

$$f_{10}(x) = \begin{cases} -\frac{\beta}{x} + 2\beta M & \text{if } x \in (0, M] \cap I, \\[2mm] \beta x & \text{if } x \in (M, \infty) \cap I, \end{cases}$$

$$\phi_{10}(x) = \begin{cases} 0 & \text{if } x \in (0, M] \cap I, \\[2mm] \ln \frac{x}{M} & \text{if } x \in (M, \infty) \cap I, \end{cases}$$

$$\psi_{10}(z) = \begin{cases} 0 & \text{if } z \in A_f(((0, M] \cap I) \times ((0, M] \cap I)), \\[2mm] \ln \frac{z}{2M^2\beta} & \text{if } x \in (M, \infty) \cap I, \end{cases}$$

where constants $\beta > 0$ and $M \in \text{Int} I$ are such that $\beta = \frac{1}{2}$ or $\beta(2M-1) = \frac{1}{2}$ depending on whether $M > 1$ or $M \leq 1$.

Now, taking into account our observations from previous sections we can state our main result.

Theorem 1. *Let I be an interval, $0 \notin I$. Then a triple (ψ, ϕ, f) of continuous functions $f : I \to \mathbb{R}$, $\phi : I \to \mathbb{R}$, and $\psi : A_f(I \times I) \to \mathbb{R}$ such that $0 \notin f(I)$ satisfies (1) if and only if there exist a $t \in I$ and real constants $a \neq 0, b$ and c such that $I = t\hat{I}$, and*

$$\psi(z) = c\hat{\psi}\left(\frac{z}{at}\right) + 2b, \qquad \phi(x) = c\hat{\phi}\left(\frac{x}{t}\right) + b, \qquad f(x) = a\hat{f}\left(\frac{x}{t}\right),$$

where $\hat{I} = I_k$ and $(\hat{\psi}, \hat{\phi}, \hat{f}) = (\psi_k, \phi_k, f_k)$ for some $k \in \{0, \ldots, 10\}$ (see the list above) or $\hat{I} = f_k(I_k)$ and $(\hat{\psi}, \hat{\phi}, \hat{f}) = (\psi_k, \phi_k \circ f_k^{-1}, f_k^{-1})$ for some $k \in \{0, \ldots, 10\}$ (provided f_k is invertible).

Similarly as in [9] and [10] we can conclude with the following remark.

Remark 1. The list of solutions given above remains unchanged if we assume that f is continuous and either ψ or ϕ is locally bounded from above or below. This follows from C. T. Ng's results from [8]. Also, if we assume that either ψ or ϕ is Lebesgue measurable and f is of class C^1 and such that for every $x \in I$ there exists a $y \in I$ such that $f'(x) \neq -f(y)/y$ then the above list of solutions remains unchanged (cf. [5]).

References

[1] N. H. Abel, *Sur les fonctions qui satisfont à l'équation $\phi x + \phi y = \psi(xfy + yfx)$*, Oeuvres Complètes de N. H. Abel rédigées par ordre du roi par B. Holmboe, Christiania 1839, tome premier, 103–110. Original version in J. reine angew. Math. **21** (1827), 386–394.

[2] J. Aczél, *Remarks and problems*, In: *The Twenty-fifth International Symposium on Functional Equations, August 16-22, 1987, Hamburg - Rissen, Germany*, Report of Meetings, Aequationes Math. **35** (1988), 116–117.

[3] J. Aczél, *The state of the second part of Hilbert's fifth problem*, Bull. Amer. Math. Soc. **20** (1989), 153–163.

[4] J. Aczél and J. K. Chung, *Integrable solutions of functional equations of a general type*, Studia Sci. Math. Hungar. **17** (1982), 51–67.

[5] A. Járai, *Regularity properties of functional equations*, Leaflets in Mathematics, Janus Pannonius University Pécs, Pécs 1996.

[6] A. Lundberg, *On the functional equation $f(\lambda(x) + g(y)) = \mu(x) + h(x + y)$*, Aequationes Math. **16** (1977), 21–30.

[7] D. Hilbert, *Mathematical Problems*, Lecture delivered before the International Congress of Mathematicians at Paris in 1900, Bull. Amer. Math. Soc. **8** (1902), 437–479.

[8] C. T. Ng, *Local boundedness and continuity for a functional equation on topological spaces*, Proc. of the Amer. Math. Soc. **39** (1973), 525–529.

[9] M. Sablik, *The continuous solution of a functional equation of Abel*, Aequationes Math. **39** (1990), 19–39.

[10] M. Sablik, *A functional equation of Abel revisited*, Abh. Math. Sem. Univ. Hamburg **64** (1994), 203–210.

V
FUNCTIONAL EQUATIONS ON ALGEBRAIC STRUCTURES

A GENERALIZATION OF D'ALEMBERT'S FUNCTIONAL EQUATION

Thomas M. K. Davison

McMaster University

davison@mcmail.cis.mcmaster.ca

Abstract   A functional equation with two unknown functions f, g is considered. When the domain of the functions is a uniquely 2-divisible abelian group, and the codomain is a field of characteristic different from 2, it is shown that there are exponential functions e, e' such that $f = e + e'$ and $g = ee'$.

Keywords: d'Alembert's functional equation, uniquely 2-divisible abelian group, exponential function, coboundary

Mathematics Subject Classification (2000): 39B52, 43A40

1.   INTRODUCTION

The d'Alembert functional equation is

$$2c(x)c(y) = c(x + y) + c(x - y). \tag{d'A}$$

The standard reference for this is Aczel [1] where it is listed on page 4 as equation (4). It is also called the cosine equation. If we put $f(x) := 2c(x)$ and $g(x) := 1$ we have a pair of functions satisfying

$$f(x)f(y) = f(x + y) + f(x - y)g(y). \tag{1}$$

Here we suppose that $f, g : A \to K$, where A is an abelian group and K is a field of characteristic not 2. It is easy to verify that if $e_1, e_2 : A \to K$ are exponential functions:

$$e_j(x + y) = e_j(x)e_j(y) \quad \text{and} \quad e_j(0) = 1,$$

then $f = e_1 + e_2$ and $g = e_1 e_2$ is a solution of equation (1). We show in Section 3 the converse of this under suitable hypotheses on A.

Z. Daróczy and Z. Páles (eds.),
Functional Equations - Results and Advances, 245–248.
© 2002 Kluwer Academic Publishers.

Theorem 1. *Assume A is a uniquely 2-divisible abelian group. If (f, g) is a solution to equation* (1) *with $f(0) = 2$ and $g(0) = 1$, then there is an extension field L of K with $[L : K] \leq 2$ and exponential functions $e_1, e_2 : A \to L$ such that $f = e_1 + e_2$ and $g = e_1 e_2$.*

In Section 2, we prove that if $g(0) = 1$ and $f(0) = 2$ then g is itself an exponential function as long as $|A/2A| \leq 2$. Then, in Section 3, using Kannappan's [2] result on the solutions of (d'A), we deduce our result.

2.   WHEN IS g AN EXPONENTIAL FUNCTION?

We say that (f, g) is an *admissible* solution of equation (1) if $f(0) = 2$ and $g(0) = 1$. *From now on we deal only with admissible solutions.* We will characterize the abelian groups A with the property that in every admissible solution (f, g) the second component, g, must be exponential.

We begin with

Proposition 1. *If $|A/2A| > 2$ there is an admissible solution (f, g) of equation* (1) *where g is not an exponential function.*

Proof. Define $f, g : A \to K$ as follows:

$$f(x) = \begin{cases} 2, & x \in 2A \\ 0, & x \notin 2A \end{cases}, \qquad g(x) = \begin{cases} 1, & x \in 2A \\ -1, & x \notin 2A \end{cases}.$$

It is easy to verify that (f, g) is an admissible solution of equation (1). Since $|A/2A| > 2$ there exist $y \notin 2A$, $z \notin 2A$ such that $y + z \notin 2A$. Then $g(y + z) = -1$ whereas $g(y)g(z) = (-1)(-1) = 1$. So g is not exponential.   □

We now proceed to show that if $|A/2A| \leq 2$ then every admissible (f, g) has g exponential. We first deduce some properties of g that are valid for every abelian group A.

Lemma 1. *For all $y \in A$, $g(y)g(-y) = 1$.*

Proof. Multiplying equation (1) by $g(-y)$ we get that, for all x, y in A,

$$f(x)f(y)g(-y) = f(x + y)g(-y) + f(x - y)g(y)g(-y). \qquad (2)$$

Putting $x = 0$ in equation (1) we see that

$$f(y) = f(-y)g(y). \qquad (3)$$

Using equation (3) we can write equation (2) as

$$f(x)f(-y) = f(x - y)g(y)g(-y) + f(x + y)g(-y).$$

But equation (1) with $-y$ replacing y yields

$$f(x)f(-y) = f(x - y) + f(x + y) g(-y).$$

Comparing the last two equations, and using the fact that f is not the zero function since $f(0) = 2 \neq 0$, we deduce the stated result.   □

Next we define the coboundary

$$G(x, y) := g(x + y) g(x)^{-1} g(y)^{-1} = g(x + y) g(-x) g(-y) \qquad (4)$$

for all $(x, y) \in A^2$. We see that G is symmetric ($G(x, y) = G(y, x)$) and $G(-y, y) = 1$ for all y. Moreover, g is exponential if, and only if, $G = 1$. The next pair of Lemmas give properties of G.

Lemma 2. *For all* $x, y, z \in A$,

(i) $f(x + y) G(x, y) = f(x + y)$

(ii) $f(x + y) f(x + y + z) G(y, z) = f(x + y) f(x + y + z).$

Proof. (i) From the x, y symmetry in equation (1) we deduce that $f(x-y)g(y) = f(y-x)g(x)$, so replacing y by $-y$ we have $f(x + y) g(-y) = f(-y - x) g(x)$.

Thus we have, using equation (3)

$$f(x + y)g(-y)g(x + y) = f(-x - y)g(x + y)g(x) = f(x + y)g(x).$$

Hence $f(x + y) g(x + y) g(-x)g(-y) = f(x + y)$, as claimed.

(ii) Since every coboundary is a cocycle, we have

$$G(x, y) G(x + y, z) = G(x, y + z) G(y, z) \qquad (5)$$

for all x, y, z in A. Multiplying both sides by $f(x + y) f(x + y + z)$ and using part (i), we deduce the stated result.   □

We remark that if f never takes the value 0, then this lemma already shows that $G = 1$, and so g is an exponential function.

Lemma 3. *If there are* y, z *in* A *such that* $G(y, z) \neq 1$, *then* $y \notin 2A$ *and* $z \notin 2A$.

Proof. Assume $G(y, z) \neq 1$. Then, from Lemma 2 (ii) $f(x + y)f(x + y + z) = 0$ for all $x \in A$. Taking $x = -y$, we see that $2f(z) = 0$, and so $f(z) = 0$. By symmetry $f(y) = 0$ too. Now, using equation (1) $f(x + y + z) f(x + y) = f(2x + 2y + z) + f(z)g(x + y)$. So $f(2x + 2y + z) = 0$ for all $x \in A$. If $z \in 2A$ choose x with $2x + z = 0$. Then

$f(2y) = 0$. But this implies $2g(y) = f(y)^2 - f(2y) = 0$, which is absurd. Hence $z \notin 2A$ and $y \notin 2A$.   □

We can now prove the main result of this section.

Proposition 2. *If* $|A/2A| \le 2$ *then* g *is an exponential function.*

Proof. Assume that $|A/2A| \le 2$. This means that $s \notin 2A$, $t \notin 2A$ implies $s+t \in 2A$. Now Lemma 3 tells us that if y or z is in 2A then $G(y, z) = 1$. So now assume $y \notin 2A$, $z \notin 2A$. Then $y + z \in 2A$. Equation (5) with $x = -y$ yields

$$G(-y, y) \, G(0, z) = G(-y, y + z) \, G(y, z)$$
$$1 \cdot 1 = 1 \cdot G(y, z).$$

Hence $G(y, z) = 1$. Thus $G(y, z) = 1$ for all y, z in A. Thus g is an exponential function.   □

3.   THE PROOF OF THEOREM 1

Kannappan's [2] solution of d'Alembert's equation (d'A) was framed for the case $K = \mathbb{C}$; however examination of the proof shows that the following is true.

If $c : A \to K$ is a solution of (d'A) with $c(0) = 1$ then there is an extension field L of K with $[L : K] \le 2$ and an exponential function $e : A \to L$ such that $2c(x) = e(x) + e(-x)$, for all $x \in A$.

Proof. By assumption $A = 2A$ so Proposition 2 applies and g is an exponential function. Define $c : A \to K$ by $c(x) = \frac{1}{2}f(2x)g(-x)$. Then c satisfies d'Alembert's equation and $c(0) = 1$. Hence there is a field L with $K \subseteq L$, $[L : K] \le 2$ and an exponential function $e : A \to L$ with

$$f(2x) = e(x)g(x) + e(-x)g(x).$$

Now let $x \in A$ be arbitrary. Since A is uniquely 2-divisible there is a unique element $\frac{x}{2} \in A$. Define $e_1(x) = e\left(\frac{x}{2}\right) g\left(\frac{x}{2}\right)$, $e_2(x) = e\left(-\frac{x}{2}\right) g\left(\frac{x}{2}\right)$. Then e_1, e_2 are exponential functions with $e_1(0) = e_2(0) = 1$. Also $e_1(x) + e_2(x) = f\left(2 \cdot \frac{x}{2}\right) = f(x)$ and $e_1(x)e_2(x) = g\left(\frac{x}{2}\right) g\left(\frac{x}{2}\right) = g(x)$. This completes the proof of the theorem.   □

References

[1] J. Aczél, *Lectures on Functional Equations and their Applications*, Academic Press, New York–London, 1966.

[2] Pl. Kannappan, *The functional equation $f(xy) + f(xy^{-1})$ $= 2f(x)f(y)$ for groups*, Proc. Amer. Math. Soc. **19** (1968), 69–74.

ABOUT A REMARKABLE FUNCTIONAL EQUATION ON SOME RESTRICTED DOMAINS

Fulvia Skof

Dipartimento di Matematica, Università di Torino,
Via Carlo Alberto 10, I-10123 Torino, Italy
fulvia.skof@unito.it

Abstract   We study the equation

$$f(x+y) + f(x-y) - 2f(x) - f(y) - f(-y) = 0$$

on restricted domains. For functions f of a real variable we give the general solution of the equation restricted to a neighbourhood of a given point $(a, b) \in \mathbb{R}^2$. In a more general functional setting, we prove an extension theorem. Finally, we state an equivalence result involving an asymptotic condition as $|x| + |y| \to \infty$.

Keywords: functional equation, equation on restricted domains, extension theorem, asymptotic condition

Mathematics Subject Classification (2000): 39B20

1.   INTRODUCTION

We shall consider the functional equation

$$f(x+y) + f(x-y) - 2f(x) - f(y) - f(-y) = A \qquad (1)$$

and the corresponding homogeneous equation

$$f(x+y) + f(x-y) - 2f(x) - f(y) - f(-y) = 0, \qquad (2)$$

[0]Research supported by Ministero dell'Università e della Ricerca Scientifica e Tecnologica of Italy.

Z. Daróczy and Z. Páles (eds.),
Functional Equations - Results and Advances, 249–262.

which are related to some of the most known equations: for instance, in the class of even functions (2) becomes the quadratic equation; in the class of odd functions (2) becomes the additive Cauchy equation, whereas (1) the Jensen one.

The general solution f of (2) was given by Gy. Szabó [6] using even and odd components of f

$$p_f(x) = \frac{f(x) + f(-x)}{2}, \qquad d_f(x) = \frac{f(x) - f(-x)}{2}. \tag{3}$$

It turns out that p_f and d_f satisfy the quadratic and the additive equation respectively; hence the general solution of (2) is given by

$$f(x) = q(x) + h(x), \tag{4}$$

where q is quadratic and h is additive.

Since the equation (1) can be transformed into an equation of the form (2) involving the new unknown function $f(x) - \frac{A}{2}$, the general solution of (1) is the abstract polynomial of the second degree

$$f(x) = q(x) + h(x) + c. \tag{5}$$

As (5) is also the general solution of the polynomial equation

$$f(x + 3y) - 3f(x + 2y) + 3f(x + y) - f(x) = 0, \tag{6}$$

we infer the equivalence of equations (1) and (6).

Our aim is to find the general solution of (2) (and consequently of (1)) on the "restricted domain" given by a bounded neighbourhood of a point $(a, b) \in \mathbb{R}^2$. The main results are Theorems 1 and 2.

Finally, we prove an equivalence result for (2) involving an asymptotic condition (Theorem 3).

2.   THE EXTENSION THEOREM

The extension result presented in this section concerns functions f having their domain in a real normed space $(X, \|\cdot\|)$ and the values $f(x)$ in a real linear space S.

We say that f is a solution of (2) on $E \subset X \times X$ if

(i) f is defined on $D_f := E_{x+y} \cup E_{x-y} \cup E_x \cup E_y \cup E_{-y} \subset X$, where

$$E_{x+y} = \{(u + v) \in X : (u, v) \in E\}$$
$$E_{x-y} = \{(u - v) \in X : (u, v) \in E\}$$
$$E_x = \{u \in X : (u, y) \in E \text{ for some } y \in X\}$$
$$E_y = \{v \in X : (x, v) \in E \text{ for some } x \in X\}$$
$$E_{-y} = \{v \in X : -v \in E_y\};$$

(ii) f satisfies (2) for all $(x, y) \in E$.

Let
$$B_r := \{x \in X : \|x\| < r\}$$
be the open ball of radius r in X. In the product space $X \times X$ the norm will be
$$\||(x, y)\|| := \|x\| + \|y\| \qquad ((x, y) \in X \times X)$$
and we use the notation
$$E(0, 0; r) := \{(x, y) \in X \times X : \||(x, y)\|| < r\}.$$

The following extension theorem, from the neighbourhood of zero to the whole space X, will be of use later for the proof of our main results.

Theorem 1. *If $f : B_r \subset (X, \| \cdot \|) \to S$ satisfies the equation (2) for every (x, y) with $\||(x, y)\|| < r$, then f may be extended from B_r over X to a function $F : X \to S$ satisfying (2) for every $(x, y) \in X \times X$. For $x \in X$, F is given by*

$$F(x) = 2^{n-1}\{(2^n + 1)f\left(\frac{x}{2^n}\right) + (2^n - 1)f\left(-\frac{x}{2^n}\right)\} \tag{7}$$

with $n = n(x) \in \mathbb{N}$ such that $\left\|\frac{x}{2^n}\right\| < \frac{r}{2}$.

Henceforth, $p_f(x)$ and $d_f(x)$ will mean, respectively, the even and odd component of f, according to formulas (3).

Proof. It is easy to prove that when
$$E = \{(x, y) \in X \times X : \||(x, y)\|| < r\}$$
$$= \{(x, y) \in X \times X : \|x\| + \|y\| < r\}$$
then $E_x = E_y = E_{-y} = B_r$ and also $E_{x+y}, E_{x-y} \subset B_r = D_f$.

Moreover, if $2u, 2v \in B_r$, then $(u, v) \in E$ and $u \pm v \in B_r = D_f$; hence, for every $(u, v) \in E(0, 0; \frac{r}{2})$ we get from (2)

$$\{p_f(u + v) + p_f(u - v) - 2p_f(u) - 2p_f(v)\} +$$
$$+ \{d_f(u + v) + d_f(u - v) - 2d_f(u)\} = 0;$$

writing the similar equation for $(-u, -v)$, by sum and difference it is immediately seen that p_f satisfies locally the quadratic equation and d_f the Jensen equation.

Now let x be an arbitrary point in X, and $n = n(x) \in \mathbb{N}$ such that $\left\|\frac{x}{2^n}\right\| < \frac{r}{2}$. Let us define $q, h : X \to S$ as follows:

$$q(x) := 2^{2n}p_f\left(\frac{x}{2^n}\right), \qquad \left\|\frac{x}{2^n}\right\| < \frac{r}{2}, \tag{8}$$

$$h(x) := 2^n d_f\left(\frac{x}{2^n}\right), \qquad \left\|\frac{x}{2^n}\right\| < \frac{r}{2}. \tag{9}$$

The values $q(x)$, $h(x)$ are independent from n (suitably large). If k is large enough and e.g. $n \geq k$ then

$$\frac{x}{2^k} = 2^{n-k}\frac{x}{2^n},$$

whence

$$p_f\left(\frac{x}{2^k}\right) = p_f\left(2^{n-k}\frac{x}{2^n}\right) = 2^{2n-2k}p_f\left(\frac{x}{2^n}\right),$$

$$2^{2k}p_f\left(\frac{x}{2^k}\right) = 2^{2n}p_f\left(\frac{x}{2^n}\right).$$

It is easily seen that $q(x)$ defined by (8) is quadratic in the whole space X.

Similarly for $h(x)$, by considering the locally additive function d_f.

Now, on the ground of (8) and (9) let us define $F : X \to S$ as follows

$$F(x) := q(x) + h(x), \qquad \text{for} \quad x \in X.$$

Obviously, F satisfies (2) for every $(x,y) \in X \times X$. The extension F is unique: Assume that F_1 is an extension of f satisfying (2) in $X \times X$; then $F_1 = q_1 + h_1$ and $q_1(x) = p_f(x)$ if $2x \in B_r$; hence $q_1\left(\frac{x}{2^n}\right) = p_f\left(\frac{x}{2^n}\right)$ and $q_1(x) = 2^{2n}q_1\left(\frac{x}{2^n}\right) = 2^{2n}p_f\left(\frac{x}{2^n}\right) = q(x)$ by (8). Similarly for $h(x)$.

We can derive from (8) and (9) the expression of $F(x)$ for $x \in X$ in terms of f as follows

$$F(x) = 2^{n-1}(2^n + 1)f\left(\frac{x}{2^n}\right) + 2^{n-1}(2^n - 1)f\left(-\frac{x}{2^n}\right),$$

for $n = n(x) \in \mathbb{N}$, $\left\|\frac{x}{2^n}\right\| < \frac{r}{2}$.
This proves the theorem.   □

3.   THE BOUNDED DOMAIN

Henceforth we shall assume that $(X, \|.\|)$ is the real space $(\mathbb{R}, |.|)$, with the view of finding the general solution of the equation (2) restricted to the neighbourhood of the given point $(a, b) \in \mathbb{R}^2$, namely to the square

$$E(a, b; r) := \{(x, y) \in \mathbb{R}^2 : \||(x, y) - (a, b)\|| < r\}$$
$$= \{(x, y) \in \mathbb{R}^2 : |x - a| + |y - b| < r\}$$

for some fixed $r > 0$; the domain $D_f \subset \mathbb{R}$ of the function f is the set

$$D_f = E_x \cup E_y \cup E_{-y} \cup E_{x+y} \cup E_{x-y},$$

which is composed by the usual projections

$$E_x := (a - r, a + r), \quad E_y := (b - r, b + r), \quad E_{-y} := (-b - r, -b + r),$$

$$E_{x+y} := (a + b - r, a + b + r), \quad E_{x-y} := (a - b - r, a - b + r).$$

Firstly we shall consider, for sake of simplicity, the case when all the above projections are pairwise disjoint, and they are also disjoint from the following opposite intervals

$$E_{-x} := (-a - r, -a + r), \quad E_{-(x+y)} := (-(a + b) - r, -(a + b) + r),$$

$$E_{-(x-y)} := (-(a - b) - r, -(a - b) + r),$$

in order to find the solution of (2) restricted to $E(a, b; r)$ in this situation. Afterwards it will be easy to adapt the result to the situation of overlapping intervals. We need the following

Lemma 1. *If* $f : D_f = E_x \cup E_y \cup E_{-y} \cup E_{x+y} \cup E_{x-y} \subset \mathbb{R} \to S$ *satisfies* (2) *in* $E(a, b; r)$, *then the function*

$$\varphi(t) := \frac{1}{2}\{f(a + t) + f(a - t) - 2f(a)\} \qquad (t \in (-r, r)) \qquad (10)$$

satisfies the quadratic equation in $E(0, 0; r) = \{(x, y) \in \mathbb{R}^2 : |x| + |y| < r\}$; *hence* φ *is the restriction to (-r,r) of the quadratic function* $\Phi : \mathbb{R} \to S$ *given by*

$$\Phi(x) = 2^{2n-1}\{f\left(a + \frac{x}{2^n}\right) + f\left(a - \frac{x}{2^n}\right) - 2f(a)\}$$

for $x \in X$ *and* $n = n(x) \in \mathbb{N}$ *such that* $\left|\frac{x}{2^n}\right| < \frac{r}{2}$.

Proof. For $(x, y) \in E(a, b; r)$ let us write $x = a + x'$, $y = b + y'$, with $(x', y') \in E(0, 0; r)$. Then, from (2) it follows

$$f(a + b + x' + y') + f(a - b + x' - y')$$
$$= 2f(a + x') + f(b + y') + f(-b - y') \qquad (11)$$

and, for $y' = 0$ and $x' = t$,

$$f(a + b + t) + f(a - b + t) = 2f(a + t) + f(b) + f(-b) \qquad (t \in (-r, r)),$$

whence, for $t \in (-r, r)$, we get

$$f(a + t) = \frac{1}{2}f(a + b + t) + \frac{1}{2}f(a - b + t) - \frac{1}{2}\{f(b) + f(-b)\}, \qquad (12)$$

By use of (11), (2) and grouping suitably the terms we get

$$f(a+t) + f(a-t) - 2f(a) =$$
$$= \frac{1}{2}\{f(a+(b+t)) + f(a-(b+t))\}$$
$$+ \frac{1}{2}\{f(a-(b-t)) + f(a+(b-t))\}$$
$$- \{f(b) + f(-b)\} - 2f(a)$$
$$= \frac{1}{2}\{2f(a) + f(b+t) + f(-b-t)\}$$
$$+ \frac{1}{2}\{2f(a) + f(b-t) + f(-b+t)\}$$
$$- \{f(b) + f(-b)\} - 2f(a),$$

whence

$$f(a+t) + f(a-t) - 2f(a) = \frac{1}{2}\{f(b-t) + f(-b+t)\}$$

$$+ \frac{1}{2}\{f(b+t) + f(-b-t)\} - \{f(b) + f(-b)\}. \tag{13}$$

Then, relatively to the function φ defined in (10) we get from (13), for $t \in (-r, r)$

$$\varphi(t) = \frac{1}{4}\{f(b-t) + f(-b+t) + f(b+t)$$

$$+ f(-b-t) - 2[f(b) + f(-b)]\}. \tag{14}$$

A direct calculation shows that φ satisfies the quadratic equation for $(x', y') \in E(0, 0; r)$. In fact, from the definition (10) we have

$$\varphi(x'+y') + \varphi(x'-y')$$
$$= \frac{1}{2}\{f(a+x'+y') + f(a-x'-y') - 2f(a) + f(a+x'-y')$$
$$+ f(a-x'+y') - 2f(a)\} =: W,$$

whence, by use of (12),

$$W = \frac{1}{4}\{f(a+b+x'+y') + f(a-b+x'+y') + f(a+b-x'-y')$$
$$+ f(a-b-x'-y') + f(a+b+x'-y') + f(a-b+x'-y')$$
$$+ f(a+b-x'+y') + f(a-b-x'+y')$$
$$- 4[f(b) + f(-b)]\} - 2f(a).$$

Applying (11) we get

$$
\begin{aligned}
W &= \frac{1}{4}\{2f(a+x') + f(b+y') + f(-b-y') + 2f(a-x') + f(b-y') \\
&\quad + f(-b+y') + 2f(a+x') + f(b-y') + f(-b+y') + 2f(a-x') \\
&\quad + f(b+y') + f(-b-y') - 4[f(b) + f(-b)]\} - 2f(a) \\
&= f(a+x') + f(a-x') - 2f(a) + \frac{1}{2}\{f(b+y') + f(b-y') \\
&\quad + f(-b+y') + f(-b-y')\} - [f(b) + f(-b)],
\end{aligned}
$$

whence, on the ground of (14),

$$
\varphi(x' + y') + \varphi(x' - y') = 2\varphi(x') + 2\varphi(y')
$$

for $(x', y') \in E(0, 0; r)$. Therefore, there exists a (unique) quadratic function $\Phi : \mathbb{R} \to S$ extending φ from $(-r, r)$ to $\mathbb{R}$; it is given by the formula

$$
\Phi(x) = 2^{2n}\varphi\left(\frac{x}{2^n}\right)
$$

for $x \in \mathbb{R}$ and $n = n(x) \in \mathbb{N}$ such that $\left|\frac{x}{2^n}\right| < \frac{r}{2}$.
This proves the lemma. $\qquad\square$

Theorem 2. *Let us assume that the set*

$$
\begin{aligned}
E(a, b; r) &= \{(x, y) \in \mathbb{R}^2 : |||(x, y) - (a, b)||| < r\} \\
&= \{(x, y) \in \mathbb{R}^2 : |x - a| + |y - b| < r\}
\end{aligned}
$$

has the usual projections $E_x, E_y, E_{-y}, E_{x+y}, E_{x-y}$ and the opposite intervals $E_{-x}, E_{-(x+y)}, E_{-(x-y)}$ all pairwise disjoint. Let S be a real linear space.

If $f : D_f = (E_x \cup E_y \cup E_{-y} \cup E_{x+y} \cup E_{x-y}) \subset \mathbb{R} \to S$ is a solution of the equation (2) on $E(a, b; r)$, then f has the following form

$$
f(t) = \begin{cases}
\Phi(t) + G(t) + \alpha, & t \in E_x = (a - r, a + r) \\
\Phi(t) + H(t) + Z(t) + \beta + \delta, & \\
& t \in E_y = (b - r, b + r) \\
\Phi(t) - H(t) + Z(t) + \beta - \delta, & \\
& t \in E_{-y} = (-b - r, -b + r) \\
\Phi(t) + G(t) + H(t) + \gamma, & \\
& t \in E_{x+y} = (a + b - r, a + b + r) \\
\Phi(t) + G(t) - H(t) + 2(\alpha + \beta) - \gamma, & \\
& t \in E_{x-y} = (a - b - r, a - b + r),
\end{cases}
\tag{15}
$$

where $\Phi : \mathbb{R} \to S$ is the quadratic function from Lemma 1, G and H: $\mathbb{R} \to S$ are additive, $Z : \mathbb{R} \to S$ is odd, and $\alpha, \beta, \gamma, \delta$ are constant vectors.

Proof. $1^{st}step$: Finding f on $E_x = (a - r, a + r)$.
   Let us define

$$z(x) := f(x) - \Phi(x) \qquad (x \in D_f), \qquad (16)$$

where Φ is the quadratic function extending φ from $(-r, r)$ to $\mathbb{R}$, according to the Lemma 1. Then z satisfies (2) in $E(a, b; r)$. Moreover, for every $(x, y') \in E(a, 0; r)$ we have $x \pm y' \in (a - r, a + r)$ and, by (16) and the properties of Φ

$$A : = z(x + y') + z(x - y')$$
$$= f(x + y') + f(x - y') - \Phi(x + y') - \Phi(x - y')$$
$$= f\left(a + (x + y' - a)\right) + f\left(a + (x - y' - a)\right)$$
$$- 2\Phi(x) - 2\Phi(y').$$

Using (2) in the form

$$f(x) = \frac{1}{2}[f(x + y) + f(x - y) - \{f(y) + f(-y)\}]$$

to calculate the first and second term (with $y = b$), we get

$$A = \frac{1}{2}\left[f\left(a + b + (x + y' - a)\right) + f\left(a - b + (x + y' - a)\right) \right.$$
$$- [f(b) + f(-b)] + f\left(a + b + (x - y' - a)\right)$$
$$\left. + f\left(a - b + (x - y' - a)\right) - \{f(b) + f(-b)\}\right] - 2\Phi(x) - 2\Phi(y')$$

or

$$A = \frac{1}{2}\left[f\left(x + (b + y')\right) + f\left(x - (b + y')\right) \right.$$
$$+ f\left(x + (b - y')\right) + f\left(x - (b - y')\right)\right]$$
$$- (f(b) + f(-b)) - 2\Phi(x) - 2\Phi(y').$$

Applying (2) again twice to find the expression in the bracket

$$A = \frac{1}{2}[2f(x) + f(b + y') + f(-b - y') + 2f(x) + f(b - y') + f(-b + y')]$$

$$-[f(b) + f(-b)] - 2\Phi(x) - 2\Phi(y')$$
$$= \frac{1}{2}[f(b + y') + f(b - y') + f(-b + y') + f(-b - y')$$
$$-2(f(b) + f(-b))] + 2f(x) - 2\Phi(x) - 2\Phi(y')$$

and by (14)

$$A = 2\varphi(y') + 2z(x) - 2\Phi(y') = 2z(x).$$

Thus we proved that $z(x + y') + z(x - y') = 2z(x)$ if $(x, y') \in E(a, 0; r)$. By obvious substitutions, we get that z satisfies the equation

$$\frac{1}{2}\{z(u) + z(v)\} = z\left(\frac{u + v}{2}\right)$$

for $(u, v) \in (a-r, a+r) \times (a-r, a+r)$; so there exist an additive $G : \mathbb{R} \to S$ and a constant $\alpha \in S$ such that $z(t) = G(t) + \alpha$ for $t \in (a - r, a + r)$: this and (16) give the form $\Phi(t) + G(t) + \alpha$ of $f(t)$ for $t \in (a - r, a + r) = E_x$.

$2^{nd} step$: Determining the auxiliary function $g : D_f \cup (-D_f) \to S$.

Since we assumed that the intervals $E_{-x} := (-a - r, -a + r)$, $E_{-(x+y)} := (-a - b - r, -a - b + r)$, $E_{-(x-y)} := (-(a - b) - r, -(a - b) + r)$ are disjoint from $D_f = E_x \cup E_y \cup E_{-y} \cup E_{x+y} \cup E_{x-y}$, we shall extend f from D_f to $D_f \cup E_{-x} \cup E_{-(x+y)} \cup E_{-(x-y)}$ as follows

$$g(t) := \begin{cases} f(t) & \text{for } t \in D_f \\ f(-t) & \text{for } t \in E_{-x} \cup E_{-(x+y)} \cup E_{-(x-y)}; \end{cases} \tag{17}$$

therefore, g is even on each of the sets

$$E_x \cup E_{-x}, \qquad E_{x+y} \cup E_{-(x+y)}, \qquad E_{x-y} \cup E_{-(x-y)},$$

and the even component of g, namely

$$p_g(t) = \frac{1}{2}\{g(t) + g(-t)\},$$

equals $f(t)$ for $t \in E_x \cup E_{x+y} \cup E_{x-y}$. This leads us to the following

Lemma 2. *The even component p_g of the function g defined by (17) satisfies the quadratic equation on the restricted domain $E(a, b; r)$.*

In fact, for $(x, y) \in E(a, b; r)$, whence $x + y \in E_{x+y}$ and $x - y \in E_{x-y}$, from (17) we get

$$p_g(x + y) + p_g(x - y) = f(x + y) + f(x - y) = 2f(x) + f(y) + f(-y)$$
$$= g(x) + g(-x) + g(y) + g(-y)$$
$$= 2p_g(x) + 2p_g(y).$$

By virtue of this Lemma, we can apply to p_g the result from [3] giving the general solution of the quadratic equation restricted to $E(a, b; r)$, namely

$$p_g(t) = \begin{cases} Q(t) + A(t) + c_1, & t \in E_x \\ Q(t) + H(t) + c_2, & t \in E_y \\ Q(t) + A(t) + H(t) + c_3, & t \in E_{x+y} \\ Q(t) + A(t) - H(t) + 2(c_1 + c_2) - c_3, & t \in E_{x-y}, \end{cases} \qquad (18)$$

where Q is a suitable quadratic function, A, H are additive and c_1, c_2, c_3 are constant.

3rd step : The general solution.

Since $p_g(t) = f(t)$ for $t \in E_x \cup E_{x+y} \cup E_{x-y}$, and from the 1st step we have

$$f(t) = \Phi(t) + G(t) + \alpha \qquad \text{for} \quad t \in E_x,$$

we can deduce from (18) the following partial result, where $Q(t) \equiv \Phi(t)$, $A(t) \equiv G(t)$, $c_1 = \alpha$ are assumed

$$f(t) = \begin{cases} \Phi(t) + G(t) + \alpha, & t \in E_x \\ \Phi(t) + G(t) + H(t) + c_3, & t \in E_{x+y} \\ \Phi(t) + G(t) - H(t) + 2(\alpha + c_2) - c_3, & t \in E_{x-y}. \end{cases} \qquad (19)$$

In order to complete the proof of the Theorem, we have now to find $f(t)$ for $t \in E_y \cup E_{-y}$, or, more exactly, its odd component

$$d_f(t) = \frac{1}{2}\{f(t) - f(-t)\} = d_g(t) \qquad \text{for} \quad t \in E_y \cup E_{-y}.$$

Since $f(t) = g(t)$ when $t \in D_f$, for $(x, y) \in E(a, b; r)$ we have from (2)

$$(p_g + d_g)(x + y) + (p_g + d_g)(x - y) = 2(p_g + d_g)(x) + 2p_g(y),$$

whence, on the ground of Lemma 2,

$$d_g(x + y) + d_g(x - y) = 2d_g(x),$$

which turns out to be the identity $0 = 0$.

In fact, by (17) g is the even extension of f from D_f to $D_f \cup E_{-x} \cup E_{-(x+y)} \cup E_{-(x-y)}$; therefore, the odd component d_g vanishes identically in the sets $E_{-x} \cup E_x$, $E_{-(x+y)} \cup E_{x+y}$, $E_{-(x-y)} \cup E_{x-y}$. For $t \in E_y \cup E_{-y}$, since $f(t) = g(t)$ and $d_f(t) = d_g(t)$, we infer that $d_f(t)$ can be an arbitrary function $\zeta(t)$ satisfying $\zeta(-t) = -\zeta(t)$ for $t \in E_y \cup E_{-y}$ and

$\zeta(t) = 0$ for $t \in \mathbb{R} \setminus (E_y \cup E_{-y})$. On account of the disjointness of the sets E_y and E_{-y}, we can write $\zeta(t)$ in the form

$$\zeta(t) = \begin{cases} Z(t) + \delta, & t \in E_y \\ Z(t) - \delta, & t \in E_{-y} \\ 0, & t \in \mathbb{R} \setminus (E_y \cup E_{-y}), \end{cases}$$

involving the restriction to $E_y \cup E_{-y}$ of an odd function $Z : \mathbb{R} \to S$ and a constant vector δ.

So, we have

$$f(t) = \begin{cases} \Phi(t) + H(t) + c_2 + Z(t) + \delta, & t \in E_y \\ \Phi(t) - H(t) + c_2 + Z(t) - \delta, & t \in E_{-y}, \end{cases} \tag{20}$$

where $Z(-y) = -Z(y)$ for $y \in E_y \cup E_{-y}$. Formulas (19), (20) complete the proof of the Theorem. $\qquad\square$

A direct check shows that the equation (2) restricted to $E(a, b; r)$ is satisfied by (15).

4.   REMARKS

Remark 1. The general solution given by (15) is related to the case where all the intervals $E_x, E_y, E_{-y}, E_{x+y}, E_{x-y}$ and $E_{-x}, E_{-(x+y)}$, $E_{-(x-y)}$ are pairwise disjoint. It can be remarked that when $a, b \neq 0$, $|a| \neq |b|$, then choosing $r > 0$ small enough the disjointness assumption can be satisfied. However, when overlappings occur, some of the expressions for $f(t)$ have to assume the same value, and therefore formulas (15) can become simpler: the odd term $Z(t)$ may be an additive function, different additive components may coincide or identically vanish, etc. For instance, if $a = b$, i.e. the domain of equation (2) is the square $E(a, a; r)$ and $E_x = E_y$, from the equation

$$G(t) + \alpha = H(t) + Z(t) + \beta + \delta \qquad (t \in E_x = E_y)$$

we find that $Z(t)$ is additive and $\delta = \alpha - \beta$; so the solution of (2) on $E(a, a; r)$ is

$$f(t) - \begin{cases} \Phi(t) + G(t) + \alpha, & t \in E_x = E_y \\ \Phi(t) + G(t) - 2H(t) + 2\beta - \alpha, & t \in E_{-y} \\ \Phi(t) + G(t) + H(t) + \gamma, & t \in E_{x+y} \\ \Phi(t) + G(t) - H(t) + 2(\alpha + \beta) - \gamma, & t \in E_{x-y}. \end{cases}$$

Remark 2. In solving either the additive Cauchy equation or the quadratic one on a bounded restricted domain, we remark the phenomenon that the solutions over the projections of the set are depending on the "restricted domain" and may contain linear combinations of functions which don't satisfy the equation related to the whole space: like constant terms in case of the restricted Cauchy equation [2], additive functions or constant terms in case of the quadratic equation [3]. Since additive, quadratic and constant functions are solutions of the equation (2) or of (1), we can infer that the solution to a restrictedly additive or quadratic equation, in front of the weakened condition, behaves locally like a solution of the general equation (2) or (1). When we are concerned with solutions of the restricted equation (2), we can observe again that solutions depend on the bounded domain $E(a, b; r)$ of the equation and are constructed by means of functions belonging to a wider class, namely the class obtained by adding to the abstract polynomials satisfying (1) an arbitrary odd function $Z(y)$ defined on the subset $E_y \cup E_{-y}$.

5.   AN ASYMPTOTIC CONDITION

Theorem 3. *Let $(S, \|.\|)$ be a Banach space and $f : \mathbb{R} \to S$ a function. Then*

$$lim_{|x|+|y| \to \infty} \|f(x+y) + f(x-y) - 2f(x) - f(y) - f(-y)\| = 0 \quad (21)$$

holds if and only if

$$f(x+y) + f(x-y) - 2f(x) - f(y) - f(-y) = 0 \quad (22)$$

for all $(x, y) \in \mathbb{R}^2$.

Proof. Clearly, the second equation implies the first. To prove the reverse implication, let us write

$$\Psi(x, y) := f(x+y) + f(x-y) - 2f(x) - f(y) - f(-y),$$

whence $\|\Psi(x, y)\| \to 0$ for $|x| + |y| \to +\infty$; $x = 0$ and $y \to \infty$ imply $f(0) = 0$. Using again the representation

$$f(t) = p_f(t) + d_f(t) \quad (23)$$

by means of even and odd components of f, let us consider

$$\Psi(x, y) = P_\Psi(x, y) + D_\Psi(x, y), \quad (24)$$

where

$$P_\Psi(x, y) = p_f(x+y) + p_f(x-y) - 2p_f(x) - 2p_f(y),$$
$$D_\Psi(x, y) = d_f(x+y) + d_f(x-y) - 2d_f(x),$$

and

$$\Psi(-x, -y) = P_\Psi(x, y) - D_\Psi(x, y). \tag{25}$$

From (24) and (25), by addition and subtraction, we get

$$\Psi(x, y) + \Psi(-x, -y) = 2P_\Psi(x, y)$$
$$\Psi(x, y) - \Psi(-x, -y) = 2D_\Psi(x, y);$$

from (21)

$$\|P_\Psi(x, y)\| \to 0, \qquad \|D_\Psi(x, y)\| \to 0 \qquad \text{for} \quad |x| + |y| \to \infty. \tag{26}$$

On the ground of known results concerning asymptotic conditions related to additive and quadratic equations, we shall easily prove that $P_\Psi(x, y)$ and $D_\Psi(x, y)$ vanish identically in $\mathbb{R}^2$.

In fact, the condition

$$\|P_\Psi(x, y)\| \to 0 \qquad \text{for} \quad |x| + |y| \to \infty$$

has been proved in 1985 [5] to be equivalent to the quadratic equation. As for $D_\Psi(x, y)$, the limit condition in (26) implies $d_f(0) = 0$, and for $x = y$

$$d_f(2x) = 2d_f(x) + o(1) \qquad \text{for} \quad |x| \to +\infty; \tag{27}$$

moreover, by the substitution $x + y = 2u$, $x - y = 2v$, it is changed into

$$\|d_f(2u) + d_f(2v) - 2d_f(u + v)\| \to 0$$

for $|u + v| + |u - v| \to \infty$ and also for $|u| + |v| \to \infty$ whence, according to (27),

$$\|d_f(u + v) - d_f(u) - d_f(v)\| \to 0 \qquad \text{for} \quad |u| + |v| \to \infty.$$

This condition has been proved in 1983 [4] to be equivalent to the additive Cauchy equation; hence $D_\Psi(x, y)$ vanishes identically too. The theorem is proved. $\qquad \square$

References

[1] J. Aczél and J. Dhombres, *Functional Equations in Several Variables*, Cambridge University Press, Cambridge, 1989.

[2] Z. Daróczy and L. Losonczi, *Über die Erweiterung der auf einer Punktmenge additiven Funktionen*, Publ. Math. Debrecen **14** (1967), 239–245.

[3] G. Peyrot and F. Skof, *Sulle funzioni localmente quadratiche nell'intorno di un punto*, Atti Accad. Sc. Torino **120** (1986), 80–92.

[4] F. Skof, *Sull'approssimazione delle applicazioni localmente δ-additive*, Atti Accad. Sc. Torino **117** (1983), 377–389.

[5] F. Skof, *Domini ristretti di validità dell'equazione funzionale quadratica e validità in grande*, Atti Accad. Sc. Torino **119** (1985), 234–244 (in particular, Theorem 4).

[6] Gy. Szabó, *Some functional equations related to quadratic functions*, Glasnik Mat. **38** (1983), 107–118.

ON DISCRETE SPECTRAL SYNTHESIS

László Székelyhidi

Institute of Mathematics and Informatics, University of Debrecen,
4010 Debrecen, Pf. 12, Hungary
szekely@math.klte.hu

Abstract   Spectral synthesis deals with the description of translation invariant function spaces over topological groups. In this work we deal with the case of discrete abelian groups. A pioneer result of M. Lefranc states that spectral synthesis holds for finitely generated discrete abelian groups. Later R. J. Elliot made an attempt to extend this result to any discrete abelian group. Unfortunately, Elliot's proof has some gaps which have not been filled yet. In this work we present a new approach to prove spectral synthesis for arbitrary discrete abelian groups.

Keywords: spectral synthesis, translation invariant function space

Mathematics Subject Classification (2000): 43A45, 39B52, 22A10

1.   INTRODUCTION

Spectral synthesis deals with the description of translation invariant function spaces over topological groups. Translation invariant function spaces appear in several different contexts: linear ordinary and partial difference and differential equations with constant coefficients, theory of group representations, classical theory of functional equations, etc. A fundamental problem is to discover the structure of such spaces of functions, or more exactly, to find an appropriate class of basic functions, the building blocks, which serve as "typical elements" of the space, a kind of basis. It turns out that these building blocks are the so-called exponential monomials. A famous and pioneer result of L. Schwartz [5]

[0]The research was supported by the Hungarian National Foundation for Scientific Research (OTKA), Grant No. T-031995 and by the Hungarian Higher Educational Research and Development Found (FKFP) Grant No. 0310/1997.

263

Z. Daróczy and Z. Páles (eds.),
Functional Equations - Results and Advances, 263–274.
© 2002 *Kluwer Academic Publishers.*

exhibits the situation. We consider the space of all complex valued continuous functions on the real line which is a locally convex topological linear space with respect to the pointwise linear operations (addition, multiplication with scalars) and to the topology of uniform convergence on compact sets. Suppose that a closed linear subspace in this space is given, which is translation invariant. This subspace may or may not contain any basic function of the above mentioned form, that is, an exponential monomial, which is the product of a power function and an exponential function. It is not even clear, if any exponential function is included in such subspaces. If so, then we say that spectral analysis holds for the subspace in question. It is not quite difficult to show, that the appearance of an exponential monomial in a subspace of this type implies that the exponential function occurring in this exponential monomial must belong to the same subspace. The complex number, characterizing this exponential function can be considered as a kind of spectral value. A fundamental consequence of L. Schwartz's result is that spectral values do exist, that means, any closed translation invariant linear subspace of the space mentioned above contains an exponential. The next step is to find the multiplicity of this spectral value, or exponential. This multiplicity refers to the highest exponent of the power function which - multiplied by the exponential function found above - belongs to the subspace. The complete description of the subspace means that all the exponential monomials corresponding to the spectral exponentials and their multiplicities characterize the subspace: their linear hull is dense in the subspace. If this happens, then we say that spectral synthesis holds for the subspace. Actually this is L. Schwartz's result: any closed translation invariant linear space of continuous functions on the reals is synthesizable from its exponential monomials.

The situation presented in this example can be generalized. Suppose, that a locally compact abelian group is given and we consider the set of all continuous complex valued functions on it, equipped with the pointwise linear operations and with the topology of uniform convergence on compact sets. In order to set up the problem of spectral analysis and spectral synthesis in this context we have to define exponential functions and exponential monomials on commutative topological groups. Continuous homomorphisms of such groups into the additive topological group of complex numbers, and into the multiplicative topological group of nonzero complex numbers are called additive, and exponential functions, respectively. An exponential monomial is a product of additive and exponential functions. Now the problem of spectral analysis, and spectral synthesis can be formulated: is it true, that any closed, translation invariant linear subspace of the space mentioned above contains an exponential function (spectral analysis), and is it true, that in any subspace of this type the linear hull of all exponential monomials is dense (spectral synthesis)? It is easy to see that we can go one step further: instead of the space of continuous functions with the given topology one can start with other important function spaces, which are translation invariant. For instance, the space of integrable functions is the natural medium of the Wiener–Tauberian theory: different versions of the Wiener–Tauberian theorem can be stated as spectral analysis theorems.

An interesting special case is presented by discrete abelian groups. Here the problem seems to be purely algebraic: all complex functions are continuous, and convergence is meant in the pointwise sense. The archetype is the additive group of integers: in this case the closed translation invariant function spaces can be characterized by systems of linear difference equations with constant coefficients. It is known that these function spaces are spanned by exponential monomials corresponding to the characteristic values of the equation, together with their multiplicities.

The next simplest case is the case of finitely generated abelian groups. As in this case structure theorem is available, it is not very surprising to have the corresponding - nontrivial - result, due to M. Lefranc [4]: on finitely generated abelian groups spectral synthesis holds for any closed translation invariant subspace.

Having these results the natural question arises: what about any discrete abelian groups? In [1] R. J. Elliot made an attempt to prove the validity of spectral synthesis on any discrete abelian group. Unfortunately, there are several gaps in his proof. In what follows we will try to offer a new approach to prove Elliot's result concerning spectral synthesis on arbitrary discrete abelian groups.

2.   NOTATION AND TERMINOLOGY

Let G be an abelian group written additively. The set of all complex valued functions on G will be denoted by $\mathcal{C}(G)$. This set equipped with the pointwise linear operations (addition and multiplication by complex numbers) and with the topology of pointwise convergence (the Thychonoff–topology) bears the structure of a locally complex topological vector space. The dual $\mathcal{M}(G)$ of this space can be identified with the space of all finitely supported complex measures on G, endowed with the pointwise linear operations. Actually, the linear functionals of $\mathcal{C}(G)$ can be realized as finitely supported functions on G. This space has a natural ring structure realized by the convolution, as multiplication. With respect to convolution $\mathcal{M}(G)$ is a commutative algebra with unit. Usually this space is considered as one equipped with the weak*-topology. Topological concepts (closed, continuous, etc.) applied to this space will always refer to this topology.

Closed translation invariant subspaces of the space $\mathcal{C}(G)$ are called *varieties*. Actually, we shall deal with *proper varieties* only: these are the ones different from the trivial subspaces, consisting the zero function only, or all complex valued functions on G.

As it is well-known from Hahn–Banach–theory, the study of closed subspaces of topological vector spaces can be converted into the study of their *annihilators*. If H is any subset of $\mathcal{C}(G)$, then its annihilator, $H^{\perp}$ is the set of all linear functionals from $\mathcal{M}(G)$ which vanish on H. Similarly, we define the annihilator of a subset M of $\mathcal{M}(G)$: this is the set of all elements of $\mathcal{C}(G)$ which are annihilated by the elements of M. This set can also be denoted by $M^{\perp}$. It is clear by the linearity and the continuity of the vector space operations, that the annihilators are closed linear spaces, independently of the structure of the original set H, or M. The Hahn–Banach–theorem enlightens the dual connection between annihilators in our situation.

Theorem 1. *Let G be an abelian group, V a variety in $\mathcal{C}(G)$, and I a closed ideal of $\mathcal{M}(G)$. Then $V^{\perp}$ is a closed ideal in $\mathcal{M}(G)$, $I^{\perp}$ is a variety in $\mathcal{C}(G)$, further $V^{\perp\perp} = V$, and $I^{\perp\perp} = I$.*

Proof. If f is an element of V and μ is an element of $V^{\perp}$, then by definition $\mu(f) = 0$, hence f belongs to $V^{\perp\perp}$. Suppose conversely, that V is a proper closed subspace of $V^{\perp\perp}$; then by the Hahn–Banach–theorem there exists a nonzero linear functional μ on $V^{\perp\perp}$ which annihilates V, that is, belongs to $V^{\perp}$, which is impossible. Similarly, we can prove the statement $I^{\perp\perp} = I$ for any closed ideal I in $\mathcal{M}(G)$. The proof of the other statements (the variety–ideal correspondence) can be found in [7]. $\qquad\square$

3.   EXPONENTIALS AND EXPONENTIAL MONOMIALS

In spectral synthesis an important role is played by *exponential functions* and *exponential monomials*. Homomorphisms of G into the additive group of complex numbers are called *additive functions*, and homomorphisms of G into the multiplicative group of nonzero complex numbers are called *exponential functions*, or simply *exponentials*. Products of additive and exponential functions are called *exponential monomials*. As the product of exponentials is an exponential too, the general form of an exponential monomial is

$$x \mapsto a_1(x)^{\alpha_1} a_2(x)^{\alpha_2} \ldots a_n(x)^{\alpha_n} m(x),$$

where $\alpha_1, \alpha_2, \ldots, \alpha_n$ are nonnegative integers, $a_1, a_2, \ldots, a_n$ are additive functions and m is an exponential. In the study of exponential monomials on an abelian group we usually fix a maximal linearly independent set of additive functions and suppose that all exponential monomials are built up from these additive functions. For instance, if $G = \mathbb{Z}^n$, then this set can be formed by the projections $x \mapsto x_i$ $(i = 1, 2 \ldots, n)$. In this case every exponential monomial can be written uniquely in the form

$$x \mapsto c \cdot x_1^{k_1} x_2^{k_2} \ldots x_n^{k_n} \exp(\lambda_1 x_1 + \lambda_2 x_2 + \cdots + \lambda_n x_n),$$

where c is a complex number, $k_1, k_2, \ldots, k_n$ are nonnegative integers, and $\lambda_1, \lambda_2, \ldots, \lambda_n$ are complex numbers.

The set of all exponentials in a given variety is called the *spectrum* of the variety, and the set of all exponential monomials in the variety is called the *spectral set* of the variety. We say that *spectral analysis holds* for a given variety if the spectrum of the variety is nonempty, and we say that *spectral synthesis holds* for a given variety if the linear hull of its spectral set is dense in the variety. In this latter case we call the variety *synthesizable*. We say that *spectral analysis holds* for a given abelian group if spectral analysis holds for any proper variety, and we say that *spectral synthesis holds* for a given abelian group if spectral synthesis holds for any proper variety.

We need the following result from [7] (Lemma 4.8 on p. 44.).

Theorem 2. *Let G be an abelian group, V a variety in $\mathcal{C}(G)$ and φ an exponential monomial of the form*

$$x \mapsto a_1(x)^{\alpha_1} a_2(x)^{\alpha_2} \ldots a_n(x)^{\alpha_n} m(x),$$

where $\alpha_1, \alpha_2, \ldots, \alpha_n$ are nonnegative integers, $a_1, a_2, \ldots, a_n$ are linearly independent additive functions and m is an exponential. If φ belongs to V, then the exponential monomials of the form

$$x \mapsto a_1(x)^{\beta_1} a_2(x)^{\beta_2} \ldots a_n(x)^{\beta_n} m(x)$$

belong to V for any nonnegative integers $\beta_i \leq \alpha_i, i = 1, 2 \ldots, n$.

This "descending property" of varieties with respect to exponential monomials is of basic importance. It implies, for instance, that if the exponential monomial

$$x \mapsto a_1(x)^{\alpha_1} a_2(x)^{\alpha_2} \ldots a_n(x)^{\alpha_n} m(x)$$

belongs to a variety with linearly independent additive functions, then the exponential m itself must belong to V.

4.   HISTORY

Here we summarize and cite the most important results and observations concerning spectral analysis and synthesis on discrete abelian groups.

As we mentioned above, apart from the classical results on finite difference equations the first general result on discrete spectral synthesis is due to M. Lefranc [4].

Theorem 3. *Spectral synthesis holds for any finitely generated abelian group.*

In 1965 R. J. Elliot ([1]) published the following general result:

Theorem 4. *Spectral synthesis holds for any abelian group.*

Unfortunately, in 1987 Z. Gajda observed that Elliot's proof has several serious gaps ([3]). In 1999 L. Székelyhidi ([6]) obtained the following result:

Theorem 5. *Spectral synthesis holds for abelian torsion groups.*

5.   SPECTRAL SYNTHESIS AND IDEALS

We recall the following concept: in a commutative ring with unit an ideal is called a *primary ideal*, if it is contained in exactly one maximal ideal. It is clear that if a variety is generated by an exponential monomial, then its annihilator is a primary ideal. Such ideals are called *fixed primary ideals*. The fixed primary ideal whose annihilator is generated by the exponential monomial φ will be denoted by I_φ.

Theorem 6. *Let G be an abelian group. Then spectral synthesis holds for a variety in $\mathcal{C}(G)$ if and only if its annihilator ideal is the intersection of fixed primary ideals.*

Proof. Let V be a synthesizable variety in $\mathcal{C}(G)$ and let E denote the set of all exponential monomials in V. We claim that

$$V^\perp = \bigcap_{\varphi \in E} I_\varphi.$$

If μ is any element of $V^\perp$ and φ is an exponential monomial in V, then μ annihilates the variety generated by φ, hence μ belongs to I_φ. Conversely, if μ belongs to I_φ for any exponential monomial φ in V, then μ vanishes on the closed linear hull of these exponential monomials, which is V, hence μ annihilates V.

On the other hand, if

$$V^\perp = \bigcap_{\varphi \in E} I_\varphi$$

holds and V is not synthesizable, then the variety, generated by all exponential monomials in V is a proper subvariety of V. Hence by the Hahn–Banach theorem there exists a μ in $\mathcal{M}(G)$ which annihilates this subvariety, but is not identically zero on V. This means, that μ does not belong to $V^\perp$, however, it belongs to $\bigcap_{\varphi \in E} I_\varphi$, which is a contradiction.

□

If G is an abelian group then $\widetilde{G}$ denotes the set of all exponentials on G. For any μ in $\mathcal{M}(G)$ we introduce the Fourier–Laplace transform of μ as the function defined on $\widetilde{G}$ by

$$\hat{\mu}(m) = \mu(\check{m})$$

for any m in $\widetilde{G}$. (The function $\check{m}$ is defined by $\check{m}(x) = m(-x)$ for all x in G.) A basic property - besides linearity and injectivity - of the Fourier–Laplace transform is expressed by the convolution–formula:

$$(\mu * \nu)\hat{} = \hat{\mu} \cdot \hat{\nu},$$

which is valid for any μ, ν in $\mathcal{M}(G)$. It follows that the Fourier–Laplace transforms for all measures in $\mathcal{M}(G)$ form a commutative algebra, isomorphic to $\mathcal{M}(G)$. This algebra will be denoted by $\mathcal{A}(G)$. It follows also, that for any ideal I in $\mathcal{M}(G)$ the Fourier–Laplace transforms of the elements of I form an ideal in the algebra $\mathcal{A}(G)$. This ideal will be denoted by $\hat{I}$. It is clear that I is maximal if and only if $\hat{I}$ is maximal, and I is primary if and only if $\hat{I}$ is primary.

6. SPECTRAL SYNTHESIS ON FREE ABELIAN GROUPS

Any free abelian group has the form $\mathbb{Z}^{\Gamma *}$, where Γ is any nonempty set. Here $*$ refers to the noncomplete, or *algebraic direct product*, that is, $\mathbb{Z}^{\Gamma *}$ is the set of all finitely supported $\mathbb{Z}$-valued functions on the set Γ. If x is an element of $\mathbb{Z}^{\Gamma *}$ and γ is in Γ, then we may write x_γ instead of $x(\gamma)$, and we call it the *γ-th coordinate* of x.

The projections $x \mapsto x_\gamma$ for γ in Γ are obviously linearly independent additive functions on $\mathbb{Z}^{\Gamma *}$. Moreover, any additive function on $\mathbb{Z}^{\Gamma *}$ is a linear combination of projections.

It is also easy to see that any exponential function on $\mathbb{Z}^{\Gamma *}$ has the form

$$x \mapsto \exp\left(\sum_{\gamma \in \Gamma} \lambda(\gamma) x_\gamma \right),$$

where $\lambda : \Gamma \to \mathbb{C}$ is an arbitrary function. Here the exponential can be identified with λ, hence the set of all exponentials on $\mathbb{Z}^{\Gamma *}$ can be identified with $\mathbb{C}^{\Gamma}$. Together with the previous remark we can see that from the point of view of spectral synthesis we can restrict ourselves to exponential monomials of the form

$$x \mapsto x_{\gamma_1}^{\alpha_1} x_{\gamma_2}^{\alpha_2} \dots x_{\gamma_n}^{\alpha_n} \exp\left(\sum_{\gamma \in \Gamma} \lambda(\gamma) x_{\gamma} \right),$$

where $\lambda : \Gamma \to \mathbb{C}$ is any function, $\gamma_1, \gamma_2, \dots, \gamma_n$ are in Γ, and $\alpha_1, \alpha_2, \dots, \alpha_n$ are nonnegative integers.

Finally, if $\mu : \mathbb{Z}^{\Gamma *} \to \mathbb{C}$ is any finitely supported function, then its Fourier–Laplace transform can be identified with the complex valued function $\hat{\mu}$ on $\mathbb{C}^{\Gamma}$ defined by

$$\hat{\mu}(\lambda) = \sum_{x \in \mathbb{Z}^{\Gamma *}} \mu(x) \exp\left(-\sum_{\gamma \in \Gamma} \lambda(\gamma) x_{\gamma} \right).$$

(The empty sum is considered to be zero.)

Observe that in the previous representation of $\hat{\mu}$ the sums are finite, as both μ and any x in $\mathbb{Z}^{\Gamma *}$ have finite support. It follows that any $\hat{\mu}$ depends on finitely many "coordinates" only. More precisely, we have the following theorem.

Theorem 7. *Let Γ be a nonempty set and μ a finitely supported measure on $\mathbb{Z}^{\Gamma *}$. Then there exists a finite subset Γ_0 in Γ with the following property: if $\lambda_1, \lambda_2 : \Gamma \to \mathbb{C}$ are functions such that $\lambda_1(\gamma) = \lambda_2(\gamma)$ holds for all γ in Γ_0, then $\hat{\mu}(\lambda_1) = \hat{\mu}(\lambda_2)$.*

Proof. Let $\Gamma_0 = \bigcup_{x \in \mathrm{supp}\,\mu} \mathrm{supp}\,x$. Suppose that $\lambda_1, \lambda_2 : \Gamma \to \mathbb{C}$ are functions such that $\lambda_1(\gamma) = \lambda_2(\gamma)$ holds for all γ in Γ. Then we have

$$\hat{\mu}(\lambda_1) = \sum_{x \in \mathbb{Z}^{\Gamma *}} \mu(x) \exp\left(-\sum_{\gamma \in \Gamma} \lambda_1(\gamma) x_{\gamma} \right)$$

$$= \sum_{x \in \mathrm{supp}\,\mu} \mu(x) \exp\left(-\sum_{\gamma \in \mathrm{supp}\,x} \lambda_1(\gamma) x_{\gamma} \right)$$

$$= \sum_{x \in \mathrm{supp}\,\mu} \mu(x) \exp\left(-\sum_{\gamma \in \mathrm{supp}\,x} \lambda_2(\gamma) x_{\gamma} \right)$$

$$= \sum_{x \in \mathbb{Z}^{\Gamma *}} \mu(x) \exp\left(-\sum_{\gamma \in \Gamma} \lambda_2(\gamma) x_{\gamma} \right) = \hat{\mu}(\lambda_2).$$

□

It is also clear that the Fourier–Laplace transform $\hat{\mu}$ of any finitely supported measure μ on $\mathbb{Z}^{\Gamma*}$ is an entire function of those coordinates on which it depends. We introduce the notation $\mathcal{A}(\Gamma)$ for the algebra of the Fourier–Laplace transforms of all finitely supported complex measures on $\mathbb{Z}^{\Gamma*}$. It is isomorphic to $\mathcal{M}(\mathbb{Z}^{\Gamma*})$ and consists of complex valued functions on $\mathbb{C}^{\Gamma}$ which depend on finitely many coordinates only (in the sense of Theorem 7.) and are entire functions of those coordinates on which they depend.

We need the following notation. Let $\alpha : \Gamma \to \mathbb{N}$ be a finitely supported function, and let $\Gamma_0 = \{\gamma_1, \gamma_2, \ldots, \gamma_n\}$ be the support of α. For any μ in $\mathcal{M}(\mathbb{Z}^{\Gamma*})$ and for any λ in $\mathbb{C}^{\Gamma}$ let

$$\partial^{\alpha}_{\Gamma_0}\hat{\mu}(\lambda) = \partial^{\alpha_1}_{\gamma_1}\partial^{\alpha_2}_{\gamma_2}\ldots\partial^{\alpha_n}_{\gamma_n}\hat{\mu}(\lambda).$$

Here $\alpha_i = \alpha(\gamma_i)$ $(i = 1, 2, \ldots, n)$. In order to explain the meaning of the right hand side of the previous equation we remark that $\hat{\mu}$ depends on finitely many, say m coordinates of λ only, hence it can be considered as a function of m complex variables. If, for instance, $\gamma_1, \gamma_2, \ldots, \gamma_n$ are included in the set of coordinates on which $\hat{\mu}$ depends, then the above differential expression has the obvious meaning. However, if some γ_i with $i = 1, 2 \ldots, n$ is not included in the set of coordinates on which $\hat{\mu}$ depends, then the right hand side of the above equation is zero.

We will also use the analoguous notation for monomials on $\mathbb{Z}^{\Gamma*}$. Namely, if $\Gamma_0 = \{\gamma_1, \gamma_2, \ldots, \gamma_n\}$ is a finite subset of Γ and $\alpha : \Gamma \to \mathbb{N}$ is a finitely supported function with support Γ_0, then we let

$$x^{\alpha}_{\Gamma_0} = x^{\alpha_1}_{\gamma_1}x^{\alpha_2}_{\gamma_2}\ldots x^{\alpha_n}_{\gamma_n}.$$

Here again $\alpha_i = \alpha(\gamma_i)$ $(i = 1, 2, \ldots, n)$.

If x is an element of $\mathbb{Z}^{\Gamma}$ and λ is an element of $\mathbb{C}^{\Gamma}$, then we let

$$(\lambda \cdot x) = \sum_{\gamma \in \Gamma} \lambda(\gamma)x_{\gamma}.$$

Theorem 8. *Let Γ be a nonempty set and let I be an ideal in $\mathcal{M}(\mathbb{Z}^{\Gamma*})$. The ideal I is an intersection of fixed primary ideals if and only if it has the following property: there exists a nonempty subset Λ of $\mathbb{C}^{\Gamma}$, for each λ in Λ there exists a nonempty family Δ_{λ} of nonempty sets, and for each Δ in Δ_{λ} there exists a nonempty set $\Omega_{\lambda, \Delta}$ of finitely supported functions $\alpha : \Gamma \to \mathbb{N}$ with support $\Omega_{\lambda, \Delta}$ such that the finitely supported measure μ belongs to I if and only if*

$$\partial^{\alpha}_{\Delta}\hat{\mu}(-\lambda) = 0$$

for all λ in Λ, Δ in Δ_λ and α in $\Omega_{\lambda,\Delta}$.

Proof. Suppose, that I is the intersection of fixed primary ideals:

$$I = \bigcap_{\varphi \in \Phi} I_\varphi.$$

Any φ in Φ has the form

$$\varphi(x) = x_\Delta^\alpha \exp(\lambda \cdot x),$$

where λ is an element of $\mathbb{C}^\Gamma$, Δ is a nonempty finite subset of Γ, and $\alpha : \Gamma \to \mathbb{N}$ is a finitely supported function with support Δ. Let Λ be the set of all functions λ, which occur in the representations of the exponential monomials φ in Φ; for each $-\lambda$ in Λ let Δ_λ be the family of all sets Δ which occur as supports of the finitely supported functions α corresponding to λ, finally for each λ in Λ and Δ in Δ_λ let $\Omega_{\lambda,\Delta}$ be the set of finitely supported functions α which occur in the representations of the exponential monomials in Φ corresponding to λ and having support Δ. Then μ belongs to I if and only if $\mu(\varphi) = 0$ for all φ in Φ. In other words, μ belongs to I if and only if

$$\mu(x_\Delta^\alpha \exp(\lambda \cdot x)) = 0$$

for all λ in Λ, Δ in Δ_λ, and α in $\Omega_{\lambda,\Delta}$. By the definition of the Fourier–Laplace transform, this is equivalent to

$$\partial_\Delta^\alpha \hat{\mu}(-\lambda) = 0.$$

The converse of the statement is also obvious.   □

For later purposes we call an ideal in $\mathcal{M}(\mathbb{Z}^{\Gamma*})$ a *differentiation ideal* if it has the property formulated in Theorem 8. Theorem 6. and Theorem 8. imply the following result:

Theorem 9. *Spectral synthesis holds for any free abelian group if and only if any ideal in $\mathcal{M}(\mathbb{Z}^{\Gamma*})$ is a differentiation ideal for any nonempty set Γ.*

7.   SPECTRAL SYNTHESIS FOR ABELIAN GROUPS

The key to the extension of spectral synthesis from free abelian groups to arbitrary abelian groups is provided by the following theorem.

Theorem 10. *If spectral synthesis holds for an abelian group then it holds for its homomorphic images, too.*

Proof. Let G be an abelian group for which spectral synthesis holds and let H be a homomorphic image of G: let $F : G \to H$ be a surjective homomorphism. If V is a variety in $\mathcal{C}(H)$ then we let

$$V_F = \{f \circ F : \quad f \in V\}.$$

Using the surjectivity of F a routine calculation shows that V_Φ is a variety in $\mathcal{C}(H)$. Let Φ be an exponential monomial in V_F of the form

$$\Phi(x) = A_1(x)^{\alpha_1} A_1(x)^{\alpha_2} \ldots A_n(x)^{\alpha_n} M(x),$$

where $A_1, A_2, \ldots, A_n$ are linearly independent additive functions on G, M is an exponential on G and $\alpha_1, \alpha_2, \ldots, \alpha_n$ are nonnegative integers. By Theorem 2. the exponential M is in V_F, too, hence $M = m \circ F$ holds for some m in V. If u, v are arbitrary in H, then $u = F(x)$ and $v = F(y)$ for some x, y in G, which implies

$$m(u + v) = m(F(x) + F(y)) = m(F(x + y)) = M(x + y)$$
$$= M(x)M(y) = m(F(x))m(F(y)) = m(u)m(v).$$

As m is never zero, hence m is an exponential in V. Theorem 2. also implies that $A_i M$ belongs to V_F ($i = 1, 2 \ldots, n$). That is, $A_i(m \circ F) = h \circ F$ with some h in V. This means that $A_i = a_i \circ F$ with some function $a_i : H \to \mathbb{C}$. If u, v are arbitrary in H, then $u = F(x)$ and $v = F(y)$ for some x, y in G, which implies

$$a_i(u + v) = a_i(F(x) + F(y)) = a_i(F(x + y)) = A_i(x + y)$$
$$= A_i(x) + A_i(y) = a_i(F(x)) + a_i(F(y))$$
$$= a_i(u) + a_i(v),$$

that is a_i is additive and $(a_i m) \circ F = A_i(m \circ F)$ belongs to V_F. If u is any element of H, then there exists x in G such that $F(x) = u$. Then we have

$$a_i m(u) = a_i m(F(x)) = (a_i m \circ F)(x)$$
$$= (h \circ F)(x) = h(F(x)) = h(u),$$

hence $a_i m = h$, which belongs to V. It follows that the exponential monomial $a_1^{\alpha_1} a_2^{\alpha_2} \ldots a_n^{\alpha_n} m$ corresponding to Φ belongs to V. If f is any element of V, then there is a net of linear combinations of exponential monomials converging pointwise on G to $f \circ F$ and by the surjectivity of F it is clear that the net of linear combinations of the corresponding exponential monomials converges pointwise on H to f. Hence spectral synthesis holds for H and the theorem is proved. $\qquad\square$

The following theorem is well-known (see [2], p. 38.).

Theorem 11. *Any abelian group is homomorphic image of a free abelian group.*

By Theorem 9. we have the following theorem.

Theorem 12. *Spectral synthesis holds for any abelian group if and only if for any nonempty set Γ all ideals in $\mathcal{A}(\Gamma)$ are differentiation ideals.*

Here we state the conjecture:

Conjecture. For any nonempty set Γ all ideals in $\mathcal{A}(\Gamma)$ are differentiation ideals.

We note that if G is a finitely generated abelian group then it is the homomorphic image of $\mathbb{Z}^n$ with some positive integer n. As it seems to be true that all ideals in $\mathcal{A}(n)$ are differentiation ideals, the results of this section may provide a new proof for Lefranc's result.

References

[1] R. J. Elliot, *Some results in spectral synthesis*, Proc. Cambridge Phil. Soc. **61** (1965), 617–620.

[2] L. Fuchs, *Abelian groups*, Akadémiai Kiadó, Budapest, 1966.

[3] Z. Gajda, (1987), private communication.

[4] M. Lefranc, *L'analyse harmonique dans $\mathbb{Z}^n$*, C. R. Acad. Sci. Paris **246** (1958), 1951–1953.

[5] L. Schwartz, *Théorie génerale des fonctions moyenne-périodiques*, Ann. of Math. **48** (1947), 857–929.

[6] L. Székelyhidi, *A Wiener Tauberian theorem on discrete abelian torsion groups*, in publication.

[7] L. Székelyhidi, *Convolution Type Functional Equations on Topological Abelian Groups*, World Scientific Publishing Co. Pte. Ltd., Singapore–New Jersey–London–Hong Kong, 1991.

HYERS THEOREM AND THE COCYCLE PROPERTY

Jacek Tabor

Institute of Mathematics, Jagiellonian University,
Reymonta 4 st., 30-059 Kraków, Poland
tabor@im.uj.edu.pl

Abstract   Using the cocycle property we generalize Hyers Theorem for the case when the target space is an arbitrary abelian group. As a corollary we obtain in particular the following result

Theorem.   *Let $\varepsilon \in [0,1)$ be arbitrary and let $(S, \cdot)$ be an amenable semigroup. Then for every $f : S \to \mathbb{C} \setminus \{0\}$ satisfying*

$$\left| \frac{f(x)f(y)}{f(xy)} - 1 \right| \le \varepsilon \qquad for \quad x, y \in S$$

there exists a unique multiplicative function $m : S \to \mathbb{C} \setminus \{0\}$ such that

$$\left| \frac{f(x)}{m(x)} - 1 \right| \le \varepsilon \qquad for \quad x \in S.$$

Keywords: local Hyers Theorem, cocycle property, stability of complex multiplication, ideally convex set

Mathematics Subject Classification (2000): 39B32, 39B52, 39B82

1.   INTRODUCTION

The field of stability of functional equations originated from the problem posed by S. Ulam of stability of homomorphisms:

Problem. *(cf. [19]) We are given a group $(G, +)$ and a metric group $(X, +, d)$. Given $\varepsilon > 0$, does there exist a $\delta > 0$ such that if $f : G \to X$*

---

[0]The research was supported by State Scientific Research Grant No. 2 P03A 021 18.

Z. Daróczy and Z. Páles (eds.),
Functional Equations - Results and Advances, 275–290.

satisfies

$$d(f(x+y), f(x) + f(y)) < \delta \qquad for \quad x, y \in G$$

then a homomorphism $a : G \to X$ *exists with*

$$d(f(x), a(x)) < \varepsilon \qquad for \quad x \in G?$$

A positive answer in the case when X is a Banach space was given by D. Hyers in [10]. Since then various authors attacked the general Ulam's problem. Probably the first results were obtained by D. Cenzer (see [1], [2]) in the case when X is the group $\mathbb{R}/\mathbb{Z}$. X is represented by the interval $[0, 1)$ with addition modulo 1, the norm and metric are defined by $\|x\| = \min\{x, 1 - x\}$, $d(x, y) = \|x - y\|$. D. Cenzer obtained the following:

Theorem C. *(cf. [1]) Let G be an abelian group and let ε be a positive real less than $\frac{1}{6}$. Then for any $f : G \to X$ satisfying*

$$\|f(x) + f(y) - f(x+y)\| \le \varepsilon \qquad for \quad x, y \in G$$

there is a unique homomorphism h mapping G into X such that $\|f(x) - h(x)\| \le \varepsilon$.

D. Cenzer shows in [2] that this result cannot be extended to the case $\varepsilon \ge \frac{1}{4}$ (the extension to $\varepsilon < \frac{1}{4}$ is contained in [7]).

Another attempt was made by Z. Moszner in [13], and further investigated by R. Ger in [6] and Józef Tabor in[17]. Z. Moszner proposed to study the Ulam's Problem in the case when $X = \mathbb{R} \setminus \{0\}$ with multiplication as a group operation. R. Ger and P. Šemrl in [7] obtained stability in the case when X is $\mathbb{C} \setminus \{0\}$.

In this paper we present a general method which enables us to generalize the previously discussed results. For the convenience of the reader we present the basic idea of the paper.

Main idea. Let G, X be abelian groups and let $V \subset X$. Let $f : G \to X$ be such that

$$Cf(x, y) := f(x) + f(y) - f(x+y) \in V \qquad for \quad x, y \in G.$$

We want to obtain an additive function $a : G \to X$ such that $f(x) - a(x) \in V$ for all $x \in G$.

We need an additional assumption that there exists a Banach space E and a injective mapping $h : V \cup (V + V) \to E$ such that $h(V)$ is a closed convex bounded subset of E and that h is locally additive (that is $h(x + y) = h(x) + h(y)$ for $x, y \in V$). This means that V has a "local structure" of a closed convex bounded set.

We define $F_h = h \circ Cf$. Then $F_h : G \times G \to h(V) \subset E$ satisfies the cocycle equation

$$F_h(x + y, z) + F_h(x, y) = F_h(x, y + z) + F_h(y, z) \qquad \text{for} \quad x, y, z \in G.$$

By the generalization of the result of L. Székelyhidi from [15] we obtain that there exists a unique function $g_h : G \to h(V)$ such that $Cg_h = F_h$. We define $g := h^{-1} \circ g_h : G \to V$. Then

$$Cg = h^{-1} \circ Cg_h = h^{-1} \circ F_h = Cf,$$

which implies that $C(f - g) = 0$. Thus $a := f - g$ is an additive function such that $f(x) - a(x) = g(x) \in V$ for all $x \in G$.

To make this simple idea rigorous in the next section we obtain some new results on the cocycle equation.

As a corollary we obtain stability of multiplication in commutative Banach algebras and of quasi-multiplicative functions (understood as in [18]).

2. COCYCLE EQUATION

Let $(S, \cdot)$ be a semigroup and let $(X, +)$ be an abelian group. A function $F : S \times S \to X$ is called a *cocycle* if for all $x, y, z \in S$

$$F(xy, z) + F(x, y) = F(x, yz) + F(y, z).$$

For a function $f : S \to X$ we define its *Cauchy difference* by the formula

$$Cf(x, y) := f(x) + f(y) - f(xy) \qquad \text{for} \quad x, y \in S.$$

It is easy to check that every Cauchy difference is a cocycle. We say that a cocycle $F : S \times S \to X$ is a *coboundary* if there exists an f with $F = Cf$.

The problem when a given cocycle is a coboundary has attracted a lot of mathematical attention. One of the first results was obtained by J. Erdős (see [4]) who showed that if S is an abelian group and X a divisible abelian group then every symmetric cocycle $F : S \times S \to X$ is a coboundary. For further references and results on this subject we refer the reader to [3].

We show that every bounded cocycle on a left (right) amenable semigroup is a coboundary of a unique bounded function. A semigroup S is called left (right) amenable if there exists a left (right) translation invariant mean on S (for more details we refer the reader to [9], [8]). We would like to mention that every abelian, finite or solvable group is amenable.

We will need the following result of L. Székelyhidi from [15] (since it is stated for right amenable semigroups, we present here the proof for left amenable semigroups).

Theorem Sz. *Let S be a left (right) amenable semigroup and let $F : S \times S \to \mathbb{R}$ be a bounded cocycle. Then there exists a bounded function $g : S \to \mathbb{R}$ such that $F = Cg$.*

Proof. Let $g(x) := m_y(F(x,y))$, where the subscript y next to m denotes that the mean m is taken with respect to the variable y.

By the cocycle property and the left invariance of the mean we have

$$
\begin{aligned}
g(u) + g(v) &= m_y(F(u,y)) + m_y(F(v,y)) \\
&= m_y(F(u,vy)) + m_y(F(v,y)) = m_y(F(u,vy) + F(v,y)) \\
&= m_y(F(uv,y) + F(u,v)) = g(uv) + F(u,v).
\end{aligned}
$$

This means that

$$
Cg(u,v) = F(u,v).
$$

$\square$

We need the following definition (see [12]).

Definition 1. Let X be a topological vector space and let V be a subset of X. We say that V is *ideally convex* if for every bounded sequence $\{x_n\}$ of elements of V and every sequence $\{\lambda_n\}$ of non-negative real numbers such that $\sum_1^\infty \lambda_n = 1$ the series $\sum_1^\infty \lambda_n x_n$ converges to an element of V.

Every sequentially complete convex set and every finite dimensional convex set is ideally convex. In the case when X is sequentially complete every open convex set is ideally convex. However, not every convex subset of an infinite dimensional Banach space is ideally convex. For more information on ideally convex sets we refer the reader to [12].

We show that we can apply Hyers method for functions with Cauchy difference contained in an ideally convex bounded set (see also [16]).

Proposition 1. *Let $(S, \cdot)$ be a semigroup, let X be a topological vector space, and let V be a bounded ideally convex subset of X.*
Then for every $F : S \times S \to X$ satisfying

$$
F(x,x) \in V \qquad \text{for} \quad x \in S,
$$

there exists a unique bounded function $f : S \to X$ such that

$$
Cf(x,x) = F(x,x) \qquad \text{for} \quad x \in S. \tag{1}
$$

Moreover, $f(S) \subset V$.

Proof. As V is ideally convex we can define

$$f(x) := \sum_{i=1}^{\infty} 2^{-i} F(x^{2^{i-1}}, x^{2^{i-1}}) \qquad \text{for} \quad x \in S.$$

Then $f(S) \subset V$, which implies that f is bounded. Moreover, for every $x \in S$ we have

$$\mathcal{C}f(x,x) = 2 \sum_{i=1}^{\infty} 2^{-i} F(x^{2^{i-1}}, x^{2^{i-1}}) - \sum_{i=1}^{\infty} 2^{-i} F(x^{2^i}, x^{2^i})$$

$$= \sum_{i=0}^{\infty} 2^{-i} F(x^{2^i}, x^{2^i}) - \sum_{i=1}^{\infty} 2^{-i} F(x^{2^i}, x^{2^i}) = F(x,x),$$

which means that f satisfies (1). Now we show the uniqueness of such f. Let f, g be bounded functions such that

$$\mathcal{C}f(x,x) = F(x,x) = \mathcal{C}g(x,x) \qquad \text{for} \quad x \in S.$$

Then

$$(f-g)(x^2) = 2(f-g)(x) \qquad \text{for} \quad x \in S,$$

which implies that for every $n \in \mathbb{N}$

$$(f-g)(x) = 2^{-n}(f-g)(x^{2^n}) \qquad \text{for} \quad x \in S.$$

As $f - g$ is bounded we obtain that $f = g$. $\qquad\square$

Now we are ready to prove the main result of this section.

Theorem 1. *Let S be a left (right) amenable semigroup, let X be a topological vector space and let V be a bounded ideally convex subset of X.*

Let $F : S \times S \to X$ be a cocycle such that

$$F(x,y) \in V \qquad \text{for} \quad x, y \in S.$$

Then there exists a unique function $f : S \to V$ such that

$$F = \mathcal{C}f. \tag{2}$$

Proof. Without loss of generality we may assume that $0 \in V$ (in the opposite case we choose an arbitrary $v_0 \in V$, define $V_0 := V - v_0$, $F_0(x,y) := F(x,y) - v_0$ and apply the results for F_0 and V_0).

By Proposition 1 we know that there exists a unique function $f : S \to V$ such that

$$\mathcal{C}f(x,x) = F(x,x) \qquad \text{for} \quad x \in S. \tag{3}$$

To complete the proof we need to show that f satisfies (2).

Let $B := V - V$ and let $X_B := \bigcup_{n \in \mathbb{N}} nB$. By $\| \cdot \|_B$ we denote the Minkowski functional of B on the space X_B, that is

$$\|x\|_B := \inf\{r \geq 0 : x \in rB\} \qquad \text{for} \quad x \in X_B.$$

Then $(X_B, \| \cdot \|_B)$ is a normed space.

Since $f(S) \subset X_B$, $F(S \times S) \subset X_B$, to prove (2) by the Hahn-Banach Theorem it is enough to verify that for every continuous linear functional $\xi : (X_B, \| \cdot \|_B) \to \mathbb{R}$

$$\xi(Cf)(x,y) = \xi(F(x,y)) \qquad \text{for} \quad x, y \in S. \tag{4}$$

Let $F_\xi = \xi \circ F$, $f_\xi = \xi \circ f$. As $F(S \times S) \subset V$ by the continuity of ξ we obtain that F_ξ is a bounded real-valued cocycle. Moreover, (3) yields that f_ξ is a bounded function such that

$$Cf_\xi(x,x) = F_\xi(x,x) \qquad \text{for} \quad x \in S. \tag{5}$$

By Theorem Sz there exists a bounded function $g : S \to \mathbb{R}$ such that

$$Cg(x,y) = F_\xi(x,y) \qquad \text{for} \quad x, y \in S, \tag{6}$$

which implies that g satisfies (5). However, by Proposition 1 we know that such a bounded function is unique and therefore $f_\xi = g$. Thus (6) yields (4), and consequently (2). $\qquad \square$

We would like to remark that in the following version of the Hyers Theorem we do not need the usual assumptions that X is locally convex or that V is closed.

Corollary 1. *Let $(S, \cdot)$ be a left (right) amenable semigroup, let $(X, +)$ be a topological vector space and let $V \subset X$ be a bounded ideally convex set. Then for every $f : S \to X$ satisfying*

$$Cf(x,y) \in V \qquad \text{for} \quad x, y \in S$$

there exists a unique additive function $a : S \to X$ such that

$$f(x) - a(x) \in V \qquad \text{for} \quad x \in S.$$

Proof. Since Cf is a cocycle by Theorem 1 there exists a unique function $\tilde{f} : S \to V$ such that

$$Cf = C\tilde{f}.$$

We put $a = f - \tilde{f}$. Then

$$Ca = Cf - C\tilde{f} = 0,$$

so a is additive. Moreover,

$$f(x) - a(x) = \tilde{f}(x) \in V \qquad \text{for} \quad x \in S.$$

The uniqueness of a satisfying the assumptions of the Corollary is equivalent to the uniqueness of $\tilde{f}$ which follows from Theorem 1. $\square$

3.   LOCAL HOMOMORPHISMS

Let $(G, +), (X, +)$ be commutative semigroups, and let $\emptyset \neq V \subset G$. We say that $h : V \cup (V + V) \to X$ is a V-*homomorphism* if

$$h(x + y) = h(x) + h(y) \qquad \text{for} \quad x, y \in V. \tag{7}$$

This definition is connected to the definition of a local homomorphism on a topological group – a local homomorphism is a U-homomorphism for a certain neighborhood U of zero.

Definition 2. Let $(G, +)$ be a commutative semigroup. We say that $V \subset G$ is a *generalized bounded ideally convex set* if there exists a topological vector space X and an injective V-homomorphism $h : V \cup (V + V) \to X$ such that $h(V)$ is a bounded ideally convex set.

By gbic(G) we denote the family of all generalized bounded ideally convex subsets of G.

From now on we assume that the mapping $h : V \cup (V + V) \to X$ denotes a V-homomorphism such that $h(V)$ is a bounded ideally convex subset of X.

Theorem 2. *Let $(S, \cdot)$ be a left (right) amenable semigroup. Let $(G, +)$ be an abelian group and let $V \in$ gbic(G). Let $F : S \times S \to G$ be a cocycle such that*

$$F(x, y) \in V \qquad \text{for} \quad x, y \in S.$$

Then there exists a unique function $f : S \to V$ such that

$$F = \mathcal{C}f. \tag{8}$$

Proof. Let $V_h = h(V), F_h = h \circ F$. As F is a cocycle by (7) for arbitrary $x, y, z \in S$ we get

$$F_h(xy, z) + F_h(x, y) = h(F(xy, z)) + h(F(x, y))$$
$$= h(F(xy, z) + F(x, y)) = h(F(x, yz) + F(y, z))$$
$$= h(F(x, yz)) + h(F(y, z)) = F_h(x, yz) + F_h(y, z).$$

This means that F_h is a cocycle. Clearly, $F_h(S \times S) \subset V_h$. By Theorem 1 we obtain that there exists a unique function $f_h : S \to V_h$ such that

$$F_h(x, y) + f_h(xy) = f_h(x) + f_h(y) \qquad \text{for} \quad x, y \in S. \tag{9}$$

Let $f := h^{-1} \circ f_h$. Then $f(S) \subset V$. By (7) and (9) we get

$$h(F(x,y) + f(xy)) = h(F(x,y)) + h(f(xy))$$
$$= F_h(x,y) + f_h(xy) = f_h(x) + f_h(y)$$
$$= h(f(x)) + h(f(y)) = h(f(x) + f(y)),$$

for all $x, y \in S$. As h is an injection this yields (8).

We prove the uniqueness part of the theorem. Let f, g be functions satisfying (8). We define $f_h = h \circ f$, $g_h = h \circ g$. By applying (8) and (7) one can easily verify that both f_h and g_h are bounded functions which satisfy (9). Making use of Theorem 1 we obtain that $f_h = g_h$. As h is injective this means that $f = g$.   □

Now we are ready to prove a generalization of the Hyers Theorem for abelian groups.

Theorem 3. *Let $(S, \cdot)$ be a left (right) amenable semigroup, let $(G, +)$ be an abelian group and let $V \in$ gbic(G).*
Let $f : S \to G$ be such that

$$\mathcal{C}f(x,y) \in V \qquad for \quad x, y \in S.$$

Then there exists a unique additive function $a : S \to G$ such that

$$f(x) - a(x) \in V \qquad for \quad x \in S.$$

Proof. Since $\mathcal{C}f$ is a cocycle by Theorem 2 we obtain that there exists a unique function $\tilde{f} : S \to V$ such that

$$\mathcal{C}f = \mathcal{C}\tilde{f}.$$

We put $a = f - \tilde{f}$. Then

$$\mathcal{C}a = \mathcal{C}f - \mathcal{C}\tilde{f} = 0,$$

which means that a is additive. Moreover,

$$f(x) - a(x) = \tilde{f}(x) \in V \qquad for \quad x \in S.$$

The uniqueness of an a satisfying the assumptions of the Theorem is equivalent to the uniqueness of $\tilde{f}$ which follows from Theorem 2.   □

The following corollary is a direct consequence of Theorem 3.

Corollary 2. *Let $(S, \cdot)$ be a left (right) amenable semigroup, and V be a bounded convex subset of $\mathbb{R}^n$. Let $f : S \to \mathbb{R}^n$ be a function such that*

$$\mathcal{C}f(x,y) \in V \qquad for \quad x, y \in S.$$

Then there exists a unique additive function $a : S \to \mathbb{R}^n$ such that

$$f(x) - a(x) \in V \qquad for \quad x \in S.$$

4. COMPLEX PLANE

For a complex number $z \in \mathbb{C} \backslash \mathbb{R}_-$ by $\arg(z)$ we mean a unique number $\phi \in (-\pi, \pi)$ such that $\cos(\phi) + i \sin(\phi) = \frac{z}{|z|}$. The main branch of the complex logarithm is defined by the formula

$$\ln(z) := \ln(|z|) + i \arg(z) \qquad \text{for} \quad z \in \mathbb{C} \backslash \mathbb{R}_-.$$

It is well known that the following hold

$$\exp(\ln(z)) = z \qquad \text{for} \quad z \in \mathbb{C} \backslash \mathbb{R}_-,$$

$$\ln(\exp(z)) = z \qquad \text{for} \quad z \in \mathbb{C} : \operatorname{Im}(z) \in (-\pi, \pi).$$

Moreover,

$$\ln(z_1 z_2) = \ln(z_1) + \ln(z_2) \qquad \text{for} \quad z_1, z_2 \in \mathbb{C}, \operatorname{Re}(z_1), \operatorname{Re}(z_2) > 0. \quad (10)$$

Let us now consider once more Theorem C of D. Cenzer quoted in the introduction. By identifying the group $\mathbb{R}/\mathbb{Z}$ with the complex unit circle with multiplication as a group operation we may reformulate it in the following way:

Theorem (Equivalent form of Theorem C). *Let $(S, \cdot)$ be an abelian group, and let $G := \{z \in \mathbb{C} \mid |z| = 1\}$. For $\varepsilon \geq 0$ we put*

$$V(\varepsilon) := \{z \in G : |arg(z)| \leq 2\pi\varepsilon\}.$$

Then for every $\varepsilon \in [0, \frac{1}{6})$ and every function $f : S \to G$ such that

$$\frac{f(x)f(y)}{f(xy)} \in V(\varepsilon) \qquad \text{for} \quad x, y \in S$$

there exists a unique multiplicative function $m : S \to G$ such that

$$\frac{f(x)}{m(x)} \in V(\varepsilon) \qquad \text{for} \quad x \in S.$$

There arises a problem if the assumption that G is a complex circle is essential. It seems natural that it should be enough to assume G to be $\mathbb{C} \backslash \{0\}$ with multiplication as a group operation. We show that this is the case. We will need the following simple lemma.

Lemma 1. *For $\varepsilon \geq 0$ we define*

$$V(\varepsilon) := \{z \in \mathbb{C} \backslash \{0\} : |z| = 1 \text{ and } |arg(z)| \leq 2\pi\varepsilon\}.$$

Then $V(\varepsilon) \in \operatorname{gbic}(\mathbb{C} \backslash \{0\})$ if and only if $\varepsilon \in [0, \frac{1}{4})$.

Proof. We consider first the case $\varepsilon \in [0, \frac{1}{4})$. The reader can easily check that the main complex argument $h : V(\varepsilon) \cup (V(\varepsilon) \cdot V(\varepsilon)) \to \mathbb{R}$ is an injective $V(\varepsilon)$-homomorphism such that $h(V(\varepsilon)) = [-\varepsilon, \varepsilon]$, which is clearly an ideally convex bounded subset of $\mathbb{R}$

Let us now assume that $\varepsilon \geq \frac{1}{4}$. Suppose that there exists a topological vector space X and an injective $V(\varepsilon)$-homomorphism $h : V(\varepsilon) \cup (V(\varepsilon) \cdot V(\varepsilon)) \to X$. Let $x_1 = i, x_2 = -i$. Then $x_1, x_2 \in V(\varepsilon)$, $x_1^2 = x_2^2 \in V(\varepsilon) \cdot V(\varepsilon)$. This implies that $2h(x_1) = h(x_1^2) = h(x_2^2) = 2h(x_2)$. Since h is injective we obtain that $x_1 = x_2$ – a contradiction. $\qquad\square$

As an immediate corollary of Lemma 1 and Theorem 3 we obtain the following result.

Corollary 3. *Let $(S, \cdot)$ be a left (right) amenable semigroup. For an $\varepsilon > 0$ we put*

$$V(\varepsilon) := \{z \in \mathbb{C} \setminus \{0\} : |z| = 1 \text{ and } |arg(z)| \leq 2\pi\varepsilon\}.$$

Then for every $\varepsilon \in [0, \frac{1}{4})$ and every function $f : S \to \mathbb{C} \setminus \{0\}$ such that

$$\frac{f(x)f(y)}{f(xy)} \in V(\varepsilon) \qquad for \quad x, y \in S$$

there exists a unique multiplicative function $m : S \to \mathbb{C} \setminus \{0\}$ such that

$$\frac{f(x)}{m(x)} \in V(\varepsilon) \qquad for \quad x \in S.$$

From the mentioned example of D. Cenzer from [2] it follows that ε cannot be taken greater then $\frac{1}{4}$. It is also suggested by the fact that $V(\varepsilon)$ is not a generalized ideally convex set for $\varepsilon \geq \frac{1}{4}$.

Now we are going to improve Theorem 3.4 from [7]. Let us first notice that the result is not completely correct – in fact there is no uniqueness since the value of approximation constant $\sqrt{1 + \frac{1}{1-\varepsilon^2} - 2\sqrt{\frac{1+\varepsilon}{1-\varepsilon}}}$ for $\varepsilon > -\frac{2}{39} \sqrt[3]{233 + 39\sqrt{33}} - \frac{32}{39} \frac{1}{\sqrt[3]{233+39\sqrt{33}}} + \frac{44}{39} \approx 0.626$ is greater than 2, and therefore we can approximate the constant one function by an arbitrary character (simply the value of approximation constant as ε goes to 1 goes to ∞, and therefore there can be no uniqueness). Moreover, the value of approximation is a complicated function of ε, while having in mind Hyers Theorem one would rather expect it to be equal to ε. We show that this is the case.

We will need the following lemma.

Lemma 2. *For $\varepsilon \geq 0$ we define*

$$W(\varepsilon) := \{z \in \mathbb{C} \setminus \{0\} : |z - 1| \leq \varepsilon\}.$$

Then $W(\varepsilon) \in \text{gbic}(G)$ if and only if $\varepsilon \in [0, 1)$.

Proof. We first show that $W(\varepsilon) \in \text{gbic}(G)$ for $\varepsilon < 1$. As a target space we take $\mathbb{C}$, and define the mapping $h : W(\varepsilon) \cup (W(\varepsilon) \cdot W(\varepsilon)) \to \mathbb{C}$ by $h(z) = \ln(z)$. Let us notice that h is well-defined since $W(\varepsilon) \cup (W(\varepsilon) \cdot W(\varepsilon)) \subset \mathbb{C} \setminus \mathbb{R}_-$. Applying (10) we obtain directly that h is a $W(\varepsilon)$-additive function. Moreover, it is clearly an injection.

To prove that $W(\varepsilon)$ is a generalized ideally convex set it is enough to show that $h(W(\varepsilon))$ is a closed convex bounded subset of $\mathbb{C}$. Clearly

$$\partial(W(\varepsilon)) = \{1 + \varepsilon e^{it} : t \in \mathbb{R}\},$$

where ∂ denotes the boundary operation. Since $\ln$ is a homeomorphism, we obtain that

$$\partial(h(W(\varepsilon))) = h(\partial(W(\varepsilon))) = \{\ln(1 + \varepsilon e^{it}) : t \in \mathbb{R}\}$$
$$= \{\tfrac{1}{2}\ln((1 + \varepsilon \cos(t))^2 + (\varepsilon \sin(t))^2) + i \arctan(\tfrac{\varepsilon \sin(t)}{1 + \varepsilon \cos(t)}) : t \in \mathbb{R}\}.$$

One can check that this is a closed curve without self-intersections. To show that $h(W(\varepsilon))$ is convex (it is clearly closed and bounded), it is enough to check that the curvature of $\partial(h(W(\varepsilon)))$ has a constant sign. The curvature of a parametric curve is given by the formula

$$K = \frac{x'_t y''_t - x''_t y'_t}{(x'^2_t + y'^2_t)^{\frac{3}{2}}}.$$

Thus to check the sign of K is is enough to calculate that

$$x'_t y''_t - x''_t y'_t = \frac{\varepsilon^2(1 + \varepsilon \cos(t))}{(1 + \varepsilon^2 + 2\varepsilon \cos(t))^2},$$

which yields that K has a constant sign.

Now we show that $W(\varepsilon)$ is not a generalized ideally convex set for $\varepsilon \geq 1$. Suppose that there exists a topological vector space X and an injective $W(\varepsilon)$-homomorphism $h : W(\varepsilon) \cup (W(\varepsilon) \cdot W(\varepsilon)) \to X$ such that $h(W)$ is an ideally convex bounded set. Then $h(1) = 0$, which by the injectivity of h implies that $h(\tfrac{1}{2}) \neq 0$. Since $\tfrac{1}{2^n} \in W(\varepsilon)$ for every $n \in \mathbb{N}$ and h is a $W(\varepsilon)$-homomorphism we obtain that $nh(\tfrac{1}{2}) = h(\tfrac{1}{2^n}) \in W(\varepsilon)$, which contradicts the assumption that $h(W(\varepsilon))$ is bounded. ⊔⊓

As a direct corollary from Lemma 2 and Theorem 3 we obtain the following important result.

Theorem 4. Let $(S, \cdot)$ be a left (right) amenable semigroup and let $\varepsilon \in [0, 1)$. Let $f : S \to \mathbb{C} \setminus \{0\}$ be such that

$$\left| \frac{f(x)f(y)}{f(xy)} - 1 \right| \le \varepsilon \qquad \text{for} \quad x, y \in S.$$

Then there exists a unique multiplicative function $m : S \to \mathbb{C} \setminus \{0\}$ such that

$$\left| \frac{f(x)}{m(x)} - 1 \right| \le \varepsilon \qquad \text{for} \quad x \in S.$$

Example 1. We show that the assumption that $\varepsilon < 1$ is essential. Let $f : \mathbb{R}_+ \to \mathbb{C} \setminus \{0\}$ be defined by the formula

$$f(x) = e^{x^2}.$$

Then trivially

$$\left| \frac{f(x)f(y)}{f(x+y)} - 1 \right| \le 1 \qquad \text{for} \quad x, y \in \mathbb{R}_+.$$

However, for every multiplicative function $m : \mathbb{R}_+ \to \mathbb{C} \setminus \{0\}$

$$\sup_{x \in \mathbb{R}_+} \left| \frac{f(x)}{m(x)} - 1 \right| = \infty,$$

because $\lim_{k \to \infty} \frac{f(k)}{m(k)} = \frac{e^{k^2}}{m(1)^k} = \infty$.

This implies that there is no stability in this case.

The notion of generalized ideally convex set helps to explain why for $\varepsilon = 1$ there is no stability. From Lemma 2 we obtain that $W(1)$ is not a bounded generalized convex set set, and as we know Hyers Theorem does not hold for unbounded sets.

The following result is a multiplicative analogue of some of results of Józef Tabor and Zs. Páles (see [18], [14]).

Corollary 4. Let $\varepsilon \in [0, 1)$ be arbitrary and let $(G, \cdot)$ be an amenable group.

Then for every $f : G \to \mathbb{C}$ satisfying

$$|f(xy) - f(x)f(y)| \le \varepsilon \min\{|f(x)f(y)|, |f(xy)|\} \qquad \text{for} \quad x, y \in G \quad (11)$$

there exists a unique multiplicative function $m : G \to \mathbb{C}$ such that

$$|f(x) - m(x)| \le \varepsilon \min\{|f(x)|, |m(x)|\} \qquad \text{for} \quad x \in G. \qquad (12)$$

Proof. Suppose that there exists an $x \in G$ such that $f(x) = 0$. Then by (11) and the fact that G is group we trivially obtain that $f \equiv 0$. The multiplicative function $m = 0$ is clearly the unique function which satisfies the assertion.

We have yet to consider the case $f(x) \neq 0$ for every $x \in G$. By Lemma 2 we obtain that the set

$$W(\varepsilon) = \{z \in \mathbb{C} \setminus \{0\} \,|\, |z - 1| \leq \varepsilon\},$$

is a generalized ideally convex bounded set with complex logarithm as an injective $W(\varepsilon)$-homomorphism. This means that $W(\varepsilon)^{-1}$ is a generalized ideally convex set with $-\ln$ as a $W(\varepsilon)^{-1}$-homomorphism. This yields that

$$W(\varepsilon) \cap W(\varepsilon)^{-1} = \{z \in \mathbb{C} \setminus \{0\} \,|\, |z - 1| \leq \varepsilon, |z^{-1} - 1| \leq \varepsilon\},$$

is a generalized bounded ideally convex set. Let $f : G \to \mathbb{C} \setminus \{0\}$ be an arbitrary function satisfying

$$\frac{f(x)f(y)}{f(xy)} \in W(\varepsilon) \cap W(\varepsilon)^{-1} \qquad \text{for} \quad x, y \in G.$$

Let us observe that the above condition is valid iff (11) holds. By Theorem 3 there exists a unique multiplicative function $m : G \to \mathbb{C} \setminus \{0\}$ such that

$$\frac{f(x)}{m(x)} \in W(\varepsilon) \cap W(\varepsilon)^{-1} \qquad \text{for} \quad x \in G,$$

which means that

$$|f(x) - m(x)| \leq \min\{|f(x)|, |m(x)|\} \qquad \text{for} \quad x \in G.$$

Now we show the uniqueness part. Suppose that there exists another multiplicative function $m1 : G \to \mathbb{C}$ which satisfies (12). Then, since $m : G \to \mathbb{C} \setminus \{0\}$ satisfying (12) is unique, we obtain that there exists an $x \in G$ such that $m1(x) = 0$. As $m1$ is multiplicative this yields consequently that $m1 \equiv 0$. However, by (12) this implies that $f \equiv 0$, a contradiction. $\qquad \square$

Remark 1. Since we use the notion of ideally convex sets all the statements of results in this section remain valid if instead of weak inequalities we take strong ones.

5.   STABILITY OF MULTIPLICATION IN COMMUTATIVE BANACH ALGEBRAS

Let $(A, \|\cdot\|)$ be a commutative Banach algebra with unit e. For $a \in A$ we define $\exp(a) := \sum_{n=0}^{\infty} \frac{1}{n!}a^n$. We will need the following inequality

$$\| \exp(x) - e\| = \| \sum_{n=1}^{\infty} \frac{1}{n!}x^n\| \leq \sum_{n=1}^{\infty} \frac{1}{n!}\|x\|^n = \exp(\|x\|) - 1. \qquad (13)$$

By $\sigma(a)$ we denote the spectrum of an element $a \in A$. The function $\ln$ is defined for all $a \in A$ such that $\sigma(a) \subset \mathbb{C} \setminus \mathbb{R}_-$ by the formula

$$\ln(a) = \frac{1}{2\pi i} \int_\gamma \frac{\ln(z)}{z - a}dz,$$

where $\gamma : [0, 1] \to \mathbb{C} \setminus \mathbb{R}_-$ is a continuously differentiable closed curve such that $\sigma(a)$ is contained inside γ.

The following three equalities are the analogues of similar ones from the previous section

$$\exp(\ln(a)) = a \quad \text{for} \quad a \in A, \sigma(a) \subset \mathbb{C} \setminus \mathbb{R}_-,$$
$$\ln(\exp(z)) = z \quad \text{for} \quad a \in A, \text{Im}(\sigma(a)) \subset (-\pi, \pi),$$
$$\ln(a_1 a_2) = \ln(a_1) + \ln(a_2) \text{ for } a_1, a_2 \in A, \text{Re}(a_1), \text{Re}(a_2) \subset (0, \infty). \qquad (14)$$

By A_* we denote the set of all invertible elements of A.

Theorem 5. *Let $(S, \cdot)$ be a left (right) amenable semigroup, let A be a commutative Banach algebra, and let $\varepsilon \in [0, \frac{1}{2})$. Let $f : S \to A_*$ be such that*

$$\left\| \frac{f(x)f(y)}{f(xy)} - e \right\| \leq \varepsilon \quad \text{for} \quad x, y \in S.$$

Then there exists a multiplicative function $m : S \to A_*$ such that

$$\left\| \frac{f(x)}{m(x)} - e \right\| \leq \frac{\varepsilon}{1 - \varepsilon} \quad \text{for} \quad x \in S. \qquad (15)$$

Moreover, if $\varepsilon < \frac{1}{3}$ then such a multiplicative function is unique.

Proof. Let

$$U(\varepsilon) := \exp(\text{cl conv} \ln(B(e, \varepsilon))),$$

where by $B(a, \varepsilon)$ we denote the closed ball centered at a and with radius ε. We show that $U(\varepsilon) \in \text{gbic}(A_*)$ for $\varepsilon \in [0, \frac{1}{2})$. Let us first check that

$$U(\varepsilon) \subset B(e, \tfrac{\varepsilon}{1-\varepsilon}). \qquad (16)$$

Let $n \in \mathbb{N}$, $a_1, \ldots, a_n \in B(0, \varepsilon)$ and $\alpha_1, \ldots, \alpha_n \geq 0$, $\sum_{i=1}^{n} \alpha_i = 1$ be arbitrarily chosen. Applying (13) we get

$$\left\| \exp\left(\sum_{i=1}^{n} \alpha_i \ln(e + a_i) \right) - e \right\| \leq \exp\left(\sum_{i=1}^{n} |\alpha_i| \cdot \| \ln(e + a_i) \| \right) - 1$$

$$\leq \exp\left(-\sum_{i=1}^{n} |\alpha_i| \ln(1 - \|a_i\|) \right) - 1 \leq \exp\left(-\sum_{i=1}^{n} |\alpha_i| \ln(1 - \varepsilon) \right) - 1$$

$$= \frac{1}{1 - \varepsilon} - 1 = \frac{\varepsilon}{1 - \varepsilon}.$$

This yields (16).

Since $\varepsilon < \frac{1}{2}$, $\frac{\varepsilon}{1-\varepsilon} < 1$ and therefore applying (16) we get that $U(\varepsilon) \subset \{a \in A_* \mid \mathrm{Re}(\sigma(a)) \subset (0, \infty)\}$. Making use (14) we obtain that $\ln : U(\varepsilon) \cup (U(\varepsilon) \cdot U(\varepsilon)) \to A$ is an injective $U(\varepsilon)$-homomorphism such that $\ln(U(\varepsilon)) = \mathrm{cl}\,\mathrm{conv}\,\ln(B(e, \varepsilon))$, which means that $U(\varepsilon) \in \mathrm{gbic}(A_*)$.

Applying Theorem 3 we obtain a unique multiplicative function $m : S \to A_*$ such that

$$\frac{f(x)}{m(x)} \in U(\varepsilon) \qquad \text{for} \quad x \in S.$$

This implies that m satisfies (15).

We show that for $\varepsilon < \frac{1}{3}$ such m is unique. As $\tilde{\varepsilon} := \frac{\varepsilon}{1-\varepsilon} < \frac{1}{2}$ we can apply the results from the first part of the proof for $\tilde{\varepsilon}$ and obtain that there exists a unique multiplicative function $m : S \to A_*$ such that

$$\frac{f(x)}{m(x)} \in U(\tilde{\varepsilon}) = U\left(\frac{\varepsilon}{1-\varepsilon} \right) \qquad \text{for} \quad x \in S.$$

Since $B(e, \frac{\varepsilon}{1-\varepsilon}) \subset U(\frac{\varepsilon}{1-\varepsilon})$, this yields in particular the uniqueness of a multiplicative function $m : S \to A_*$ satisfying

$$\frac{f(x)}{m(x)} \in B(e, \frac{\varepsilon}{1-\varepsilon}) \qquad \text{for} \quad x \in S.$$

$\square$

References

[1] D. Cenzer, *The stability problem for transformations of the circle*, Proc. Royal Soc. Edinburgh **84A** (1979), 279–281.

[2] D. Cenzer, *The stability problem: New results and counterexamples*, Letters in Mathematical Physics **10** (1985), 155–160.

[3] T. M. Davison, B. R. Ebanks, *Cocycles on cancellative semigroups*, Publ. Math. Debrecen **26** (1995), 137–147.

[4] J. Erdős, *A remark on the paper "On some functional equations" by S. Kurepa*, Glasnik Mat.-Fiz Astronom **14** (1959), 213–216.

[5] G. L. Forti, *Hyers-Ulam stability of functional equations in several variables*, Aequationes Math. **50** (1995), 143-190.

[6] R. Ger, *Superstability is not natural*, Annales de l'Ecole Normale Supérieure à Carcovie 159, Travaux Mathematiques **XIII** (1993), 109–125.

[7] R. Ger, P. Šemrl, *The stability of the exponential equation*, Proc. Amer. Math. Soc. **124** (1996), 779–787.

[8] P. Greenleaf, *Invariant Means on Topological Groups*, Van Nostrand, Reinhold Company 1969.

[9] E. Hewitt, K. Ross, *Abstract Harmonic Analysis*, vol. 1, Academic Press, New York 1963.

[10] D. H. Hyers, *On the stability of the linear functional equation*, Proc. Natl. Acad. Sci. USA **16** (1941), 222–224.

[11] M. Kuczma, *An Introduction to the Theory of Functional Equations and Inequalities*, Państwowe Wydawnictwo Naukowe & Uniwersytet Śląski, Warszawa–Kraków–Katowice, 1985.

[12] E. A. Lifshitz, *Ideally convex sets*, Funktsional. Anal. i Prilozhen. **4** (1970), 76–77 (Russian).

[13] Z. Moszner, *Sur la stabilité de l'equation d'homomorphisme*, Aequationes Math. **29** (1985), 290–306.

[14] Zs. Páles, *Generalized stability of the Cauchy functional equation*, Aequationes Math. **56** (1998), 222–232.

[15] L. Székelyhidi, *Stability properties of functional equations in several variables*, Publ. Math. Debrecen **47** (1995), 95–100.

[16] Jacek Tabor, *Ideally convex sets and Hyers theorem*, Funkcialaj Ekvacioj **43** (2000), 121–125.

[17] Józef Tabor, *Approximate endomorphisms of the complex plane*, J. Natural Geom. **1** (1992), 71–86.

[18] Józef Tabor, *Quasi-additive functions*, Aequationes Math. **39** (1990), 179–197.

[19] S. Ulam, *A Collection of Mathematical Problems*, Interscience Publ., New York, 1960.

VI

FUNCTIONAL EQUATIONS IN FUNCTIONAL ANALYSIS

MAPPINGS WHOSE DERIVATIVES ARE ISOMETRIES

John A. Baker

Department of Pure Mathematics, University of Waterloo,
Waterloo, Ontario N2L 3G1, Canada

jabaker@math.uwaterloo.ca

Abstract   If X is a strictly convex real Banach space, U is a region in X and f is a continuously differentiable mapping of U into X such that its derivative, at each point of U, is an isometry then f is the restriction to U of an affine isometry of X onto itself.

Keywords: isometry, differentiable, functional equation

Mathematics Subject Classification (2000): 39B52, 51F99, 46B04

It is well known that if U is a region (non-empty open connected set) in $\mathbb{C}$ and f is an analytic function from U into $\mathbb{C}$ such that $|f'(z)| = 1$ for all z in U then there exist $c \in \mathbb{C}$ and $\theta \in \mathbb{R}$ such that $f(z) = c + e^{i\theta}z$ for all $z \in U$. Our main aim is to prove the following generalization of this assertion. If X is a strictly convex real Banach space, f is a C^1 (continuously differentiable) mapping of a region U in X into X such that $Df(x)$, the Fréchet derivative of f at x, is an isometry of X onto X for each x in U, then f is an isometry and Df is constant.

1.   ISOMETRIES

Throughout this paper X and Y denote real normed linear spaces. If $\phi \neq S \subseteq X$ and $f : S \to Y$ then we say that f is an *isometry* (or f is *isometric*) provided $\|f(x) - f(y)\| = \|x - y\|$ for all $x, y \in S$. Mazur and Ulam [6] showed that if f is an isometry of X *onto* Y then f is affine, i.e. there exist c in Y and a *linear* isometry, L, of X onto Y such that $f(x) = c + Lx$ for all $x \in X$.

293

Z. Daróczy and Z. Páles (eds.),
Functional Equations - Results and Advances, 293–296.
© 2002 *Kluwer Academic Publishers.*

We say that Y is *strictly convex* provided

$$a, b \in Y, \quad \|a\| = \|b\| = 1 \quad \text{and} \quad a \neq b \quad \text{imply} \quad \|a + b\| < 2.$$

The main result of [1] asserts that if f is an isometry of X into Y (not necessarily onto) and if Y is strictly convex then f is affine. The proof was based on the elementary

Lemma. *If Y is a strictly convex normed linear space, $x, y, z \in Y$ and*

$$\|x - z\| = \|y - z\| = \frac{\|x - y\|}{2}$$

then $z = (x + y)/2$.

For $a \in X$ and $r > 0$ let

$$B(a, r) = \{x \in X : \|x - a\| < r\}\text{—the } \textit{ball} \text{ with center } a \text{ and radius } r.$$

2.   FURTHER MOTIVATION

If U is a region in X, f is a differentiable mapping of U into Y and $a \in U$ then the (Fréchet) derivative of f at a will be denoted by $Df(a)$; in case $X = Y = \mathbb{R}^n$ (with the Euclidean norm) the matrix of $Df(a)$ (with respect to the usual basis of $\mathbb{R}^n$) will be denoted by $\nabla f(a)$. Recall that a linear mapping of $\mathbb{R}^n$ to itself is isometric if and only if its matrix is orthogonal.

Suppose that U is a region in $\mathbb{C}$ and $f : U \to \mathbb{C}$. If we ignore multiplication and identify $\mathbb{C}$ with $\mathbb{R}^2$ then, according to the Cauchy-Riemann equations, f is analytic on U if and only if it is Fréchet differentiable on U and, for each (x, y) in U, there exist $r \geq 0$ and $\theta \in \mathbb{R}$ such that

$$\nabla f(x, y) = r \begin{pmatrix} \cos \theta & -\sin \theta \\ \sin \theta & \cos \theta \end{pmatrix}$$

—in complex notation, $f'(x + iy) = re^{i\theta}$. Thus the introductory remark may be restated as follows: if f is a differentiable mapping of a region U in $\mathbb{R}^2$ into $\mathbb{R}^2$ such that $\nabla f(x, y)$ is a proper orthogonal matrix for each (x, y) in U then ∇f is constant and f is isometric.

3.   THE MAIN RESULT

Our main result is the theorem below which generalizes the last assertion. We will need the following well-known proposition, a proof of which may be found on page 218 of [5].

Proposition. *If U is a region in X, f is a C^1 map of U into Y and, for some $\lambda \geq 0$, $\|Df(x)\| \leq \lambda$ for all $x \in U$ then $\|f(x) - f(y)\| \leq \lambda \|x - y\|$*

whenever the line segment $[x; y] := \{(1-t)x + ty : 0 \le t \le 1\}$ *is contained in U.*

Note that if T is a linear isometry of X onto Y then $\|T\| = \|T^{-1}\| = 1$.

Theorem. *Suppose that U is a region in a strictly convex real Banach space X and f is a C^1 mapping of U into X such that, for each $x \in U$, $Df(x)$ is an isometry of X onto itself. Then f is the restriction to U of an isometry of X onto X and Df is constant.*

Proof. Since $\|Df(x)\| = 1$ for all $x \in U$, by the Proposition,

$$\|f(x) - f(y)\| \le \|x - y\| \qquad \text{whenever} \quad [x; y] \subseteq U. \qquad (1)$$

Let $a \in U$. According to the inverse function theorem (see [5], page 221) there exist open subsets Ω and V of X such that $a \in \Omega \subset U$, $f(\Omega) = V$, f is one-to-one on Ω and, if g denotes the restriction of f to Ω, g^{-1} is C^1 on V and

$$(Dg^{-1})(g(u)) = [Dg(u)]^{-1} = [Df(u)]^{-1} \qquad \text{for} \quad u \in \Omega.$$

Hence $\|Dg^{-1}(v)\| = 1$ for all $v \in V$ and thus $\|g^{-1}(v) - g^{-1}(w)\| \le \|v - w\|$ whenever $[v; w] \subset V$. That is

$$\|x - y\| \le \|f(x) - f(y)\| \qquad \text{whenever} \quad [f(x); f(y)] \subseteq V. \qquad (2)$$

Now choose $\rho > 0$ such that $B(f(a), \rho) \subseteq V$ and choose $r > 0$ such that $B(a, r) \subseteq \Omega$ and $f(B(a, r)) \subseteq B(f(a), \rho)$. It follows from (1) and (2) that $\|f(x) - f(y)\| = \|x - y\|$ for $x, y \in B(a, r)$.

Thus if $x, y \in B(a, r)$ then

$$\left\| f\left(\frac{x+y}{2}\right) - f(x) \right\| = \left\| \frac{x+y}{2} - x \right\| = \frac{\|x-y\|}{2} = \frac{\|f(x) - f(y)\|}{2}$$

and, similarly,

$$\left\| f\left(\frac{x+y}{2}\right) - f(y) \right\| = \frac{\|f(x) - f(y)\|}{2}.$$

Hence, by the Lemma,

$$f\left(\frac{x+y}{2}\right) = \frac{f(x) + f(y)}{2} \qquad \text{for} \quad x, y \in B(a, r).$$

According to Theorem 1 of [7], there is a unique $c_a \in X$ and a unique linear map L_a of X into itself such that

$$f(x) = c_a + L_a x \qquad \text{for all} \quad x \in B(a, r).$$

Now this is so for every $a \in U$. Since U is open and connected, it is not difficult to see that, for every $a, b \in U$, $c_a = c_b$ and $L_a = L_b$. Thus there is a unique $c \in X$ and a unique linear $L : X \to X$ such that

$$f(x) = c + Lx \qquad \text{for all} \quad x \in U.$$

It follows that $Df(x) = L$ for all $x \in U$ so that L is isometric. Thus f has a unique extension to an isometry of X onto itself and Df is constant. $\qquad\qquad\qquad\qquad\qquad\qquad\qquad\qquad\qquad\qquad\square$

Since linear isometries of $\mathbb{R}^n$ are those determined by orthogonal matrices, we have the

Corollary. *Suppose U is a region in $\mathbb{R}^n$, f is a C^1 map of U into $\mathbb{R}^n$ and, for each $x \in U$, $\nabla f(x)$ is an orthogonal matrix. Then there exists a $c \in \mathbb{R}^n$ and a real orthogonal $n \times n$ matrix Q such that $f(x) = c + Qx$ for all $x \in U$.*

For $n = 2$ the Corollary follows easily from the above observations concerning analytic functions; for $n \geq 3$, if f is assumed to be of class C^3, it is a special case of a beautiful theorem of Liouville concerning conformal mappings—see e.g. [2] and §9.5.4 of [3]. This theorem has been generalized, in the realm of finite dimensional Jordan algebras and C^4 maps, by W. Bertram; see Theorem 2.1.1, page 13, of [4].

References

[1] J. A. Baker, *Isometries in normed spaces*, Amer. Math. Monthly **78** (1971), 655–658.

[2] W. Benz, *Der Satz von Liouville über räumliche orthogonaltreue (winkeltreue) Abbildungen für beliebige Signatur*, Aequationes Math. **21** (1980), 257–282.

[3] M. Berger, *Geometry*, Springer-Verlag, Berlin, 1987.

[4] W. Bertram, *Jordan algebras and conformal geometry*, Positivity in Lie Theory; Open Problems, Walter de Gruyter, Berlin, New York, 1998.

[5] S. Lang, *Analysis I*, Addison-Wesley, Reading, Mass., 1968.

[6] S. Mazur and S. Ulam, *Sur les transformations isométriques d'espaces vectoriels normes*, C. R. Acad. Sci. Paris **194** (1932), 946–948.

[7] F. Radó and J. A. Baker, *Pexider's equation and aggregation of allocations*, Aequationes Math. **32** (1987), 227–239.

LOCALIZABLE FUNCTIONALS

Bruce Ebanks

Department of Mathematics and Statistics, Mississippi State University,
Mississippi State, MS 39762, U.S.A

ebanks@math.msstate.edu

Abstract   Let (X, Σ, μ) be a σ-finite measure space, and let F be a space of nonnegative Σ-measurable functions. Conditions under which a nonnegative-valued functional M on F has a representation of the form

$$M(f) = \int_X (\phi \circ f) d\mu \qquad (f \in F),$$

are presented. That is, functionals of this form are characterized. Also, some applications to the measurement of fuzziness are presented.

Keywords: functional, measure, fuzzy set, measurable function

Mathematics Subject Classification (2000): 39B52, 28C05, 28A20, 28A25, 04A72

1.   INTRODUCTION

The motivating question of this article is the following. How can one characterize those functionals M of the form

$$M(f) = \int_X \phi \circ f \, d\mu \qquad (1)$$

on some given space of measurable functions in a meaningful way? More precisely, let (X, Σ, μ) be a measure space, and let $\mathcal{F}$ be a space of non-negative Σ-measurable functions $f : X \to \mathbb{R}_+ = [0, +\infty)$ (or $[0, +\infty]$). We seek conditions on a functional $M : \mathcal{F} \to \mathbb{R}_+$ which guarantee the existence of some $\phi : \mathbb{R}_+ \to \mathbb{R}_+$ such that (1) holds.

This question grew out of a study of measures of fuzziness on a space $\mathcal{F}$ of fuzzy sets. A fuzzy set can be considered to be a function $f : X \to [0, 1]$, where for each $x \in X$ the value $f(x)$ is regarded as the degree

Z. Daróczy and Z. Páles (eds.),
Functional Equations - Results and Advances, 297–304.
© 2002 Kluwer Academic Publishers.

to which x "belongs to" the fuzzy set. (Thus $f(x) = 0$ (resp. 1) if x is definitely outside (resp. inside) the fuzzy set, while a value of $f(x)$ strictly between 0 and 1 indicates a "degree of partial inclusion" of x in the fuzzy set.)

1.1 ASSUMPTIONS ABOUT THE FUNCTIONAL M

Given this context, we shall make the following assumptions about our functional M throughout this article. First, if $f(x) = 0$ except on a set of measure zero, then we may regard f as being essentially an ordinary set (i.e. a classical characteristic function). That is,

$$M(f) = 0 \quad \text{whenever} \quad f = 0 \quad \mu - a.e. \quad (f \in \mathcal{F}). \tag{2}$$

Second, if f_n is a sequence of fuzzy sets converging pointwise in a monotone nondecreasing manner to a fuzzy set f, then the fuzziness of f should be the limit of the fuzziness of f_n as $n \to \infty$. That is,

$$M(f_n) \to M(f) \text{ whenever } f_n \nearrow f \text{ pointwise}$$
$$(f, f_n \in \mathcal{F}; n = 1, 2, ...). \tag{3}$$

Third, we assume that the difference in the measures of fuzziness of two fuzzy sets is determined locally on the set of points where the two fuzzy sets differ. That is, the functional M is **localizable**:

There exists a map $\Delta : \mathcal{F} \to \mathbb{R}_+$ for which
$$M(f) - M(f\mathcal{X}_{X \setminus D}) = \Delta(f\mathcal{X}_D) \quad (f \in \mathcal{F}, D \in \Sigma). \tag{4}$$

Here $\mathcal{X}_S$ is the characteristic function of set S. (That is, $\mathcal{X}_S(x) = 1$ if $x \in S$, $\mathcal{X}_S(x) = 0$ if $x \in X \setminus S$.)

Fourth, we assume a sort of μ-invariance of the functional M; specifically,

$$M(c\mathcal{X}_A) = M(c\mathcal{X}_B) \text{ whenever } \mu(A) = \mu(B), \ c \geq 0 \quad (A, B \in \Sigma). \tag{5}$$

This establishes a link between the measure of fuzziness of a fuzzy set f and the measure of the set of points where f is "fuzzy".

It should be noted here that results will be proved based on only the properties (2)-(5) as formulated, without reference to the fuzzy set context. Hence, these results have a broader applicability. In particular, some results will be established for spaces $\mathcal{F}$ of functions which are not necessarily uniformly bounded.

1.2 ASSUMPTIONS ABOUT THE MEASURE SPACE (X, Σ, μ)

Before proceeding further, we must introduce an assumption about our measure space. We shall say that (X, Σ, μ) is *additively suitable* if it satisfies the following two hypotheses.

(A1) Given any $x, y \in \mu(\Sigma)$ with $x + y \in \mu(\Sigma)$, there exist disjoint $A, B \in \Sigma$ such that $x = \mu(A), y = \mu(B)$.

(A2) Any map $\alpha : \mu(\Sigma) \to \mathbb{R}_+$ which is additive on the set $T = \{(x, y) \mid x, y, x + y \in \mu(\Sigma)\}$ is linear on $\mu(\Sigma)$.

The first hypothesis is clear, but the second calls for some explanation. A map $\alpha : \mu(\Sigma) \to \mathbb{R}_+$ is said to be *additive on the set* $T = \{(x, y) \mid x, y, x + y \in \mu(\Sigma)\}$ if

$$\alpha(x + y) = \alpha(x) + \alpha(y) \qquad ((x, y) \in T).$$

A map $\alpha : \mu(\Sigma) \to \mathbb{R}_+$ is *linear on* $\mu(\Sigma)$ if there exists a constant $\gamma \in \mathbb{R}_+$ such that

$$\alpha(x) = \gamma x \qquad (x \in \mu(\Sigma)).$$

As an example of a situation in which hypothesis (A2) is not satisfied, suppose $\mu(\Sigma) = \mathbb{N}_0 + \mathbb{N}_0\sqrt{2} = \{m + n\sqrt{2} \mid m, n \in \mathbb{N}_0\}$, where $\mathbb{N}_0$ is the set of nonnegative integers. There the map $\alpha : \mu(\Sigma) \to \mathbb{R}_+$ defined by

$$\alpha(m + n\sqrt{2}) = m + n \qquad (m, n \in \mathbb{N}_0),$$

is additive on $\mu(\Sigma) \times \mu(\Sigma)$, but it is not linear on $\mu(\Sigma)$.

For an example in which (A1) is not satisfied, consider a discrete measure space with $X = \{x_1, x_2, x_3, x_4\}$, where $\mu(\{x_1\}) = 1$, $\mu(\{x_2\}) = \mu(\{x_3\}) = 2$, and $\mu(\{x_4\}) = 7$. Then $x, y, x + y \in \mu(\Sigma)$ for $x = 3$ and $y = 4$, but $\mu(A) = 3$ only if $A = \{x_1, x_2\}$ or $A = \{x_1, x_3\}$, and $\mu(B) = 4$ only if $B = \{x_2, x_3\}$. There do not exist *disjoint* $A, B \in \Sigma$ with $x = \mu(A), y = \mu(B)$.

Typical kinds of sets $S = \mu(\Sigma)$ which guarantee the fulfillment of hypothesis (A2) of additive suitability are given in the following proposition.

Proposition 1. *Let p be a positive real number, suppose that either $G = \mathbb{R}$ or G is an additive subgroup of $\mathbb{Q}$, and put $S = (G \cap I)p$, where I may be $[0, 1]$, $[0, 1)$, or $\mathbb{R}_+$. Then every map $\alpha : S \to \mathbb{R}_+$ which is additive on $T = \{(x, y) \mid x, y, x + y \in S\}$ is linear on S.*

Proof. First, suppose that G is an additive subgroup of $\mathbb{Q}$. It is well-known that additive maps are homogeneous of degree 1 over $\mathbb{Q}$. Since every $x \in S$ can be written as $x = rp$ for $r \in G \cap I$, we have

$$\alpha(x) = \alpha(rp) = r\alpha(p) = \frac{\alpha(p)}{p}x \qquad (x \in S).$$

This shows that α is linear on S.

Now suppose that $G = \mathbb{R}$, so that $S = [0, p]$, $[0, p)$, or $\mathbb{R}_+$. If $S = [0, p)$ or $[0, p]$, then any map α additive on T has a unique extension to a map $\bar{\alpha} : \mathbb{R} \to \mathbb{R}$ additive on $\mathbb{R}_+ \times \mathbb{R}_+$ (even on $\mathbb{R} \times \mathbb{R}!$). (See e.g. [1].) Hence, we may suppose that $S = \mathbb{R}_+$ without loss of generality. But any additive map on $\mathbb{R}_+ \times \mathbb{R}_+$ which is nonnegative on an interval must be linear on $\mathbb{R}_+$. $\square$

We shall assume for the main results in this article that (X, Σ, μ) is an additively suitable σ-finite measure space.

Also, we assume tacitly throughout that $\mathcal{F}$ contains all *simple* non-negative Σ-measurable functions. That is, $\Sigma_k c_k \mathcal{X}_{E_k}$ belongs to $\mathcal{F}$ for all finite sequences $\{c_k\}$ of nonnegative reals in $\{f(x) \mid x \in X, f \in \mathcal{F}\}$, and for all corresponding sequences $\{E_k\}$ of Σ-measurable, pairwise disjoint $E_k \in \Sigma$.

2.   FIRST RESULTS

Proposition 2. *Suppose* (X, Σ, μ) *is* σ-finite *and that* $M : \mathcal{F} \to \mathbb{R}_+$ *satisfies* (2)-(4). *Then* $M = \Delta$, *and for each* $f \in \mathcal{F}$ *there is a nonnegative* Σ-measurable $h_f : X \to \mathbb{R}_+$ *such that*

$$M(f\mathcal{X}_E) = \Delta(f\mathcal{X}_E) = \int_E h_f \, d\mu \qquad (E \in \Sigma). \tag{6}$$

Moreover, $h_f(x) = 0$ *for almost every* x *such that* $f(x) = 0$.

Proof. To see that $M = \Delta$, just put $D = X$ in (4) and use (2). Now fix an $f \in \mathcal{F}$. Choose arbitrarily a pairwise disjoint sequence $\{A_k\}$ in Σ. Define

$$f_n := f\mathcal{X}_{\bigcup_{k=1}^{n} A_k} \qquad (n = 1, 2, ...) \qquad \text{and} \qquad f_\infty := f\mathcal{X}_{\bigcup_{k=1}^{\infty} A_k}.$$

Then on the one hand we have

$$M(f_\infty) = \Delta(f_\infty) = \Delta(f\mathcal{X}_{\bigcup_{k=1}^{\infty} A_k}). \tag{7}$$

On the other hand, since

$$M(f_k) - M(f_{k-1}) = \Delta(f_k \mathcal{X}_{A_k}) = \Delta(f \mathcal{X}_{A_k})$$

follows from (4), we have

$$
\begin{aligned}
M(f_n) &= \sum_{k=1}^{n-1} [M(f_{k+1}) - M(f_k)] + M(f_1) \\
&= \sum_{k=1}^{n-1} \Delta(f \mathcal{X}_{A_{k+1}}) + \Delta(f \mathcal{X}_{A_1}) = \sum_{k=1}^{n} \Delta(f \mathcal{X}_{A_k}).
\end{aligned}
\tag{8}
$$

Now, since $f_n \nearrow f_\infty$, it follows from (3) that $M(f_n) \to M(f_\infty)$. Hence, comparing (7) and (8), we see that

$$\sum_{k=1}^{\infty} \Delta(f \mathcal{X}_{A_k}) = \Delta(f \mathcal{X}_{\bigcup_{k=1}^{\infty} A_k}).$$

Defining $\Phi_f : \Sigma \to \mathbb{R}_+$ by

$$\Phi_f(E) := \Delta(f \mathcal{X}_E) \qquad (E \in \Sigma), \tag{9}$$

this means that Φ_f is countably additive. Moreover,

$$\Phi_f(\emptyset) = \Delta(0) = M(0) = 0,$$

so Φ_f is a measure on (X, Σ).

Furthermore, it is easy to see that Φ_f is absolutely continuous with respect to μ. Indeed, suppose $\mu(Z) = 0$ for some $Z \in \Sigma$. Then

$$\Phi_f(Z) = \Delta(f \mathcal{X}_Z) = M(f \mathcal{X}_Z) = 0$$

by (2), since $f \mathcal{X}_Z = 0$ a.e.

Therefore, since (X, Σ, μ) is σ-finite, an application of the Radon-Nikodym Theorem yields the existence of a nonnegative Σ-measurable h_f on X for which

$$\Phi_f(E) = \int_E h_f \, d\mu \qquad (E \in \Sigma).$$

Together with (9) and the fact that $M = \Delta$, this establishes (6).

Finally, for a given $f \in \mathcal{F}$, let $N = \{x \mid f(x) = 0\}$. Now

$$0 = M(0) = M(f \mathcal{X}_N) = \int_N h_f \, d\mu.$$

Since $h_f \geq 0$, this means that $h_f = 0$ a.e. on N.   □

Next, we add hypothesis (5).

Proposition 3. *Suppose* (X, Σ, μ) *is* σ-*finite and additively suitable, and that* $M : \mathcal{F} \to \mathbb{R}_+$ *satisfies* (2)-(5). *Then there is a left-continuous map* $\phi : \mathbb{R}_+ \to \mathbb{R}_+$ *such that* $\phi(0) = 0$ *and* (1) *holds for all simple* $f \in \mathcal{F}$.

Proof. By Proposition 2, to each $f \in \mathcal{F}$ there corresponds a nonnegative Σ-measurable h_f such that (6) holds. Taking $f(x) = c$ for all $x \in X$, where c is any nonnegative constant, we deduce from (6) that

$$M(c\mathcal{X}_E) = \int_E h_c \, d\mu \qquad (E \in \Sigma, \; c \in \mathbb{R}_+). \tag{10}$$

Now, by hypothesis (5), we know also that the quantity $M(c\mathcal{X}_E)$, for any given $c \geq 0$, depends on E only through its measure $\mu(E)$. Fixing $c \in \mathbb{R}_+$, we may define $\alpha_c : \mu(E) \to \mathbb{R}_+$ by

$$\alpha_c(\mu(E)) := \int_E h_c \, d\mu \qquad (E \in \Sigma). \tag{11}$$

Let $T = \{(x, y) \mid x, y, x + y \in \mu(E)\}$. Since (X, Σ, μ) is additively suitable, for any $(x, y) \in T$ there exist disjoint $A, B \in \Sigma$ for which $x = \mu(A)$ and $y = \mu(B)$. Thus we have

$$\alpha_c(x + y) = \alpha_c(\mu(A \cup B)) = \int_{A \cup B} h_c \, d\mu = \int_A h_c \, d\mu + \int_B h_c \, d\mu$$
$$= \alpha_c(\mu(A)) + \alpha_c(\mu(B)) = \alpha_c(x) + \alpha_c(y).$$

That is, α_c is additive on T. Hence, again by the fact that (X, Σ, μ) is additively suitable, α_c must be linear on $\mu(E)$. That is, there is a constant γ_c (depending on c) such that

$$\alpha_c(x) = \gamma_c x \qquad (x \in \mu(\Sigma)).$$

Freeing $c \in \mathbb{R}_+$, we have shown that

$$\alpha_c(x) = \phi(c)x, \; x \in \mu(\Sigma) \qquad (c \in \mathbb{R}_+),$$

for some map $\phi : \mathbb{R}_+ \to \mathbb{R}_+$. Combining this with (11) and (10), we have found that

$$M(c\mathcal{X}_E) = \phi(c)\mu(E) \qquad (c \in \mathbb{R}_+, \; E \in \Sigma). \tag{12}$$

Moreover, $\phi(0) = 0$ by (2), and ϕ is left-continuous by (3).

Finally, consider a simple nonnegative Σ-measurable $f = \sum_{k=1}^{n} c_k \mathcal{X}_{E_k}$ in $\mathcal{F}$. Applying (4) a total of $n-1$ times, and recalling the fact that $\Delta = M$, we find that

$$
M(f) = M\left(\sum_{k=1}^{n} c_k \mathcal{X}_{E_k}\right) = M\left(\sum_{k=1}^{n-1} c_k \mathcal{X}_{E_k}\right) + M(c_n \mathcal{X}_{E_n})
$$
$$
= \ldots = \sum_{k=1}^{n} M(c_k \mathcal{X}_{E_k}).
$$

With (12) and the fact that $\phi(0) = 0$, we have therefore

$$
M\left(\sum_{k=1}^{n} c_k \mathcal{X}_{E_k}\right) = \sum_{k=1}^{n} \phi(c_k)\mu(E_k) = \sum_{k=1}^{n} \int_{E_k} \phi(c_k \mathcal{X}_{E_k}(x))\, d\mu
$$
$$
= \sum_{k=1}^{n} \int_{E_k} \phi\left(\sum_{j=1}^{n} c_j \mathcal{X}_{E_j}(x)\right) d\mu
$$
$$
= \int_{X} \phi \circ \left(\sum_{j=1}^{n} c_j \mathcal{X}_{E_j}\right) d\mu,
$$

which is (1) for simple $f \in \mathcal{F}$. $\qquad\square$

3. CHARACTERIZATION THEOREMS

Assuming the hypotheses of Proposition 3, can we obtain (1) for all $f \in \mathcal{F}$? For any $f \in \mathcal{F}$, we can choose a sequence $\{f_k\}$ of simple $f_k \in \mathcal{F}$ such that $f_k \nearrow f$. Since ϕ is left-continuous, we then have $\phi \circ f_k \to \phi \circ f$ and we know that $\phi \circ f$ is nonnegative and Σ-measurable. We know also that $M(f_k) \to M(f)$ by (3), and we have established that

$$
M(f_k) = \int_{X} \phi \circ f_k\, d\mu.
$$

Thus (1) will be proved if we can establish that

$$
\int_{X} \phi \circ f_k\, d\mu \longrightarrow \int_{X} \phi \circ f\, d\mu. \tag{13}
$$

We now prove two characterization theorems, using the Monotone and Dominated Convergence Theorems, respectively.

Theorem 1. *Suppose (X, Σ, μ) is σ-finite and additively suitable. Then a functional $M : \mathcal{F} \to \mathbb{R}_+$ satisfies properties (2)–(5) and*

$$
c \mapsto M(c\mathcal{X}_E) \text{ is nondecreasing on } \mathbb{R}_+ \text{ for some } E \in \Sigma,
$$
$$
0 < \mu(E) < +\infty, \tag{14}
$$

if and only if M has the form (1) for some left-continuous nondecreasing $\phi : \mathbb{R}_+ \to \mathbb{R}_+$ with $\phi(0) = 0$.

Proof. In one direction, it is a straightforward verification. For the other, suppose (2)-(5) and (14) for M. By Proposition 3, we have (1) for all simple $f \in \mathcal{F}$. In particular, we have (12):

$$M(c\mathcal{X}_E) = \phi(c)\mu(E) \qquad (c \in \mathbb{R}_+),$$

valid for the set E in (14). Hence ϕ is nondecreasing, so $\phi \circ f_k \nearrow \phi \circ f$ whenever $f_k \nearrow f$. By the Monotone Convergence Theorem, we have (13), and this suffices to establish (1) by the discussion leading up to the statement of this theorem. $\qquad\square$

Theorem 2. *Let (X, Σ, μ) be σ-finite and additively suitable, and suppose $\mathcal{F}$ is uniformly bounded, say $f(X) \subset [0, K]$ for all $f \in \mathcal{F}$. Then a functional $M : \mathcal{F} \to \mathbb{R}_+$ satisfies (2)-(5) and*

$$\{M(c\mathcal{X}_E) \mid c \in [0, K]\} \text{ is bounded for some } E \in \Sigma,$$
$$0 < \mu(E) < +\infty, \tag{15}$$

if and only if M has the form (1) for some left-continuous bounded $\phi : [0, K] \to \mathbb{R}_+$ with $\phi(0) = 0$.

Proof. Make the obvious changes in the proof of Theorem 1. The major change is the use of Lebesgue's Dominated Convergence Theorem instead of monotone convergence. Boundedness of ϕ follows from (15) and (12). $\qquad\square$

Taking $K = 1$ in Theorem 2, we have the following consequence.

Corollary 1. *Let (X, Σ, μ) be σ-finite and additively suitable, and suppose $\mathcal{F}$ is a collection of fuzzy sets containing all $[0,1]$-valued Σ-measurable simple functions. Then a functional $M : \mathcal{F} \to \mathbb{R}_+$ satisfies (2)-(5) and (15) with $K = 1$ if and only if M has the form (1) with a left-continuous bounded $\phi : [0,1] \to \mathbb{R}_+$ satisfying $\phi(0) = 0$.*

References

[1] J. Aczél, *A Short Course on Functional Equations*, D. Reidel, 1987, pp. 16–21.

[2] R. L. Wheeden and A. Zygmund, *Measure and Integral, an Introduction to Real Analysis*, Marcel Dekker, 1977.

JORDAN MAPS ON STANDARD OPERATOR ALGEBRAS

Lajos Molnár

Institute of Mathematics and Informatics, University of Debrecen,
4010 Debrecen, P.O.Box 12, Hungary
molnarl@math.klte.hu

Abstract   Jordan isomorphisms of rings are defined by two equations. The first one is the equation of additivity while the second one concerns multiplicativity with respect to the so-called Jordan product. In this paper we present results showing that on standard operator algebras over spaces with dimension at least 2, the bijective solutions of that second equation are automatically additive.

Keywords: standard operator algebra, Jordan homomorphism

Mathematics Subject Classification (2000): 47B49, 39B52

1.   INTRODUCTION AND STATEMENT OF THE RESULTS

It is an interesting problem to study the interrelation between the multiplicative and the additive structures of a ring. The first quite surprising result on how the multiplicative structure of a ring determines its additive structure is due to Martindale [5]. In [5, Corollary] he proved that every bijective multiplicative map from a prime ring containing a nontrivial idempotent onto an arbitrary ring is necessarily additive and, hence, it is a ring isomorphism. This result has been utilized by Šemrl in [9] to describe the form of the semigroup isomorphisms of standard operator algebras on Banach spaces.

---

[0]This research was supported by the Hungarian National Foundation for Scientific Research (OTKA), Grant No. T030082, T031995, and by the Ministry of Education, Hungary, Reg. No. FKFP 0349/2000.

Z. Daróczy and Z. Páles (eds.),
Functional Equations - Results and Advances, 305–320.
© 2002 Kluwer Academic Publishers.

Beyond ring homomorphisms there is another very important class of transformations between rings. These are the Jordan homomorphisms. The Jordan structure of associative rings has been studied by many people in ring theory. Moreover, Jordan operator algebras have serious applications in the mathematical foundations of quantum mechanics. If $\mathcal{R}, \mathcal{R}'$ are rings and $\phi : \mathcal{R} \to \mathcal{R}'$ is a transformation, then ϕ is called a Jordan homomorphism if it is additive and satisfies

$$\phi(A^2) = \phi(A)^2 \qquad (A \in \mathcal{R}). \tag{1}$$

If $\mathcal{R}'$ is 2-torsion free, then, under the assumption of additivity, (1) is equivalent to

$$\phi(AB + BA) = \phi(A)\phi(B) + \phi(B)\phi(A) \qquad (A, B \in \mathcal{R}). \tag{2}$$

Clearly, every ring homomorphism is a Jordan homomorphism and the same is true for ring antihomomorphisms (the transformation $\phi : \mathcal{R} \to \mathcal{R}'$ is called a ring antihomomorphism if ϕ is additive and satisfies $\phi(AB) = \phi(B)\phi(A)$ for all $A, B \in \mathcal{R}$). In algebras, it seems more frequent that instead of (2), one considers the equation

$$\phi((1/2)(AB + BA)) = (1/2)(\phi(A)\phi(B) + \phi(B)\phi(A)) \qquad (A, B \in \mathcal{R}). \tag{3}$$

Under the assumption of additivity these two equations are obviously equivalent.

The additivity of bijective maps ϕ between von Neumann algebras without commutative direct summand satisfying the equation (3) and

$$\phi(A^*) = \phi(A)^*$$

was studied in [3] and [4]. The aim of this paper is to investigate similar problems on standard operator algebras. From our present point of view the main difference between standard operator algebras and von Neumann algebras is that standard operator algebras need not have unit and they are not necessarily closed in any operator topology.

In what follows we shall study the equations (2) and (3). Both equations play important role; the first one is because of ring theory and the second one is because of the applications of Jordan operator algebras in mathematical physics. Our main results describe the form of the bijective solutions of the considered equations. It will turn out that all such solutions are automatically additive. We refer to our recent papers [7], [6] for some other results of similar spirit.

We now summarize the results of the paper. In what follows we consider all linear spaces over the complex field. If X is a Banach space, then we denote by $B(X)$ and $F(X)$ the algebra of all bounded linear operators and the ideal of all bounded linear finite rank operators on X, respectively. A subalgebra of $B(X)$ is called a standard operator algebra if it contains $F(X)$. The dual space of X is denoted by X' and A' stands for the Banach space adjoint of the operator $A \in B(X)$.

Our first result describes the form of the bijective solutions of (3) on standard operator algebras.

Theorem 1. *Let* X, Y *be Banach spaces,* $\dim X > 1$, *and let* $\mathcal{A} \subset B(X)$, $\mathcal{B} \subset B(Y)$ *be standard operator algebras. Suppose that* $\phi : \mathcal{A} \to \mathcal{B}$ *is a bijective transformation satisfying*

$$\phi((1/2)(AB + BA)) = (1/2)(\phi(A)\phi(B) + \phi(B)\phi(A)) \qquad (4)$$

for every $A, B \in \mathcal{A}$.

If X *is infinite dimensional, then we have the following possibilities:*

(i) *there exists an invertible bounded linear operator* $T : X \to Y$ *such that*

$$\phi(A) = TAT^{-1} \qquad (A \in \mathcal{A});$$

(ii) *there exists an invertible bounded conjugate-linear operator* $T : X \to Y$ *such that*

$$\phi(A) = TAT^{-1} \qquad (A \in \mathcal{A});$$

(iii) *there exists an invertible bounded linear operator* $T : X' \to Y$ *such that*

$$\phi(A) = TA'T^{-1} \qquad (A \in \mathcal{A});$$

(iv) *there exists an invertible bounded conjugate-linear operator* $T : X' \to Y$ *such that*

$$\phi(A) = TA'T^{-1} \qquad (A \in \mathcal{A}).$$

If X *is finite dimensional, then we have* $\dim X = \dim Y$. *So, our transformation* ϕ *can be supposed to act on the matrix algebra* $M_n(\mathbb{C})$. *In this case we have the following possibilities:*

(v) *there exist a ring automorphism* h *of* $\mathbb{C}$ *and an invertible matrix* $T \in M_n(\mathbb{C})$ *such that*

$$\phi(A) = Th(A)T^{-1} \qquad (A \in M_n(\mathbb{C}));$$

(vi) *there exist a ring automorphism* h *of* $\mathbb{C}$ *and an invertible matrix* $T \in M_n(\mathbb{C})$ *such that*

$$\phi(A) = Th(A)^t T^{-1} \qquad (A \in M_n(\mathbb{C})).$$

Here, ^{t} stands for the transpose and $h(A)$ denotes the matrix obtained from A by applying h on every entry of it.

From the theorem above we easily have the following corollary. If H is a Hilbert space and $A \in B(H)$, then A^* denotes the Hilbert space adjoint of A.

Corollary 1. Let H, K be Hilbert spaces, $\dim H > 1$, and let $\mathcal{A} \subset B(H)$, $\mathcal{B} \subset B(K)$ be standard operator algebras which are closed under taking adjoints. Suppose that $\phi : \mathcal{A} \to \mathcal{B}$ is a bijective transformation satisfying

$$\phi(A^*) = \phi(A)^*$$
$$\phi((1/2)(AB + BA)) = (1/2)(\phi(A)\phi(B) + \phi(B)\phi(A)) \qquad (5)$$

for every $A, B \in \mathcal{A}$.

Then we have the following possibilities:

(i) there exists a unitary operator $U : H \to K$ such that

$$\phi(A) = UAU^* \qquad (A \in \mathcal{A});$$

(ii) there exists an antiunitary operator $U : H \to K$ such that

$$\phi(A) = UAU^* \qquad (A \in \mathcal{A});$$

(iii) there exists a unitary operator $U : H \to K$ such that

$$\phi(A) = UA^*U^* \qquad (A \in \mathcal{A});$$

(iv) there exists an antiunitary operator $U : H \to K$ such that

$$\phi(A) = UA^*U^* \qquad (A \in \mathcal{A}).$$

Unfortunately, we do not have a result concerning the equation (2) in the Banach space setting. However, we have the following result describing the self-adjoint solutions of that equation on standard operator algebras over Hilbert spaces. The restriction to self-adjoint solutions is very natural in operator theory where they usually consider transformations on operator algebras which preserve adjoints.

Theorem 2. Let H, K be Hilbert spaces, $\dim H > 1$, and let $\mathcal{A} \subset B(H)$, $\mathcal{B} \subset B(K)$ be standard operator algebras which are closed under taking adjoints. Let $\phi : \mathcal{A} \to \mathcal{B}$ be a bijective transformation satisfying

$$\phi(A^*) = \phi(A)^*$$
$$\phi(AB + BA) = \phi(A)\phi(B) + \phi(B)\phi(A)$$

for every $A, B \in \mathcal{A}$. Then we have the same possibilities for ϕ as in Corollary 1.

Although we do not have a result on the equation (2) in the Banach space setting, we suspect that its solutions are the same as those ones listed in Theorem 1. Unfortunately, this is only a conjecture which is left as an open problem.

Finally, we point out the fact that in all of our statements we have supposed that the underlying spaces are at least 2-dimensional. In fact, this assumption is necessary to put as it turns out from the following example. Over 1-dimensional spaces, standard operator algebras are trivially identified with the complex field $\mathbb{C}$. Now, consider a bijective additive function $a : \mathbb{R} \to \mathbb{R}$ for which $a(1) = 1$ and $a(\log_2 3) \neq \log_2 3$. Such a function exists since 1 and $\log_2 3$ are linearly independent in the linear space $\mathbb{R}$ over the field of rationals. Let $f :]0, +\infty[\to]0, +\infty[$ be defined by

$$f(t) = 2^{a(\log_2 t)}.$$

Define the function $h : \mathbb{C} \to \mathbb{C}$ by

$$h(z) = \begin{cases} 0, & \text{if } z = 0; \\ f(|z|)\frac{z}{|z|}, & \text{if } z \neq 0. \end{cases}$$

It is easy to see that $h : \mathbb{C} \to \mathbb{C}$ is a bijective multiplicative function, $h(2) = 2$ and $h(\bar{z}) = \overline{h(z)}$ ($z \in \mathbb{C}$). On the other hand, h is not additive, since $h(3) \neq h(1) + h(2)$. This function serves as a counterexample for all of our results above after omitting the assumption on dimension.

2.   PROOFS

This section is devoted to the proofs of our results.

If X is a Banach space, then the operator $P \in B(X)$ is an idempotent if $P^2 = P$. There is a partial ordering between idempotents. If $P, Q \in B(X)$ are idempotents, then we write $P \leq Q$ if $PQ = QP = P$. The idempotents $P, Q \in B(X)$ are said to be orthogonal if $PQ = QP = 0$.

Let $x \in X$ and $f \in X'$ be nonzero. The rank-1 operator $x \otimes f$ is defined by

$$(x \otimes f)(z) = f(z)x \qquad (z \in X).$$

It is trivial to see that $x \otimes f$ is an idempotent if and only if $f(x) = 1$. Conversely, every rank-1 idempotent can be written in this form.

We begin with the proof of our first result.

Proof of Theorem 1. First observe that $\phi(0) = 0$. Indeed, if $A \in \mathcal{A}$ is such that $\phi(A) = 0$, then we have

$$\phi(0) = \phi((1/2)(A0 + 0A)) = (1/2)(\phi(A)\phi(0) + \phi(0)\phi(A)) = 0.$$

We deduce from (4) that ϕ preserves the idempotents. Since ϕ^{-1} has the same properties as ϕ, it follows that ϕ preserves the idempotents in both directions. If the idempotents $P, Q \in \mathcal{A}$ are orthogonal, then we have

$$0 = \phi(0) = \phi((1/2)(PQ + QP)) = (1/2)(\phi(P)\phi(Q) + \phi(Q)\phi(P)).$$

Multiplying this equality by $\phi(Q)$ from the left and from the right respectively, we have $\phi(Q)\phi(P)\phi(Q) = \phi(P)\phi(Q)$ and $\phi(Q)\phi(P)\phi(Q) = \phi(Q)\phi(P)$. This implies that

$$0 = \phi(Q)\phi(P)\phi(Q) = \phi(P)\phi(Q) = \phi(Q)\phi(P).$$

Therefore, ϕ preserves the orthogonality between idempotents in both directions.

We assert that ϕ preserves the partial order $\leq$ between the idempotents. If $P, Q \in \mathcal{A}$ are idempotents and $P \leq Q$, then we obtain

$$\phi(P) = \phi((1/2)(PQ + QP)) = (1/2)(\phi(P)\phi(Q) + \phi(Q)\phi(P)).$$

Multiplying this equality by $\phi(Q)$ from the left and from the right respectively, we get that $\phi(Q)\phi(P)\phi(Q) = \phi(Q)\phi(P)$ and $\phi(Q)\phi(P)\phi(Q) = \phi(P)\phi(Q)$. This implies that $\phi(P) = \phi(P)\phi(Q) = \phi(Q)\phi(P)$ and hence we have $\phi(P) \leq \phi(Q)$.

It is easy to see that an idempotent $P \in \mathcal{A}$ is of rank n if and only if there is a system $P_1, \ldots, P_n \in \mathcal{A}$ of pairwise orthogonal nonzero idempotents for which $P_k \leq P$ $(k = 1, \ldots, n)$, but there is no such system of $n+1$ members. It now follows that ϕ preserves the rank of idempotents.

Let $P, Q \in \mathcal{A}$ be orthogonal finite rank idempotents. We know that $\phi(P), \phi(Q)$ are orthogonal finite rank idempotents. As ϕ preserves the order, we have $\phi(P), \phi(Q) \leq \phi(P+Q)$ implying $\phi(P) + \phi(Q) \leq \phi(P+Q)$. Since ϕ preserves also the rank of idempotents, it follows that $\phi(P) + \phi(Q) = \phi(P + Q)$. This means that ϕ is orthoadditive on the set of all finite rank idempotents in $\mathcal{A}$.

Let $P_1, \ldots, P_n \in \mathcal{A}$ be pairwise orthogonal finite rank idempotents and $\lambda_1, \ldots, \lambda_n \in \mathbb{C}$. Using the orthoadditivity of ϕ we have

$$
\begin{aligned}
\phi\Big(\sum_k \lambda_k P_k\Big) &= \phi\Big(\frac{1}{2}\Big(\Big(\sum_k \lambda_k P_k\Big)\Big(\sum_l P_l\Big) + \Big(\sum_l P_l\Big)\Big(\sum_k \lambda_k P_k\Big)\Big)\Big) \\
&= \frac{1}{2}\Big(\phi\Big(\sum_k \lambda_k P_k\Big)\phi\Big(\sum_l P_l\Big) + \phi\Big(\sum_l P_l\Big)\phi\Big(\sum_k \lambda_k P_k\Big)\Big) \\
&= \frac{1}{2}\Big(\phi\Big(\sum_k \lambda_k P_k\Big)\sum_l \phi(P_l) + \sum_l \phi(P_l)\phi\Big(\sum_k \lambda_k P_k\Big)\Big) \\
&= \sum_l \frac{1}{2}\Big(\phi\Big(\sum_k \lambda_k P_k\Big)\phi(P_l) + \phi(P_l)\phi\Big(\sum_k \lambda_k P_k\Big)\Big) \\
&= \sum_l \phi\Big(\frac{1}{2}\Big(\Big(\sum_k \lambda_k P_k\Big)P_l + P_l\Big(\sum_k \lambda_k P_k\Big)\Big)\Big) \\
&= \sum_l \phi(\lambda_l P_l).
\end{aligned}
$$

$$(6)$$

Next we prove that $\phi(-P) = -\phi(P)$ for every finite rank idempotent $P \in \mathcal{A}$. Let P be of rank 1. We have

$$
\begin{aligned}
\phi(\lambda P) &= \phi((1/2)((\lambda P)P + P(\lambda P))) \\
&= (1/2)(\phi(\lambda P)\phi(P) + \phi(P)\phi(\lambda P)).
\end{aligned}
$$

Multiplying this equality by $\phi(P)$ from the left and from the right respectively, we have $\phi(P)\phi(\lambda P)\phi(P) = \phi(P)\phi(\lambda P)$ and $\phi(P)\phi(\lambda P)\phi(P) = \phi(\lambda P)\phi(P)$. It follows that

$$
\phi(\lambda P) = \phi(P)\phi(\lambda P)\phi(P).
$$

Since $\phi(P)$ is of rank 1, it follows from the equality above that

$$
\phi(\lambda P) = \mu\phi(P) \tag{7}
$$

for some scalar $\mu \in \mathbb{C}$. So, we obtain that $\phi(-P) = c\phi(P)$ for some scalar $c \in \mathbb{C}$. Since

$$
c^2\phi(P) = (c\phi(P))^2 = \phi(-P)^2 = \phi((-P)^2) = \phi(P),
$$

we have $c = \pm 1$. By the injectivity of ϕ we get $\phi(-P) = -\phi(P)$. Using (6) we deduce that

$$
\phi(-P) = -\phi(P) \tag{8}
$$

for every finite rank idempotent $P \in \mathcal{A}$.

For any $A, B \in \mathcal{A}$ we write

$$A \circ B = (1/2)(AB + BA).$$

With this notation the equation (4) can be rewritten as

$$\phi(A \circ B) = \phi(A) \circ \phi(B) \qquad (A, B \in \mathcal{A}).$$

Let $T \in F(X)$ be arbitrary and let $P \in F(X)$ be an idempotent. Choose a finite rank idempotent $Q \in \mathcal{A}$ for which $QT = TQ = T$ and $QP = PQ = P$. Such a Q can be constructed in the following way. Let $S \in F(X)$. Pick a finite rank idempotent Q_S^l with range containing the range of S. We have $Q_S^l S = S$. Next, pick a finite dimensional subspace M of X whose direct sum with the kernel N of S is X. Consider the idempotent Q_S^r with range M corresponding to the direct sum $M \oplus N = X$. We have $S Q_S^r = S$. Finally, as the partially ordered set of all finite rank idempotents on X is cofinal (see, for example, [8, Lemma]), we can choose a finite rank idempotent Q for which $Q_T^l, Q_T^r, Q_P^l, Q_P^r \le Q$. It is easy to check that Q has the desired properties. Now, it requires only trivial computation to verify that

$$(2P - Q) \circ (T \circ P) = PTP. \tag{9}$$

It follows that

$$\phi(2P - Q) \circ (\phi(T) \circ \phi(P)) = \phi(PTP).$$

We prove that $\phi(2P - Q) = 2\phi(P) - \phi(Q)$. Indeed, since $Q - P$ is an idempotent which is orthogonal to P, by (6) and (8) we can compute

$$\begin{aligned}
\phi(2P - Q) &= \phi(P - (Q - P)) = \phi(P) + \phi(-(Q - P)) \\
&= \phi(P) - \phi(Q - P) = \phi(P) - (\phi(Q) - \phi(P)) \\
&= 2\phi(P) - \phi(Q).
\end{aligned}$$

So, we have

$$(2\phi(P) - \phi(Q)) \circ (\phi(T) \circ \phi(P)) = \phi(PTP).$$

We assert that $\phi(Q)\phi(T)\phi(Q) = \phi(T)$ and $\phi(Q)\phi(P)\phi(Q) = \phi(P)$. In fact, these follow from the equalities

$$\phi(T) = (1/2)(\phi(T)\phi(Q) + \phi(Q)\phi(T))$$

and

$$\phi(P) = (1/2)(\phi(P)\phi(Q) + \phi(Q)\phi(P))$$

after mutliplying them by $\phi(Q)$ from the left and from the right, respectively. Similarly as in the case of (9), one can now easily check that

$$(2\phi(P) - \phi(Q)) \circ (\phi(T) \circ \phi(P)) = \phi(P)\phi(T)\phi(P).$$

Therefore, we have $\phi(PTP) = \phi(P)\phi(T)\phi(P)$. We note that in this part of the proof we have used an idea similar to what was followed in the proof of [3, Lemma 1.6].

In the next section of the proof we apply some ideas from the proof of [6, Theorem]. Fix a rank-1 idempotent $P \in \mathcal{A}$. By (7), there is a function $h_P : \mathbb{C} \to \mathbb{C}$ such that

$$\phi(\lambda P) = h_P(\lambda)\phi(P) \qquad (\lambda \in \mathbb{C}).$$

We show that h_P does not depend on P. If $Q \in \mathcal{A}$ is another rank-1 idempotent not orthogonal to P, then we compute

$$\phi((1/2)((\lambda P)Q + Q(\lambda P))) = (1/2)(h_P(\lambda)\phi(P)\phi(Q) + h_P(\lambda)\phi(Q)\phi(P))$$
$$= h_P(\lambda)(1/2)(\phi(P)\phi(Q) + \phi(Q)\phi(P)).$$

We similarly have

$$\phi((1/2)(P(\lambda Q) + (\lambda Q)P)) = h_Q(\lambda)(1/2)(\phi(P)\phi(Q) + \phi(Q)\phi(P)).$$

Since $\phi(P)\phi(Q) + \phi(Q)\phi(P) \neq 0$ ($\phi(P)$ is not orthogonal to $\phi(Q)$), it follows that $h_P = h_Q$. If Q is orthogonal to P, then we can choose a rank-1 idempotent $R \in \mathcal{A}$ such that R is not orthogonal to P and not orthogonal to Q. We have $h_P = h_R = h_Q$. Therefore, there is a function $h : \mathbb{C} \to \mathbb{C}$ such that

$$\phi(\lambda P) = h(\lambda)\phi(P) \tag{10}$$

for every $\lambda \in \mathbb{C}$ and every rank-1 idempotent $P \in \mathcal{A}$.

We assert that $\phi(\lambda A) = h(\lambda)\phi(A)$ for every $A \in F(X)$. If A is a finite rank idempotent, then this follows from (10) and (6). If $A \in F(X)$ is arbitrary, then there is a finite rank idempotent P such that $PA = AP = A$. We compute

$$\phi(\lambda A) = \phi((1/2)(A(\lambda P) + (\lambda P)A))$$
$$= (1/2)(\phi(A)h(\lambda)\phi(P) + h(\lambda)\phi(P)\phi(A))$$
$$= h(\lambda)\phi(A).$$

We next prove that h is multiplicative. Let $P \in \mathcal{A}$ be a nonzero finite rank idempotent. We have

$$h(\lambda\mu)\phi(P) = \phi(\lambda\mu P) = \phi((1/2)((\lambda P)(\mu P) + (\mu P)(\lambda P)))$$
$$= (1/2)(h(\lambda)\phi(P)h(\mu)\phi(P) + h(\mu)\phi(P)h(\lambda)\phi(P))$$
$$= h(\lambda)h(\mu)\phi(P)$$

and this shows that h is multiplicative.

We prove that h is additive. Let $x, y \in X$ be linearly independent vectors, and choose linear functionals $f, g \in X'$ such that $f(x) = 1, f(y) = 0$ and $g(x) = 0, g(y) = 1$. Let $\lambda, \mu \in \mathbb{C}$ be such that $\lambda + \mu = 1$. Define $R = (\lambda x + \mu y) \otimes (f + g)$, $P = x \otimes f$, $Q = y \otimes g$. Clearly, R, P, Q are rank-1 idempotents and P is orthogonal to Q. By what we already know, we deduce

$$
\begin{aligned}
h(\lambda + \mu)\phi(R) &= \phi((\lambda + \mu)R) = \phi(R(P + Q)R) \\
&= \phi(R)\phi(P + Q)\phi(R) = \phi(R)\phi(P)\phi(R) + \phi(R)\phi(Q)\phi(R) \\
&= \phi(RPR) + \phi(RQR) = \phi(\lambda R) + \phi(\mu R) \\
&= (h(\lambda) + h(\mu))\phi(R).
\end{aligned}
$$

By the multiplicativity of h, we have $h(\lambda + \mu) = h(\lambda) + h(\mu)$ whenever $\lambda + \mu \neq 0$. To see the additivity of h, it remains to prove that $h(-\lambda) = -h(\lambda)$. Since h is multiplicative, it follows that $h(-\lambda)^2 = h(\lambda^2) = h(\lambda)^2$. By the injectivity of h we have the desired equality $h(-\lambda) = -h(\lambda)$.

We now verify that ϕ is additive on $F(X)$. Let $A, B \in F(X)$ be arbitrary and pick any rank-1 idempotent $P \in F(X)$. Choose $x \in X, f \in X'$ such that $P = x \otimes f$. We compute

$$
\begin{aligned}
\phi(P)\phi(A + B)\phi(P) &= \phi(P(A + B)P) = \phi(f((A + B)x)P) \\
&= h(f((A + B)x))\phi(P) \\
&= h(f(Ax))\phi(P) + h(f(Bx))\phi(P) \\
&= \phi(f(Ax)P) + \phi(f(Bx)P) = \phi(PAP) + \phi(PBP) \\
&= \phi(P)\phi(A)\phi(P) + \phi(P)\phi(B)\phi(P) \\
&= \phi(P)(\phi(A) + \phi(B))\phi(P).
\end{aligned}
$$

Since this holds true for every rank-1 idempotent P on X, we easily obtain that $\phi(A + B) = \phi(A) + \phi(B)$. Consequently, $\phi : F(X) \to F(Y)$ is an additive bijection satisfying (4).

Since the algebra $F(X)$ (as well as every standard operator algebra) is prime (this means that for every $A, B \in F(X)$, the equality $AF(X)B = \{0\}$ implies $A = 0$ or $B = 0$), we can apply a result of Herstein [2] to obtain that ϕ is necessarily a ring isomorphism or a ring antiisomorphism of $F(X)$. In the isomorphic case we can apply the result in [9] and obtain the desired form of ϕ on $F(X)$. In the finite dimensional case we are done since in that case any standard operator algebra coincides with $F(X)$. Observe that the argument given in [9] for the finite dimensional case can be changed to give the antiisomorphic part of our result in the finite dimensional case. So, let us assume that X is infinite dimensional and that ϕ is a ring isomorphism. By [9] there is a bounded invertible either linear or conjugate-linear operator $T : X \to Y$ such that

$$\phi(A) = TAT^{-1} \qquad (A \in F(X)).$$

If $A \in \mathcal{A}$ is arbitrary, then for every finite rank idempotent $P \in F(X)$ we have

$$(1/2)T(AP + PA)T^{-1} = \phi((1/2)(AP + PA))$$
$$= (1/2)(\phi(A)\phi(P) + \phi(P)\phi(A))$$
$$= (1/2)(\phi(A)TPT^{-1} + TPT^{-1}\phi(A)).$$

Multiplying this equality by T^{-1} from the left and by T from the right, we get

$$AP + PA = T^{-1}\phi(A)TP + PT^{-1}\phi(A)T.$$

Now, multiplying this equality by P from both sides, we arrive at

$$PAP = PT^{-1}\phi(A)TP.$$

Since $P \in \mathcal{A}$ was an arbitrary finite rank idempotent, it follows that $A = T^{-1}\phi(A)T$ $(A \in \mathcal{A})$. Therefore, we have $\phi(A) = TAT^{-1}$ $(A \in \mathcal{A})$.

Suppose finally that X is infinite dimensional and ϕ is a ring antiisomorpism of $F(X)$. Performing trivial modifications in the proofs of [9, Theorem] and [1, Proposition 3.1], one can verify that there is a bounded invertible either linear or conjugate-linear operator $T : X' \to Y$ such that

$$\phi(A) = TA'T^{-1} \qquad (A \in F(X)).$$

Similarly to the isomorphic case, we can arrive at the equality

$$P'A' + A'P' = (AP + PA)' = T^{-1}\phi(A)TP' + P'T^{-1}\phi(A)T.$$

As P' is an idempotent, multiplying this equality by P' from both sides, we deduce

$$P'A'P' = P'T^{-1}\phi(A)TP'.$$

Since, as we learn from [1, Proposition 3.1], in the antiisomorphic case X, Y are reflexive, it follows that P' runs through the set of all finite rank idempotents in $B(X')$ as P runs through the set of all finite rank idempotents in $B(X)$. So, just as in the isomorphic case we can infer that

$$\psi(A) = TA'T^{-1} \qquad (A \in \mathcal{A}).$$

This completes the proof of the theorem. $\qquad\qquad\qquad\qquad\square$

It is now easy to prove Corollary 1. We recall that the self-adjoint idempotents in $B(H)$ are called projections.

Proof of Corollary 1. Clearly, Theorem 1 can be applied. According to that result, we have several possibilities concerning the form of ϕ. We give the proof in the case of only one such possibility. The other cases can be handled in a quite similar way. Suppose that H is infinite dimensional. By Theorem 1, we have, for example, a bounded linear operator $T : H \to K$ such that

$$\phi(A) = TAT^{-1} \qquad (A \in \mathcal{A}).$$

Pick an arbitrary rank-1 projection $P \in \mathcal{A}$. By the self-adjointness of ϕ we have

$$(T^{-1})^* PT^* = (TPT^{-1})^* = TPT^{-1}.$$

Since this holds for every rank-1 projection P on H we easily obtain that the vectors $(T^{-1})^* x$, Tx are linearly dependent for every $x \in H$. It needs only an elementary linear algebraic argument to show that in this case T^{-1*} and T are necessarily linearly dependent, that is, we have $T^{-1} = \lambda T^*$ for some $\lambda \in \mathbb{C}$. On the other hand, it follows from (5) that ϕ sends projections to projections. This implies that the scalar λ above is necessarily positive. Denote $U = \sqrt{\lambda} T$. We infer that $U : H \to K$ is an invertible bounded linear operator with $U^{-1} = U^*$. This gives us that U is unitary.

As for the case when H is finite dimensional, we recall the well-known fact that if $h : \mathbb{C} \to \mathbb{C}$ is a ring automorphism of $\mathbb{C}$ for which $h(\overline{\lambda}) = \overline{h(\lambda)}$ ($\lambda \in \mathbb{C}$), then h is either the identity or the conjugation. $\square$

The proof of Theorem 2 will rest on the following lemmas. Recall that an operator $A \in B(H)$ is said to be positive if $\langle Ax, x \rangle \geq 0$ holds for every $x \in H$. In this case we write $A \geq 0$.

Lemma 1. *Let H be a Hilbert space and $A, B \in B(H)$. Suppose that A is positive and $AB + BA = 0$. Then we have $AB = BA = 0$.*

Proof. Since $AB = -BA$, we obtain

$$A^2 B = A(AB) = A(-BA) = (-AB)A = (BA)A = BA^2.$$

That is, A^2 commutes with B. It is well-known that if a positive operator T commutes with an operator, then the same holds true for the positive square root of T. In fact, this follows from the fact that the square root of T is the norm limit of polynomials of T. Therefore, we get that A commutes with B which gives us that $AB = BA = 0$. $\square$

In what follows let $\overline{\mathrm{rng}}\,A$ denote the closure of the range of the operator $A \in B(H)$.

Lemma 2. *Let $\mathcal{A}$ be a standard operator algebra on a Hilbert space. Let $A, B \in \mathcal{A}$ be self-adjoint. Then we have $\overline{\mathrm{rng}}\,A \subset \overline{\mathrm{rng}}\,B$ if and only if for every positive operator $C \in \mathcal{A}$ with $BC = 0$ it follows that $AC = 0$.*

Proof. All we have to do is to note that $\mathcal{A}$ contains all projections of rank 1 and that the condition $\overline{\mathrm{rng}}\,A \subset \overline{\mathrm{rng}}\,B$ is equivalent to the condition that $\ker B \subset \ker A$. $\square$

As for the proof of our next lemma we recall the following useful notation. If $x, y \in H$, then $x \otimes y$ stands for the operator defined by

$$(x \otimes y)(z) = \langle z, y \rangle x \qquad (z \in H).$$

Lemma 3. *Let H be a Hilbert space. If $A \in B(H)$ is such that $TA + AT \geq 0$ holds for every $0 \leq T \in F(H)$, then A is a nonnegative scalar multiple of the identity.*

Proof. First observe that A is positive. Indeed, for every finite rank projection P on H we have $PA + AP \geq 0$. Considering an increasing net of finite rank projections weakly converging to the identity, we obtain that $A + A \geq 0$ and this implies $A \geq 0$.

If $0 \neq x \in H$ is arbitrary, then we have

$$x \otimes Ax + Ax \otimes x \geq 0.$$

It follows from this inequality that for any $y \in H$ we have

$$\langle y, Ax \rangle \langle x, y \rangle + \langle y, x \rangle \langle Ax, y \rangle \geq 0,$$

which implies that

$$\mathrm{Re}(\langle y, Ax \rangle \langle x, y \rangle) \geq 0. \tag{11}$$

We can write $Ax = \lambda x + x^{\perp}$, where $\lambda \in \mathbb{C}$ and $x^{\perp} \in H$ is a vector orthogonal to x. Define $y = \mu x + x^{\perp}$ for an arbitrary $\mu \in \mathbb{C}$. It follows from (11) that

$$\mathrm{Re}(\mu \bar{\lambda} \|x\|^2 + \|x^{\perp}\|^2) \bar{\mu} \geq 0.$$

This implies that

$$|\mu|^2 \, \mathrm{Re}\,\bar{\lambda} \|x\|^2 + \|x^{\perp}\|^2 \, \mathrm{Re}\,\bar{\mu} \geq 0$$

holds for every $\mu \in \mathbb{C}$. It is easy to see that we necessarily have $\|x^{\perp}\|^2 = 0$.

The above observation yields that for every $x \in H$, the vectors Ax and x are linearly dependent. As we have mentined in the proof of Corollary 1, such a local linear dependence implies global linear dependence. Therefore, it follows that A is a scalar multiple of the identity. It is clear that the scalar in question is nonnegative.   □

Lemma 4. *Let $n \in \mathbb{N}$, $n > 1$. Suppose that $\psi : M_n(\mathbb{C}) \to M_n(\mathbb{C})$ is a bijective transformation for which*

$$\psi(A^*) = \psi(A)^*$$
$$\psi(AB + BA) = \psi(A)\psi(B) + \psi(B)\psi(A)$$

holds for every $A, B \in M_n(\mathbb{C})$. Then ψ satisfies (5).

Proof. First observe that ψ preserves positivity in both directions. Indeed, if $A \in M_n(\mathbb{C})$ is positive, then there is a positive $B \in M_n(\mathbb{C})$ such that $2B^2 = A$. We have

$$\psi(A) = \psi(B^*B + BB^*) = \psi(B)^*\psi(B) + \psi(B)\psi(B)^* \geq 0.$$

Since ψ^{-1} has the same properties as ψ, we get that ψ preserves positivity in both directions.

Let $A \in M_n(\mathbb{C})$ be positive. We have

$$\psi(A)\psi(I) + \psi(I)\psi(A) = \psi(2A) \geq 0.$$

Since $\psi(A)$ runs through the positive elements of $M_n(\mathbb{C})$, by Lemma 3 we infer that $\psi(I)$ is a positive scalar multiple of the identity. Denote $\psi(I) = \lambda I$. Consider the transformation $\tilde{\psi} : M_n(\mathbb{C}) \to M_n(\mathbb{C})$ defined by

$$\tilde{\psi}(A) = (1/\lambda)\psi(A) \qquad (A \in M_n(\mathbb{C})).$$

Since we have

$$\psi(A) = \psi(I(A/2) + (A/2)I) = 2\lambda\psi(A/2),$$

one can easily check that $\tilde{\psi}$ satisfies

$$\tilde{\psi}((1/2)(AB + BA)) = (1/2)(\tilde{\psi}(A)\tilde{\psi}(B) + \tilde{\psi}(B)\tilde{\psi}(A)).$$

By Theorem 1, $\tilde{\psi}$ is additive. It follows from the definition of $\tilde{\psi}$ that ψ is also additive which plainly implies the assertion of the lemma.   □

We are now in a position to prove our final result.

Proof of Theorem 2. Just as in the proof of Theorem 1 one can prove that $\phi(0) = 0$.

We next show that ϕ preserves the positive elements in both directions. This can be done quite similarly to the first part of the proof of Lemma 4.

Let $A \in \mathcal{A}$ be positive and $B \in \mathcal{A}$ be arbitrary. Suppose that $AB = BA = 0$. We have

$$0 = \phi(0) = \phi(AB + BA) = \phi(A)\phi(B) + \phi(B)\phi(A).$$

Since $\phi(A)$ is positive, it follows from Lemma 1 that $\phi(A)\phi(B) = \phi(B)\phi(A) = 0$. As ϕ^{-1} has the same properties as ϕ, we find that for any two self-adjoint operators $A, B \in \mathcal{A}$ one of them being positive we have $AB = 0$ if and only if $\phi(A)\phi(B) = 0$. From Lemma 2 we deduce that for any two self-adjoint operators $A, B \in \mathcal{A}$ we have $\overline{\mathrm{rng}}\, A \subset \overline{\mathrm{rng}}\, B$ if and only if $\overline{\mathrm{rng}}\, \phi(A) \subset \overline{\mathrm{rng}}\, \phi(B)$.

Let n be a positive integer. Using the spectral theorem, one can easily verify the following characterization of positive rank-n operators. The positive operator $A \in \mathcal{A}$ is of rank n if and only if there exists a system $A_1, \ldots, A_n \in \mathcal{A}$ of nonzero positive operators such that $\overline{\mathrm{rng}}\, A_k \subset \overline{\mathrm{rng}}\, A$ $(k = 1, \ldots, n)$, $A_k A_l = 0$, $(k \neq l)$ but there is no such system of $n+1$ members. By this characterization, ϕ preserves the positive rank-n operators in both directions.

Let $A \in \mathcal{A}$ be a positive rank-n operator. Then $\phi(A)$ is also positive and is of rank n. Let $B \in \mathcal{A}$ be any operator acting on $H_0 = \overline{\mathrm{rng}}\, A$. We mean by this that B maps H_0 into itself and B is zero on $H_0^\perp$. One can easily verify that $\overline{\mathrm{rng}}\, (B^*B + BB^*) \subset H_0$. Denote $C = \phi(B)$ and $K_0 = \overline{\mathrm{rng}}\, \phi(A)$. It follows that $\overline{\mathrm{rng}}\, (C^*C + CC^*) \subset K_0$. If $k \in K_0^\perp$, then we have $k \in \ker(C^*C + CC^*)$. Since $\langle C^*Ck, k \rangle + \langle CC^*k, k \rangle = 0$, we obtain that $\langle C^*Ck, k \rangle = 0$ and $\langle CC^*k, k \rangle = 0$. It follows that $Ck = 0$ and $C^*k = 0$. Consequently, we get that $C(K_0^\perp) = \{0\}$ and $C(K_0) \subset K_0$. Therefore, we have proved that if $B \in \mathcal{A}$ acts on $\overline{\mathrm{rng}}\, A$, then $\phi(B) \in \mathcal{B}$ acts on $\overline{\mathrm{rng}}\, \phi(A)$.

The argument above gives us that ϕ sends finite rank operators to finite rank operators. Indeed, any finite rank operator can be considered as an operator acting on the range of a positive finite rank operator. Since ϕ^{-1} has the same properties as ϕ, we obtain that ϕ maps $F(H)$ onto $F(H)$ and, identifying the operator algebra over H_0 and K_0 with $M_n(\mathbb{C})$, the map ϕ induces a bijective transformation $\psi : M_n(\mathbb{C}) \to M_n(\mathbb{C})$ for which

$$\psi(T^*) = \psi(T)^*$$
$$\psi(TS + ST) = \psi(T)\psi(S) + \psi(S)\psi(T)$$

for every $T, S \in M_n(\mathbb{C})$. Lemma 4 tells us that ψ satisfies (5) on $M_n(\mathbb{C})$. Since A was arbitrary, it follows that ϕ fulfils (5) on $F(H)$. Now, referring to Corollary 1 we have the form of ϕ on $F(H)$ which can be shown to be valid on the whole $\mathcal{A}$ in a way very similar to the last part of the proof of Theorem 1. $\square$

References

[1] M. Brešar and P. Šemrl, *Mappings which preserve idempotents, local automorphisms, and local derivations*, Canad. J. Math. **45** (1993), 483–496.

[2] I. N. Herstein, *Jordan homomorphisms*, Trans. Amer. Math. Soc. **81** (1956), 331–341.

[3] J. Hakeda, *Additivity of Jordan *-maps on AW*-algebras*, Proc. Amer. Math. Soc. **96** (1986), 413–420.

[4] J. Hakeda and K. Saitô, *Additivity of Jordan *-maps on operator algebras*, J. Math. Soc. Japan **38** (1986), 403–408.

[5] W. S. Martindale III, *When are multiplicative mappings additive?*, Proc. Amer. Math. Soc. **21** (1969), 695–698.

[6] L. Molnár, *On isomorphisms of standard operator algebras*, Studia Math. **142** (2000), 295–302.

[7] L. Molnár, **-semigroup endomorphisms of B(H)*, in I. Gohberg (Edt.), *Operator Theory: Advances and Applications*, Proceedings of the Memorial Conference for Béla Szőkefalvi-Nagy, Szeged, 1999, Birkhäuser, to appear.

[8] L. Molnár, *Orthogonality preserving transformations on indefinite inner product spaces: generalization of Uhlhorn's version of Wigner's theorem*, preprint.

[9] P. Šemrl, *Isomorphisms of standard operator algebras*, Proc. Amer. Math. Soc. **123** (1995), 1851–1855.

BISYMMETRY AND ASSOCIATIVITY TYPE EQUATIONS ON QUASIGROUPS

ON THE FUNCTIONAL EQUATION $S_1(x,y) = S_2(x, T(N(x), y))$

Claudi Alsina

Secció de Matemàtiques i Informàtica, Universitat Politècnica de Catalunya
Avda. Diagonal 649, 08028 Barcelona, Spain
alsina@ea.upc.es

Enric Trillas

Depto. de Inteligencia Artificial, Universidad Politécnica de Madrid.
Campus Montegancedo. 28660 Boadilla del Monte, Spain
trillas@fi.upm.es

Abstract   Motivated by some functional models arising in Fuzzy Logic, we study the functional equation $S_1(x,y) = S_2(x, T(N(x), y))$, where S_1, S_2 are continuous t-conorms, T is a continuous t-norm and N is a strong negation. In doing this, some interesting methods for solving this equation are introduced.

Keywords: associative function, t-norm, t-conorm, strong negation, fuzzy logic

Mathematics Subject Classification (2000): 39B40, 03E72

1.   INTRODUCTION

Our aim in this paper is to deal with the functional equation of Pexider type

$$S_1(x,y) = S_2(x, T(N(x), y)), \qquad (1)$$

where S_2 is a non-strict Archimedean t-conorm, N is a strong negation, S_1 is a continuous t-conorm, and T is a continuous t-norm. This functional equation plays a crucial role in Fuzzy Logic either in generalizing the boolean property $(a \cdot b')' = b + (a' \cdot b')$ or in modelling implication functions satisfying the property $p \to q = p \to p \land q$ (see [5], [6], where

323

Z. Daróczy and Z. Páles (eds.),
Functional Equations - Results and Advances, 323–334.
© 2002 Kluwer Academic Publishers.

some preliminar results were stated). Note that the substitution $y = 1$ into (1) yields $S_2(x, N(x)) = 1$ so clearly S_2 is in the class of non-strict Archimedean t-conorms. But since S_1 and T as associative functions may have various possible representations, the main difficulty that we face in this paper is to solve (1) in all possible classes of functions for the binary operations S_1 and T.

2.   PRELIMINARIES

Following [9], we state the following:

Definition 1. A *t-norm* is a two-place function T from $[0, 1]^2$ into $[0, 1]$ such that the following conditions are satisfied for all x, x', y, y' and z in $[0,1]$:

  (i) Associativity: $T(x, T(y, z)) = T(T(x, y), z)$;
 (ii) Commutativity: $T(x, y) = T(y, x)$;
(iii) Monotonicity: $T(x, y) \leq T(x', y')$ whenever $x \leq x'$ and $y \leq y'$;
 (iv) Unit element: $T(x, 1) = T(1, x) = x$;
  (v) Null element: $T(x, 0) = T(0, x) = 0$.

Note that (v) follows from (iii) and (iv) and that with continuity conditions, (ii) follows from the other conditions.

The most celebrated t-norms are $\text{Min}(x, y) = \text{Minimum}\{x, y\}$, $\text{Prod}(x, y) = x \cdot y$, $W(x, y) = \text{Max}(x + y - 1, 0)$.

Definition 2. A *strict involution* or *strong negation* on $[0,1]$ is a function N from $[0,1]$ onto $[0,1]$ which is strictly decreasing, $N(0) = 1$, $N(1) = 0$ and $N \circ N = j$, where j denotes the identity function on $[0,1]$.

The classical strong negation is $1 - j$, i.e., $(1 - j)(x) = 1 - x$. A representation for strong negations in the form $N(x) = g^{-1}(1 - g(x))$ was given in [10].

Definition 3. A *t-conorm* is a binary operation S on $[0,1]$ such that $S^*(x, y) = 1 - S(1 - x, 1 - y)$ is a t-norm.

Thus, if N is a strong negation and T is a t-norm, $S(x, y) = N(T(N(x), N(y)))$ is a t-conorm.

Let us quote a representation theorem for continuous t-norms in its latest version [7]:

Theorem 1. *Let T be a two-place function from $[0, 1]^2$ into $[0, 1]$ such that:*

(i) $T(x, 0) = T(0, x) = 0$,

(ii) $T(1, 1) = 1$,

(iii) T is associative,

(iv) T is jointly continuous.

Then T admits one of the following representations:

(a) $T(x, y) = \text{Min}(x, y)$;

(b) $T(x, y) = t^{(-1)}(t(x) + t(y))$, where t is a continuous and strictly decreasing function from $[0, 1]$ into $\mathbb{R}^+$, with $t(1) = 0$ and $t^{(-1)}$ is the pseudo-inverse of t;

(c) There exists a countable collection $\{[a_n, b_n]\}$ of non-overlapping, closed, non-degenerate subintervals of $[0, 1]$ and a collection of t-norms T_n each of them representable in the form (b) such that

$$T(x, y) = \begin{cases} a_n + (b_n - a_n) T_n \left(\frac{x - a_n}{b_n - a_n}, \frac{y - a_n}{b_n - a_n} \right), & \text{if } (x, y) \in [a_n, b_n]^2 \\ & \text{for some } n, \\ \text{Min}(x, y), & \text{otherwise.} \end{cases}$$

The previous theorem yields a corresponding representation for all continuous t-conorms. In 1979, M. J. Frank proved a remarkable result. Frank's result concerns the study of which continuous t-norms T and t-conorms S may satisfy the functional equation

$$T(x, y) + S(x, y) = x + y. \tag{*}$$

Theorem 2. *A continuous t-norm T and a t-conorm S satisfy equation (*) if and only if the couple (T, S) has one of the following forms:*

(i) $T_0(x, y) = \text{Min}(x, y)$, $S_0(x, y) = \text{Max}(x, y)$;

(ii) $T_1(x, y) = \text{Prod}(x, y)$, $S_1(x, y) = \text{Prod}^*(x, y)$;

(iii) $T_\infty(x, y) = W(x, y)$; $S_\infty(x, y) = W^*(x, y)$;

(iv) $T_\lambda(x, y) = \log_\lambda[1 + (\lambda^x - 1)(\lambda^y - 1)/(\lambda - 1)]$, $0 < \lambda < \infty$, $\lambda \neq 1$, $S_\lambda(x, y) = T_\lambda^*(x, y)$;

(v) T is representable as an ordinal sum of t-norms each of which is a member of the family $T_\lambda (0 < \lambda \leq \infty)$, and $S(x, y) = x + y - T(x, y)$.

3.   STUDY OF THE FUNCTIONAL EQUATION (1)

Our aim here is to solve (1) when S_1 is a continuous t-conorm, S_2 is a continuous t-conorm, T is a continuous t-norm and N is a strong negation. As we have seen before it follows from (1) that $S_2(N(x), x) = 1$

so S_2 will be a non-strict Archimedean t-conorm representable in the form

$$S_2(x,y) = s_2^{(-1)}(s_2(x) + s_2(y)), \tag{2}$$

where $s_2 : [0,1] \to [0,1]$ is continuous, strictly increasing, $s_2(0) = 0$, $s_2(1) = 1$ and $s_2^{(-1)}(x) = s_2^{-1}(x)$ for x in [0,1], $s_2^{(-1)}(x) = 1$ for $x \geq 1$. Thus, S_2 has its associated strong negation

$$N_2(x) = s_2^{-1}(1 - s_2(x)), \tag{3}$$

and $S_2(x,y) = 1$ if and only if $y \geq N_2(x)$. In particular, since $S_2(x, N(x)) = 1$, we need to have $N \geq N_2$. Note that (1) is a functional equation of Pexider type with two variables x, y and four unknown functions S_2, S_1, T, N.

Lemma 1. *If S_2 and N_2 are given by (2) and (3), respectively, N is a strong negation, T is a continuous t-norm, S_1 is a continuous t-conorm with $S_1(x,y) < 1$ whenever $x,y \neq 1$, and (1) holds, then necessarily $N = N_2$ and T must have one of the following forms:*

  (i) $T = s_2^{-1} \circ \mathrm{Prod} \circ s_2 \times s_2;$
  (ii) $T = s_2^{-1} \circ T_\lambda \circ s_2 \times s_2$, *where T_λ belongs to Frank's family;*
  (iii) $T = s_2^{(-1)} \circ W \circ s_2 \times s_2;$
  (iv) $T = \mathrm{Min}$ *or T is an ordinal sum of Archimedean t-norms of the above type.*

Proof. The commutativity of S_1 and (1) yield at once

$$S_1(x,y) = S_2(x, T(N(x), y)) = S_2(y, T(N(y), x)) = S_1(y, x),$$

and since we are assuming $S_1(x,y) < 1$ for $x, y \neq 1$ we deduce, using (2), that

$$s_2(x) + s_2(T(N(x), y)) = s_2(y) + s_2(T(N(y), x)). \tag{4}$$

Introduce the new variables $u = s_2(y)$ and $v = s_2(N(x))$ into (4). Then, for u, v in (0,1) arbitrary, we obtain

$$\begin{aligned}
\left(s_2 \circ N \circ s_2^{-1}\right)(v) + s_2\left(T(s_2^{-1}(v), s_2^{-1}(u))\right) \\
= u + s_2\left(T(N(s_2^{-1}(u)), N(s_2^{-1}(v)))\right).
\end{aligned} \tag{5}$$

Let $f : [0,1] \to [0,1]$ be defined by

$$f(t) = (s_2 \circ N \circ s_2^{-1})(t). \tag{6}$$

Then f is strictly decreasing continuous and $f(0) = 1$, $f(1) = 0$. Consider now the t-norm:

$$G(u, v) = s_2 \left(T(s_2^{-1}(u), s_2^{-1}(v)) \right)$$

and the t-conorm

$$H(u, v) = (s_2 \circ N)[T((N \circ s_2^{-1}(u), (N \circ s_2^{-1})(v))].$$

By virtue of the equation (5)

$$f(v) + G(u, v) = u + H(u, v),$$

therefore, changing the roles of u, v, we must have also

$$f(u) + G(u, v) = v + H(u, v),$$

whence substraction of the last equalities yield $f(v) - f(u) = u - v$, i.e., $f(v) + v = f(u) + u$, and $f(x) + x$ is a constant function: $f(x) + x = k$. The condition $f(0) = 1$ yields $k = 1$, i.e., $f(x) = 1 - x$, whence by (6), $N(x) = s_2^{-1}(1 - s_2(x)) = N_2(x)$.

Now with $N = N_2$ we go back to (4) to get

$$s_2(x) + s_2(T(s_2^{-1}(1 - s_2(x)), y)) = s_2(y) + s_2(T(s_2^{-1}(1 - s_2(y)), x)),$$

and the substitutions $u = s_2(x)$, $v = 1 - s_2(y)$ yield

$$u + s_2[T(s_2^{-1}(1 - u), s_2^{-1}(1 - v))] = 1 - v + s_2[T(s_2^{-1}(v), s_2^{-1}(u))].$$

Thus the t-norm $T_2(u, v) = s_2 \left[T(s_2^{-1}(u), s_2^{-1}(v)) \right]$ and its associated t-conorm

$$T_2^*(u, v) = 1 - T_2(1 - u, 1 - v)$$

would satisfy Frank's equation

$$T_2(u, v) + T_2^*(u, v) = u + v,$$

i.e., T_2 is completely determined by Theorem 2 and so is T. The lemma follows. □

In the next lemmas we will consider the possible representations for the continuous t-conorm S_1. We begin with the case $S_1 = $ Max where we can apply the previous lemma.

Lemma 2. *Let S_2 and N_2 be given by (2) and (3), respectively. Let N be a strong negation and let T be a continuous t-norm satisfying*

$$S_2(x, T(N(x), y)) = Max(x, y),$$

for all x, y in $[0,1]$. This is possible if and only if T is a non-strict Archimedean t-norm with additive generator $t(x) = 1 - s_2(x)$, i.e., $T = s_2^{(-1)} \circ W \circ s_2 \times s_2$ and $N = N_2$.

Proof. By Lemma 1 we know $N = N_2$. So, for all x, y such that $x, y < 1$, we will have by virtue of (2), (3) and the above equation $S_2(x, T(N_2(x), y)) = \text{Max}(x, y) < 1$, i.e.,

$$s_2(\text{Max}(x, y)) = s_2(x) + s_2(T(N_2(x), y)),$$

whence

$$\begin{aligned} s_2\left(T(a, b)\right) &= s_2(\text{Max}(N_2(a), b)) - s_2(N_2(a)) \\ &= \text{Max}\left(1 - s_2(a), s_2(b)\right) + s_2(a) - 1 \\ &= \text{Max}\left(s_2(a) + s_2(b) - 1, 0\right), \end{aligned}$$

so $T = s_2^{(-1)} \circ W \circ s_2 \times s_2$. The converse is immediate. □

Now we will consider the case where S_1 is a strict t-conorm, i.e., S^* is a strict t-norm.

Lemma 3. *If S_2 and N_2 are given by (2) and (3), respectively, N is a strong negation, T is a continuous t-norm, S_1 is a strict t-conorm and (1) holds, then $N = N_2$ and there are two types of solutions either*
   (a) $T = s_2^{-1} \circ \text{Prod} \circ s_2 \times s_2$; $S_1 = s_2^{-1} \circ \text{Prod}^* \circ s_2 \times s_2$,
or
   (b) $T = s_2^{-1} \circ T_\lambda \circ s_2 \times s_2$; $S_1 = s_2^{-1} \circ T_{1/\lambda}^* \circ s_2 \times s_2$,
where T_λ belongs to Frank's family (see Theorem 2).

Proof. If S_1 is a strict t-conorm, and $S_1(x, y) < 1$ whenever $x \cdot y \neq 1$, we can apply again Lemma 1, so $N = N_2$ and the possibles forms of the t-norm T are determined. But we need to check which one can be consistent with the fact that (1) holds for a strict t-conorm S_1. In cases (i) and (ii) of Lemma 1, we obtain (a) and (b) as claimed in the statement of this lemma. What we need to show now is that cases (iii) and (iv) are not possible. Indeed, in case (iii) $T = s_2^{(-1)} \circ W \circ s_2 \times s_2$ and we would obtain, by Lemma 2, $S_1 = \text{Max}$ contradicting the assumption that S_1 must be a strict t-conorm.

Finally, in case (iv) there would exist for T at least an idempotent element u_0 in $(0,1)$, so we would have for all y

$$\begin{aligned} S_1(N_2(u_0), y) &= S_2\left(N_2(u_0), T(u_0, y)\right) = S_2(N_2(u_0), \text{Min}(u_0, y)) \\ &= \text{Min}\left(S_2(N_2(u_0), u_0), S_2(N_2(u_0), y)\right) \\ &= S_2(N_2(u_0), y), \end{aligned}$$

which is impossible because S_1 is strict but S_2 is not.   □

In order to deal with S_1 as an ordinal sum of Archimedean t-conorms we prove first the following

Lemma 4. *If S_2 and N_2 are given by (2) and (3), respectively, T is a continuous t-norm, N is a strong negation, S_1 is an ordinal sum of Archimedean t-conorms and (1) holds, then $S_1(x,y) < 1$ for all $x, y \neq 1$.*

Proof. Under the above conditions consider the set $O = \{(x,y) \in [0,1]^2 \mid S_1(x,y) = 1\}$. Let us suppose that there exists a point $(x_0, y_0) \in O$ with $x_0, y_0 \neq 1$. Since S_1 is an ordinal sum, then S_1 takes the value 1 in a region of the form

$$P = \{(x,y) \mid x \in [b, 1], \, y \geq g(x)\} \subset O,$$

for some b in (0,1) idempotent element of S_1 and for some function $g : [b, 1] \to [b, 1]$ continuous, strictly decreasing, $g(b) = 1$, $g(1) = b$ and $g = g^{-1}$.

When (x,y) moves in P, the corresponding point $(N(x), y)$ moves in

$$Q = \{(a, b) \mid 0 \leq a \leq N(b), \, b \geq g(N(a))\}$$

and in this case $(x,y) \in P$ and (1) imply $1 = S_1(x,y) = S_2(x, T(N(x), y))$, i.e., $T(N(x), y) \geq N_2(x)$, inequality which implies in particular that T cannot vanish in Q.

Next note that we must have $S_1 = \text{Max}$ on $[b, 1] \times [0, b] \cup [0, b] \times [b, 1]$, whence, for $0 \leq y \leq b \leq x \leq 1$,

$$x = \text{Max}(x,y) = S_1(x,y) = S_2(x, T(N(x), y)),$$

and necessarily $T(N(x), y) = 0$, but when (x,y) moves in $[b, 1] \times [0, b]$, the points $(N(x), y))$ move in $K = [0, N(b)] \times [0, b]$, i.e., $T \equiv 0$ on K. Recalling the representation theorem, T cannot be Min or a strict t-norm because T vanishes on K, T cannot be a non-strict Archimedean t-norm because T does not vanish in Q and T cannot be an ordinal sum because its vanishing Archimedean component would include Q in its zero set and therefore its zero set would cut Q effectively, which is not possible. We get a contradiction and the lemma is proved.   □

Lemma 5. *If S_2 and N_2 are given by (2) and (3), respectively, N is a strong negation, and S_1 is an ordinal sum of Archimedean t-conorms, then there exists no continuous t-norm T satisfying (1).*

Proof. Assume that (1) would hold for some continuous t-norm T. Then (1) and the commutativity of S_1 would yield

$$S_1(x,y) = S_2(x, T(N_2(x), y)) = S_2(y, T(N_2(y), x)),$$

for all x, y in $[0,1]$, however, by the previous Lemma 4, we know that $S_1(x, y) < 1$ whenever $x, y \neq 1$, so we can apply Lemma 1, $N = N_2$ and T becomes completely determined. Thus our last job is to check again which of the t-norms T described in Lemma 1 can satisfy (1) when S_1 is an ordinal sum. Cases (i) and (ii) of Lemma 2 would yield by (1) that S_1 must be strict, so they are not possible. Case (iii) would yield (as seen in Lemma 3) that $S_1(N_2(x), y) = \text{Max}(N_2(x), y)$ and $S_1 = \text{Max}$ which is not possible. Finally case (iv) (following the reasoning used at the end of Lemma 3) would imply

$$S_1(N_2(u_0), y) = S_2(N_2(u_0), y)$$

for any idempotent element u_0 in $(0,1)$ of T. But this last equality is impossible: any vertical section of an ordinal sum of a t-conorm cannot be the vertical section of a non-strict Archimedean t-conorm.   □

Finally, we turn our attention to (1) when S_1 is a non-strict Archimedean t-conorm representable in the form

$$S_1(x, y) = s_1^{(-1)}(s_1(x) + s_1(y)) \tag{7}$$

with associated strong negation

$$N_1(x) = s_1^{-1}(1 - s_1(x)), \tag{8}$$

where $s_1 : [0, 1] \to [0, 1]$ is continuous, strictly increasing, $s_1(0) = 0$, $s_1(1) = 1$. This case becomes complicated because (1) links four unknown functions S_1, S_2, N, T but behind it there are three strong negations N_1, N_2 and N which satisfy the relation $S_2(x, N(x)) = 1$, i.e., $N \geq N_2$ but we may have other complicated relationships, e.g., N_1 may be non-comparable with N, etc. The most general relation of N, N_1, and N_2 is given by the following

Lemma 6. *Let N be a strong negation, let T be a continuous t-norm and let S_2, N_2, S_1, N, be given by* (2), (3), (7), *and* (8), *respectively. If* (1) *holds then, for all x in $[0,1]$,*

$$T(N(x), N_1(x)) = N_2(x). \tag{9}$$

Proof. For $y < N_1(x) = s_1^{-1}(1 - s_1(x))$, i.e., $s_1(x) + s_1(y) < 1$, we have by (1)

$$s_1^{-1}(s_1(x) + s_1(y)) = s_2^{-1}(s_2(x) + s_2(T(N(x), y))).$$

Thus, for $y = N_1(x) - \epsilon$ with $\epsilon > 0$ arbitrarily small,

$$\left(s_2 \circ s_1^{-1}\right)(s_1(x) + s_1(N_1(x) - \epsilon)) = s_2(x) + s_2\left(T(N(x), N_1(x) - \epsilon)\right),$$

and letting ϵ tend to zero, we conclude, using the continuity of the previous functions, that

$$1 = s_2(x) + s_2(T(N(x), N_1(x))),$$

so (9) has been proved.   □

Introducing the functions $f_1, f_2 : [0, 1] \to [0, 1]$

$$f_1(x) = N_1(N(x)) \qquad f_2(x) = N_2(N(x)),$$

we have a couple of continuous strictly increasing functions such that $f_i(0) = 0$, $f_i(1) = 1$, $i = 1, 2$,

$$T(x, f_1(x)) = f_2(x) \leq \operatorname{Min}(x, f_1(x)). \tag{10}$$

Now we deal with S_1 being a non-strict and Archimedean t-conorm, we will distinguish the possible forms of T.

Lemma 7. *If S_2, N_2, S_1, and N_1 are given by (2), (3), (7), and (8), respectively and $T = \operatorname{Min}$, then (1) holds if and only if $S_1 = S_2$, and $N(x) \geq N_2(x) = N_1(x)$ for all x in [0,1].*

Proof. We note that if (1) holds for $T = \operatorname{Min}$, then $S_1(x,y) = S_2(x, \operatorname{Min}(N(x), y)) = \operatorname{Min}(S_2(x, N(x)), S_2(x, y)) = \operatorname{Min}(1, S_2(x, y)) = S_2(x, y)$, so $S_1 = S_2$ and therefore $N_1 = N_2$. The fact $N_2 \leq N$ has already been seen in the first paragraph of page 4. The converse is immediate.   □

Let us note that condition $T(N(x), N_1(x)) = N_2(x)$ yields (10) and since f_1 and f_2 are continuous strictly increasing functions from [0,1] onto [0,1] with $f_1(0) = f_2(0) = 0$ and $f_1(1) = f_2(1) = 1$, we deduce that T cannot vanish in $(0, 1)^2$ so, for non-strict Archimedean t-conorms S_1, S_2, there is no continuous t-norm T satisfying (1) which is Archimedean and non-strict. In the case $N_1 = N_2$ or $N = N_2$ one sees that strict t-norms T satisfying (1) do not exist.

Finally, we consider the case where T can be an ordinal sum of Archimedean t-norms. Then equation (1) does not determine T. The following example shows how one can have ordinal sums of an arbitrary countable set of Archimedean t-norms satisfying (1).

Example 1. Consider $S_1 = S_2 = W^*$, with $s_1(x) = s_2(x) = x$, $N_1(x) = N_2(x) = 1 - x$. Take the strong negation $N > 1 - j$ given by

$$N(x) = \begin{cases} (1 + \sqrt{1 - 4x^2})/2, & \text{if } 0 \leq x \leq 1/2, \\ \sqrt{x(1-x)}, & \text{if } 1/2 \leq x \leq 1. \end{cases}$$

Let T be an ordinal sum of Archimedean t-norm such that $T(x,y) =$ Min(x,y) at least for all (x,y) such that $y \geq 1-N(x)$ or $y \leq (1-N)^{-1}(x)$ and with all the squares along the diagonal (where T is Archimedean) located in the region $\{(x,y) \mid (1-N)^{-1}(x) \leq y \leq 1 - N(x)\}$. Then (1) holds independently of the Archimedean components of T.

In fact, by construction, $T(N(x), 1-x) = 1-x$ so (9) is satisfied. We need to check that $W^*(x,y) = W^*(x, T(N(x), y))$, i.e.,

$$\text{Min}(x + y, 1) = \text{Min}(x + T(N(x), y), 1). \tag{11}$$

When $x+y \leq 1$ we have $y \leq 1-x \leq N(x)$, so $N(y) \geq x$ and $(1-N)(y) = 1 - N(y) \leq 1 - x \leq N(x)$, and, therefore, $y \leq (1-N)^{-1}(N(x))$. Thus $T(N(x), y) = \text{Min}(N(x), y) = y$ and (11) follows. When $x + y > 1$ then $y > 1 - x$ and, since $T(N(x), 1-x) = 1-x$, we obtain $x + T(N(x), y) \geq x + T(N(x), 1-x) = x + 1 - x = 1$ and (11) holds.

In the very special case $N = N_2$ one can have the following result:

Lemma 8. *Let S_2 and N_2 be given by (2) and (3); let S_1 and N_1 be given by (7) and (8) and assume that $N = N_2$. Then (1) holds if and only if $S_1 = S_2$ is an arbitrary non-strict Archimedean t-conorm, $N_1 = N_2 = N$ is its associated strong negation and $T = $ Min.*

Proof. It is obvious that if $T = $ Min, $S_1 = S_2$, $N_1 = N_2 = N$ then (1) holds. Conversely, assume that (1) holds and $N = N_2$, i.e., $f_2(x) = x$, therefore, $T(x, f_1(x)) = x \leq f_1(x)$ so $T(x, f_1(x)) = x = \text{Min}(f_1(x), x)$. If $N_1 = N = N_2$, then $f_1(x) = x$, $T(x,x) = x$ for all x and by Theorem 1, $T = $ Min, implying $S_1 = S_2$. So let us consider the possibility that $N_1 > N$, i.e., $f_1(x) \geq x$ for all x but $f_1(x_0) > x_0$ for some x_0, then we know that $T = $ Min on the region $B = [0,1]^2 \backslash A$, where $A = \{(x,y) \in [0,1]^2 \mid f_1^{-1}(x) \leq y \leq f_1(x)\}$, so we need to face the possibility that T is a special ordinal sum with $T = $ Min at least on B. Let $C = \{(x,y) \mid T(x,y) = \text{Min}(x,y)\}$. Then $B \subsetneq C$ and if $[0,1]\backslash E(T) = \bigcup_{j\in J} (a_j, b_j)$ we need to have $[0,1]^2 \backslash C = \bigcup_{j \in J}(a_j, b_j)^2$, where J is at most countable and $E(T)$ denotes the closed set of idempotent elements of T. Pick up one of these boxes $[a_{j_0}, b_{j_0}]^2$ and consider, for $\epsilon > 0$, points of the form

$$(x_0, y_\epsilon) = \left(\frac{a_{j_0} + b_{j_0}}{2}, b_{j_0} + \epsilon \right) \in C.$$

We can choose values $\epsilon > 0$ such that

$$x_0 = \frac{a_{j_0} + b_{j_0}}{2} < y_\epsilon = b_{j_0} + \epsilon < f_1 \left(\frac{a_{j_0} + b_{j_0}}{2} \right) = f_1(x_0) \tag{12}$$

then $T(x_0, y_\epsilon) = \mathrm{Min}(x_0, y_\epsilon) = x_0$, $y_\epsilon \geq x_0 = N_2(N_2(x_0))$, i.e., $S_2(N_2(x_0), y_\epsilon) = 1$ and, consequently,

$$1 = S_2(N_2(x_0), x_0) = S_2(N_2(x_0), \mathrm{Min}(x_0, y_\epsilon))$$
$$= S_2(N_2(x_0), T(x_0, y_\epsilon)) = S_1(N_2(x_0), y_\epsilon)$$

i.e., with $N = N_2$, $y_\epsilon \geq N_1(N(x_0)) = f_1(x_0)$ contradicting (12). $\square$

From all the above lemmas our main result follows at once.

Theorem 3. *Let S_2 be a non-strict Archimedean t-conorm given by (2) with additive generator s_2 and let N_2 be its associated strong negation (3). Let N be a strong negation, T a continuous t-norm, and S_1 a continuous t-conorm. Then, the solutions of (1) can be described as follows*

(i) *If $S_1 = \mathrm{Max}$, then $T = s_2^{(-1)} \circ W \circ s_2 \times s_2$, $N = N_2$;*

(ii) *If S_1 is strict, then either $S_1 = s_2^{-1} \circ \mathrm{Prod}^* \circ s_2 \times s_2$, $T = s_2^{-1} \circ \mathrm{Prod} \circ s_2 \times s_2$, $N = N_2$, or $S_1 = s_2^{-1} \circ T_{1/\lambda}^* \circ s_2 \times s_2$, $T = s_2^{-1} \circ T_\lambda \circ s_2 \times s_2$; $N = N_2$;*

(iii) *In the case that S_1 is a non-strict Archimedean t-conorm and T is an ordinal sum of Archimedean t-norms then (1) can have solutions, where T is not determined by S_1 and S_2 and there are cases where (1) has no Archimedean t-norms T as solutions. When $T = \mathrm{Min}$ then (1) yields $S_1 = S_2$ and $N_1 = N_2 \leq N$. Moreover, if $N = N_2$, necessarily $T = \mathrm{Min}$, $S_1 = S_2$, and $N_1 = N_2$.*

4.   ACKNOWLEDGEMENT

The authors want to thank the referees for their interesting remarks.

References

[1] J. Aczél, *Lectures on Functional Equations and Their Applications*, Academic Press, New York–London, 1966.

[2] C. Alsina, E. Trillas, and L. Valverde, *On some logical connectives for Fuzzy Set Theory*, J. Math. Anal. Appl. **93** (1983), 15–26.

[3] C. Alsina, *As you like them: connectives in Fuzzy Logic*, Proc. IS-MVL 96, Santiago de Compostela, 1996, pp. 1–7.

[4] C. Alsina, *On connectives in Fuzzy Logic satisfying the condition $S(T_1(x, y), T_2(x, N(y))) = x$*, Proc. FUZZ'IEEE-97, Barcelona, 1997, pp. 149–153.

[5] C. Alsina and E. Trillas, *On (S, N)-Implications in Fuzzy Logic consistent with T-conjunctions*, Proc. 1999 EUSFLAT-ESTYLF Joint Conf., Univ. Illes Balears, Palma, (1999), pp. 425–428.

[6] C. Alsina and E. Trillas, *On (S, N)-implications in fuzzy logic consistent with T-conjunctions*, Proc. EUSFLAT Conf., 1999, pp. 425–428.

[7] C. Alsina, M. J. Frank, and B. Schweizer, *A primer of t-norms. Associative functions on real intervals*, in preparation.

[8] J. C. Fodor and M. Roubens, *Fuzzy preference modelling and multicriteria decision support*, Kluwer, Dordrecht, 1994.

[9] B. Schweizer and A. Sklar, *Probabilistic Metric Spaces*, Elsevier North-Holland, New York, 1983.

[10] E. Trillas, *Sobre funciones de negación en la teoría de los subconjuntos difusos*, (in spanish), Stochastica **3** (1979), 47-60. [Reprinted (english version) in *Advances in Fuzzy Logic*, (eds. by S. Barro et al.), Publicacions Universidade de Santiago de Compostela, 1998, pp. 31–43.]

[11] E. Trillas and L. Valverde, *On implication and indistinguishability in the setting of Fuzzy Logic*, in *Management Decision Support System using Fuzzy Set and Possibility Theory*, (eds. by R. R. Yager and J. Kacpryck), North-Holland, 1985, pp. 198–212.

GENERALIZED ASSOCIATIVITY ON RECTANGULAR QUASIGROUPS

Aleksandar Krapež

Matematički Institut SANU,

Beograd, Yugoslavia

sasa@mi.sanu.ac.yu

Dedicated to the memory of Professor Dr Aleksandar Kron

Abstract   A type of groupoid called a *rectangular quasigroup* is defined as a direct product of a left zero semigroup, a quasigroup and a right zero semigroup. We give three different axiom systems for these groupoids. Some important properties of rectangular quasigroups are derived, the solvability of the word problem among them.

Finally, it is proved that under a special condition, all four rectangular quasigroups satisfying the generalized associativity equation are isotopic to the same rectangular group. This result generalizes the Four Quasigroups Theorem but is incomparable to J. Aczél's generalization given in his book [1].

Keywords: rectangular quasigroup, axiomatization, variety, word problem, functional equation of generalized associativity

Mathematics Subject Classification (2000): 39B52, 20N02, 03C05, 08A50

1.   INTRODUCTION

One of the prominent structures in the semigroup theory is the so called *rectangular group* (see for example [6]). It is a direct product of a left zero semigroup (satisfying the identity $xy = x$), a group and a right

---

[0]This paper is financed by the Ministry of Science and Technology of R. Serbia through project 04M03.

335

Z. Daróczy and Z. Páles (eds.),
Functional Equations - Results and Advances, 335–349.
© 2002 Kluwer Academic Publishers.

zero semigroup (which satisfies the identity $xy = y$). A common generalization of a rectangular group and a quasigroup called a *rectangular quasigroup* is defined.

Three different axiom systems are given for these groupoids proving that the class of all rectangular quasigroups is a variety, although in a language extended by the two division operations \ and /.

The functional equation of generalized associativity on rectangular quasigroups is solved in some special cases.

2.   RECTANGULAR QUASIGROUPS

Of the several possible ways to define rectangular quasigroups we choose the following:

Definition 1. Groupoid S is a *rectangular quasigroup* iff it is isomorphic to the direct product of a left zero semigroup, a quasigroup and a right zero semigroup.

We are now faced with the problem of the axiomatization of the class of all rectangular quasigroups. For that we have at our disposal the standard method of R. A. Knoebel [5]. Adjusting the types of the left/right zero semigroups to that of (equational) quasigroups, we define $x \backslash y = x/y = xy$ in both of them. Different definition of \ and / would affect the form of the axioms for rectangular quasigroups. As it is, the resulting axiom system (K) consists of the following 20 identities:

$$(xy) \backslash (xy)(tt/t) = x \backslash x(tt/t) \tag{1}$$

$$(x \backslash y) \backslash (x \backslash y)(tt/t) = x \backslash x(tt/t) \tag{2}$$

$$(x/y) \backslash (x/y)(tt/t) = x \backslash x(tt/t) \tag{3}$$

$$t \backslash t((x \backslash xy)t/t) = t \backslash t(yt/t) \tag{4}$$

$$t \backslash t((x(x \backslash y) \cdot t)/t) = t \backslash t(yt/t) \tag{5}$$

$$t \backslash t((xy/y)t/t) = t \backslash t(xt/t) \tag{6}$$

$$t \backslash t(((x/y)y \cdot t)/t) = t \backslash t(xt/t) \tag{7}$$

$$t \backslash t((t \cdot xy)/(xy)) = t \backslash t(ty/y) \tag{8}$$

$$t \backslash t(t(x \backslash y)/(x \backslash y)) = t \backslash t(ty/y) \tag{9}$$

$$t \backslash t(t(x/y)/(x/y)) = t \backslash t(ty/y) \tag{10}$$

$$(x \backslash x(tt/t)) \backslash (x \backslash x(tt/t))((t \backslash t(xt/t))(t \backslash t(tx/x))/(t \backslash t(tx/x))) = x \tag{11}$$

$$(x \backslash x(tt/t))(y \backslash y(tt/t)) \backslash ((x \backslash x(tt/t))(y \backslash y(tt/t)) \cdot (tt/t)) \\ = (xy) \backslash (xy)(tt/t) \tag{12}$$

$$((x \backslash x(tt/t)) \backslash (y \backslash y(tt/t))) \backslash (((x \backslash x(tt/t)) \backslash (y \backslash y(tt/t))) \cdot (tt/t)) \\ = (x \backslash y) \backslash (x \backslash y)(tt/t) \tag{13}$$

$$((x \backslash x(tt/t))/(y \backslash y(tt/t))) \backslash (((x \backslash x(tt/t))/(y \backslash y(tt/t))) \cdot (tt/t)) \\ = (x/y) \backslash (x/y)(tt/t) \tag{14}$$

$$t \backslash t(((t \backslash t(xt/t)) \cdot (t \backslash t(yt/t)))t/t) = t \backslash t((xy \cdot t)/t) \tag{15}$$

$$t \backslash t(((t \backslash t(xt/t)) \backslash (t \backslash t(yt/t)))t/t) = t \backslash t((x \backslash y)t/t) \tag{16}$$

$$t \backslash t(((t \backslash t(xt/t))/(t \backslash t(yt/t)))t/t) = t \backslash t((x/y)t/t) \tag{17}$$

$$t \backslash t((t \cdot ((t \backslash t(tx/x)) \cdot (t \backslash t(ty/y))))/((t \backslash t(tx/x)) \cdot (t \backslash t(ty/y)))) \\ = t \backslash t((t \cdot xy)/(xy)) \tag{18}$$

$$t \backslash t((t \cdot ((t \backslash t(tx/x)) \backslash (t \backslash t(ty/y))))/((t \backslash t(tx/x)) \backslash (t \backslash t(ty/y)))) \\ = t \backslash t((t \cdot xy)/(xy)) \tag{19}$$

$$t \backslash t((t \cdot ((t \backslash t(tx/x))/(t \backslash t(ty/y))))/((t \backslash t(tx/x))/(t \backslash t(ty/y)))) \\ = t \backslash t((t \cdot xy)/(xy)). \tag{20}$$

Following Knoebel we define:

Definition 2. $p(x, y, z) = x \backslash x(yz/z)$, $p_1(x, t) = p(x, t, t) = x \backslash x(tt/t)$, $p_2(x, t) = p(t, x, t) = t \backslash t(xt/t)$, $p_3(x, t) = p(t, t, x) = t \backslash t(tx/x)$.

Using these functions we can simplify the form of the axioms of the system (K). For example axiom (1) becomes $p_1(xy, t) = p_1(x, t)$, axiom (11) becomes $p(p_1(x, t), p_2(x, t), p_3(x, t)) = x$ etc.

Theorem 1. *The formulas* (12)–(14) *and* (18)–(20) *follow from* (1)–(3) *and* (8)–(10).

Proof. If we denote any of the operations $\cdot, \backslash, /$ by $*$, then the formulas (12)–(14) may be written as $p_1(p_1(x,t) * p_1(y,t),t) = p_1(x * y,t)$ and formulas (18)–(20) as $p_3(p_3(x,t) * p_3(y,t),t) = p_3(xy,t)$.

The proof of (12)–(14):

$$
\begin{aligned}
p_1(p_1(x,t) * p_1(y,t),t) &= p_1(p_1(x,t),t) &&\text{(by (1), (2) or (3))}\\
&= p_1(x\backslash x(tt/t),t)\\
&= p_1(x,t) &&\text{(by (2))}\\
&= p_1(x * y,t) &&\text{(by (1), (2) or (3)).}
\end{aligned}
$$

Analogously, (18)–(20) can be proved.   □

Even in the shorter version the axiom system (K) is not very elegant. Therefore we propose another one, called $(\square Q)$, which consists of the following identities:

$$x\backslash xx = x \tag{21}$$

$$xx/x = x \tag{22}$$

$$x(x\backslash y) = x\backslash xy \tag{23}$$

$$(x/y)y = xy/y \tag{24}$$

$$xy\backslash(xy \cdot z) = x\backslash xz \tag{25}$$

$$(x\backslash y)\backslash(x\backslash y)z = x\backslash xz \tag{26}$$

$$(x/y)\backslash(x/y)z = x\backslash xz \tag{27}$$

$$x(y\backslash yz) = xz \tag{28}$$

$$(xy/y)z = xz \tag{29}$$

$$(x \cdot yz)/yz = xz/z \tag{30}$$

$$x(y\backslash z)/(y\backslash z) = xz/z \tag{31}$$

$$x(y/z)/(y/z) = xz/z \tag{32}$$

$$x\backslash x(yz/z) = (x\backslash xy)z/z \tag{33}$$

$$(x\backslash xy)z = x\backslash(x \cdot yz) \tag{34}$$

$$x(yz/z) = (xy \cdot z)/z. \tag{35}$$

Yet another axiom system (G) is given below:

Define relations $\mathcal{H}$ and $\mathcal{P}$ by: $x\mathcal{H}y \leftrightarrow x/(y\backslash x) = y$, $x\mathcal{P}y \leftrightarrow x/(y\backslash x) = x$. By Δ and $\square$ we denote the diagonal i.e. the equality relation and the full relation respectively. The axioms of (G) are:

$$\mathcal{H} \text{ is a congruence} \tag{36}$$

$$\mathcal{P} \text{ is a congruence} \tag{37}$$

$$\mathcal{H} \cap \mathcal{P} = \Delta \tag{38}$$

$$\mathcal{H}\mathcal{P} = \mathcal{P}\mathcal{H} \tag{39}$$

$$\mathcal{H} \vee \mathcal{P} = \square \tag{40}$$

$$xx \; \mathcal{H} \; x \tag{41}$$

$$x \cdot yz \; \mathcal{H} \; xz \tag{42}$$

$$xy \cdot z \; \mathcal{H} \; xz \tag{43}$$

$$x\backslash y \; \mathcal{H} \; xy \tag{44}$$

$$x/y \; \mathcal{H} \; xy \tag{45}$$

$$x\backslash xy \; \mathcal{P} \; y \tag{46}$$

$$x(x\backslash y) \; \mathcal{P} \; y \tag{47}$$

$$xy/y \; \mathcal{P} \; x \tag{48}$$

$$(x/y)y \; \mathcal{P} \; x. \tag{49}$$

In Theorem 2 below we shall prove that the axiom systems $(\square Q)$ and (G) are equivalent to (K).

We use the following notation: $\mathcal{L}$ for the class of all left zero semigroups, $\mathcal{R}$ for the class of all right zero semigroups, $\mathcal{B}$ for the class of all rectangular bands, $\mathcal{Q}$ for the class of all quasigroups and $[\square Q]$ for the class of all rectangular quasigroups.

Theorem 2. *The following conditions for the groupoid S are equivalent:*

a) S *is a rectangular quasigroup*

b) $S \simeq L \times Q \times R, \quad L \in \mathcal{L}, Q \in \mathcal{Q}, R \in \mathcal{R}$

c) $S \simeq B \times Q, \quad B \in \mathcal{B}, Q \in \mathcal{Q}$

d) S *satisfies* (K)

e) S *satisfies* $(\square Q)$

f) S *satisfies* (G).

Proof. a) and b) are equivalent by the definition of rectangular quasigroup. Every rectangular band is isomorphic to the direct product of a left and a right zero semigroup and therefore b) and c) are equivalent. As previously noted the equivalence of b) and d) follows from the Theorem of Knoebel [5].

d) $\Rightarrow$ e) requires just tedious checking. We prove several axioms of the system $(\square Q)$.
The proof of formula (21):

$$\begin{aligned}
x\backslash xx &= p(p_1(x\backslash xx, t), p_2(x\backslash xx, t), p_3(x\backslash xx, t)) && \text{(by (11))} \\
&= p(p_1(x, t), p_2(x, t), p_3(xx, t)) && \text{(by (2), (4), and (9))} \\
&= p(p_1(x, t), p_2(x, t), p_3(x, t)) && \text{(by (8))} \\
&= x && \text{(by (11))}.
\end{aligned}$$

Similarly we can prove (22).

The proof of formula (23):

$$
\begin{aligned}
x(x\backslash y) &= p(p_1(x(x\backslash y),t), p_2(x(x\backslash y),t), p_3(x(x\backslash y),t)) && \text{(by (11))}\\
&= p(p_1(x,t), p_2(y,t), p_3(x\backslash y,t)) && \text{(by (1), (5), (8))}\\
&= p(p_1(x,t), p_2(y,t), p_3(y,t)) && \text{(by (9))}\\
&= p(p_1(x,t), p_2(y,t), p_3(xy,t)) && \text{(by (8))}\\
&= p(p_1(x\backslash xy,t), p_2(x\backslash xy,t), p_3(x\backslash xy,t)) && \text{(by (2), (4), (9))}\\
&= x\backslash xy.
\end{aligned}
$$

We can prove the formula $(x*y)\backslash(x*y)t = x\backslash xt$ (which represents any of the axioms (25), (26), (27)) substituting t for tt/t in (1), (2) or (3). The proof of formula (L): $p_2(x(y\backslash yz),t) = p_2(xz,t)$:

$$
\begin{aligned}
p_2(x(y\backslash yz),t) &= p_2(p_2(x,t)\cdot p_2(y\backslash yz,t),t) && \text{(by (15))}\\
&= p_2(p_2(x,t)\cdot p_2(z,t),t) && \text{(by (4))}\\
&= p_2(xz,t) && \text{(by (15))}.
\end{aligned}
$$

The proof of formula (28):

$$
\begin{aligned}
x(y\backslash yz) &= p(p_1(x(y\backslash yz),t), p_2(x(y\backslash yz),t), p_3(x(y\backslash yz),t)) && \text{(by (11))}\\
&= p(p_1(x,t), p_2(xz,t), p_3(y\backslash yz,t)) && \text{(by (1), (L), (8))}\\
&= p(p_1(xz,t), p_2(xz,t), p_3(yz,t)) && \text{(by (1), (9))}\\
&= p(p_1(xz,t), p_2(xz,t), p_3(z,t)) && \text{(by (8))}\\
&= p(p_1(xz,t), p_2(xz,t), p_3(xz,t)) && \text{(by (8))}\\
&= xz && \text{(by (11))}.
\end{aligned}
$$

The proof of formula (30):

$$
\begin{aligned}
(t\cdot xy)/xy &= p(p_1((t\cdot xy)/xy,t), p_2((t\cdot xy)/xy,t), p_3((t\cdot xy)/xy,t)) \\
&\hspace{7cm} \text{(by (11))}\\
&= p(p_1(t\cdot xy,t), p_2(t,t), p_3(xy,t)) && \text{(by (3), (6), (10))}\\
&= p(p_1(t,t), p_2(ty/y,t), p_3(y,t)) && \text{(by (1), (6), (8))}\\
&= p(p_1(ty,t), p_2(ty/y,t), p_3(ty/y,t)) && \text{(by (1), (10))}\\
&= p(p_1(ty/y,t), p_2(ty/y,t), p_3(ty/y,t)) && \text{(by (3))}\\
&= ty/y && \text{(by (11))}.
\end{aligned}
$$

The proof of formula (33):

$$x\backslash x(yz/z) = p(p_1(x\backslash x(yz/z), t), p_2(x\backslash x(yz/z), t), p_3(x\backslash x(yz/z), t))$$
$$\text{(by (11))}$$
$$= p(p_1(x, t), p_2(yz/z, t), p_3(x(yz/z), t)) \quad \text{(by (2), (4), (9))}$$
$$= p(p_1(x\backslash xy, t), p_2(y, t), p_3(yz/z, t)) \quad \text{(by (2), (6), (8))}$$
$$= p(p_1((x\backslash xy)z, t), p_2(x\backslash xy, t), p_3(z, t)) \quad \text{(by (1), (4), (10))}$$
$$= p(p_1((x\backslash xy)z/z, t), p_2((x\backslash xy)z/z, t), p_3((x\backslash xy)z/z, t))$$
$$\text{(by (3), (6), (10))}$$
$$= (x\backslash xy)z/z \quad \text{(by (11))}.$$

The proof of formula (34):

$$(x\backslash xy)z = p(p_1((x\backslash xy)z, t), p_2((x\backslash xy)z, t), p_3((x\backslash xy)z, t)) \quad \text{(by (11))}$$
$$= p(p_1(x\backslash xy, t), p_2(p_2(x\backslash xy, t)p_2(z, t), t), p_3(z, t))$$
$$\text{(by (1), (15), (8))}$$
$$= p(p_1(x, t), p_2(p_2(y, t)p_2(z, t), t), p_3(z, t)) \quad \text{(by (1), (4))}$$
$$= p(p_1(x, t), p_2(yz, t), p_3(z, t)) \quad \text{(by (15))}.$$

Also:

$$x\backslash(x \cdot yz) = p(p_1(x\backslash(x \cdot yz), t), p_2(x\backslash(x \cdot yz), t), p_3(x\backslash(x \cdot yz), t))$$
$$\text{(by (11))}$$
$$= p(p_1(x, t), p_2(yz, t), p_3(x \cdot yz, t)) \quad \text{(by (1), (4), (9))}$$
$$= p(p_1(x, t), p_2(yz, t), p_3(yz, t)) \quad \text{(by (8))}$$
$$= p(p_1(x, t), p_2(yz, t), p_3(z, t)) \quad \text{(by (8))},$$

and consequently $(x\backslash xy)z = x\backslash(x \cdot yz)$.

e) $\Rightarrow$ d) is also straightforward. Again we prove just some of the formulas.

Formula $(x*y)\backslash(x*y)(tt/t) = x\backslash x(tt/t)$ represents any of the formulas (1), (2), (3). We can prove it from (25), (26) or (27), using (22).

The proof of formula (4):

Changing variables in (28), we get: $t(x\backslash xy) = ty$.

Therefore $(t\backslash t(x\backslash xy))t/t = (t\backslash ty)t/t$ which yields (4) by applying (33) on both sides.

Formulas (8), (9) and (10) follow easily from (21), (30), (31) and (32).

It is easy to prove that $p_1(x, t) = x\backslash xt$ and $p_3(x, t) = tx/x$.

The proof of formula (11):

$$p(p_1(x,t), p_2(x,t), p_3(x,t)) = p_1(x,t) \backslash p_1(x,t)(p_2(x,t)p_3(x,t)/p_3(x,t))$$

$$\text{(by definition)}$$

$$= (x\backslash xt)\backslash(x\backslash xt)(p_2(x,t)p_3(x,t)/p_3(x,t))$$

$$= x\backslash x(p_2(x,t)p_3(x,t)/p_3(x,t)) \quad \text{(by (26))}$$

$$= x\backslash x(p_2(x,t)(tx/x)/(tx/x))$$

$$= x\backslash x(p_2(x,t)x/x)) \quad \text{(by (32))}$$

$$= x\backslash x((t\backslash t(xt/t))x/x)$$

$$= x\backslash x((xt/t)x/x) \quad \text{(by (4))}$$

$$= x\backslash x(xx/x) \quad \text{(by (6))}$$

$$= x\backslash xx \quad \text{(by (22))}$$

$$= x \quad \text{(by (21))}.$$

The proof of formula (15):

$$p_2(p_2(x,t)p_2(y,t),t)) = t\backslash t((p_2(x,t)p_2(y,t)\cdot t)/t)$$

$$= t\backslash t(p_2(x,t)(p_2(y,t)t/t)) \quad \text{(by (35))}$$

$$= (t\backslash tp_2(x,t))((t\backslash t(yt/t))t/t) \quad \text{(by (34))}$$

$$= (t\backslash t(t\backslash t(xt/t)))(((t\backslash ty)t/t)t/t) \quad \text{(by (33))}$$

$$= (t\backslash t(xt/t))((t\backslash ty)t/t) \quad \text{(by (28), (29))}$$

$$= ((t\backslash tx)t/t)(t\backslash t(yt/t)) \quad \text{(by (33))}$$

$$= (t\backslash tx)(yt/t) \quad \text{(by (29), (28))}$$

$$= t\backslash(t\cdot x(yt/t)) \quad \text{(by (34))}$$

$$= t\backslash t((xy\cdot t)/t) \quad \text{(by (35))}$$

$$= p_2(xy,t).$$

c) and *f)* are equivalent by the theorems 19.2 and 19.3 of G. Birkhoff (see [4], pp 119–120). $S/\mathcal{H}$ is a rectangular band by (41)–(43) and $S/\mathcal{P}$ is a quasigroup by (46)–(49). $\qquad\qquad\Box$

Definition 3. *head(t) (tail(t))* is the first (last) variable of the term t.

Theorem 3. $u = v$ *is true in all rectangular quasigroups iff head(u) = head(v), tail(u) = tail(v) and* $u = v$ *is true in all quasigroups.*

Using the result of T. Evans [3] we get:

Corollary 1. *The word problem for rectangular quasigroups is solvable.*

Also:

Corollary 2. *Let* $*$ *and* $\circ$ *be any of the operations* $\cdot, \backslash, /$. *If* $u = v$ *is true in all quasigroups then* $x*(u \circ y) = x*(v \circ y)$ *and* $(x*u) \circ y = (x*v) \circ y$ *is true in all rectangular quasigroups.*

Corollary 3. *If* $u = v$ *is true in all quasigroups then* $x/(u \backslash x) = (x/v) \backslash x$ *is true in all rectangular quasigroups.*

Definition 4. This is just a reminder of the standard notation.
$Sub(S)$ is the lattice of subalgebras of S.
$Sub^0(S)$ is the lattice of subalgebras of S with the empty set added as the smallest element (used when two subalgebras have an empty intersection).
2 is the two element lattice.
2^S is the lattice of all subsets of S (including the empty set $\emptyset$).
$Con(S)$ – the lattice of congruences of S.
$Eq(S)$ – the lattice of equivalences of S.
$Hom(S,T)$ – the set of homomorphisms from S to T.
$End(S)$ – the monoid of endomorphisms of S.
$Aut(S)$ – the group of automorphisms of S.
$Var(\mathcal{K})$ is the lattice of varieties of a class $\mathcal{K}$ of algebras.

Corollary 4. *For all* $L, M \in \mathcal{L}$; $Q, Q' \in \mathcal{Q}$ *and* $R, N \in \mathcal{R}$:

a) $Sub^0(L \times Q \times R) \simeq ((2^L \setminus \{\emptyset\}) \times Sub(Q) \times (2^R \setminus \{\emptyset\})) \cup \{(\emptyset, \emptyset, \emptyset)\}$.
b) $Con(L \times Q \times R) \simeq Eq(L) \times Con(Q) \times Eq(R)$.
c) $Hom(L \times Q \times R, M \times Q' \times N) = M^L \times Hom(Q, Q') \times N^R$.
d) $End(L \times Q \times R) \simeq L^L \times End(Q) \times R^R$.
e) $Aut(L \times Q \times R) \simeq S_{|L|} \times Aut(Q) \times S_{|R|}$.

Corollary 5.

$$FreeRectangularQuasigroup(n) \simeq L_n \times FreeQuasigroup(n) \times R_n.$$

The notions $FreeRectangularQuasigroup(n)$ and $FreeQuasigroup(n)$ are just what the names indicate: appropriate free structures with n generators. $L_n(R_n)$ is the unique n-element left (right) zero semigroup – which also happen to be free.

Corollary 6. $Var([\square Q]) \simeq \mathbf{2} \times Var(\mathcal{Q}) \times \mathbf{2}$.

Theorem 4. *The equation* $ax = b$ *is consistent on a rectangular quasigroup* S *iff* $a(a \backslash b) = b$ *and then its general solution is given by:* $x = p \backslash p(a \backslash b)$, $p \in S$.
Consistent equation $ax = b$ *has exactly as many solutions as each of the equations* $ax = a, bx = b$.

Of course, the dual theorem is also true.

Axiom systems (K) and (G) are dependent and probably is $(\square Q)$. Therefore:

Problem 1. *Give an independent axiom system for the class of all rectangular quasigroups.*

3.   GENERALIZED ASSOCIATIVITY ON RECTANGULAR QUASIGROUPS

Aczél, Belousov and Hosszú solved in 1960 the generalized associativity equation on quasigroups (see [2]):

Theorem 5. (Four Quasigroups Theorem) *If the four quasigroups A, B, C, D (on a set $S \neq \emptyset$) satisfy the generalized associativity equation:*

$$A(x, B(y, z)) = C(D(x, y), z) \tag{50}$$

then they are all isotopic to the same group. The general solution of (50) is given by:

$$\begin{cases} A(x, y) = A_1 x \cdot A_2 y, \\ B(x, y) = A_2^{-1}(A_2 B_1 x \cdot A_2 B_2 y), \\ C(x, y) = C_1 x \cdot C_2 y, \\ D(x, y) = C_1^{-1}(C_1 D_1 x \cdot C_1 D_2 y), \end{cases} \tag{51}$$

where $\cdot$ is an arbitrary group on S and $A_1, A_2, B_1, B_2, C_1, C_2, D_1, D_2$ are arbitrary permutations of S satisfying:

$$\begin{cases} A_1 = C_1 D_1 \\ A_2 B_1 = C_1 D_2 \\ A_2 B_2 = C_2. \end{cases} \tag{52}$$

Since then, this remarkable result has been reproved, generalized and cited many times. One of the generalizations is the following theorem of Aczél (see [1], p. 311).

Theorem 6. *If there exist a constant a, so that each individual equation*

$$f_1(s) \equiv F(s, a) = t, \qquad g_1(s) \equiv G(s, a) = t, \qquad g_2(s) \equiv G(a, s) = t,$$

$$h_2(s) \equiv H(a, s) = t, \qquad k_2(s) \equiv K(a, s) = t,$$

can be solved uniquely with respect to s on a set Q, then on Q, every solution of functional equation

$$F[G(x, y), z] = H[x, K(y, z)] \tag{53}$$

is of the form

$$F(x, y) = f_1(x) \circ h_2[k_2(y)] = f(x) \circ g(y), \qquad (54)$$

$$G(x, y) = f_1^{-1}\{f_1[g_1(x)] \circ f_1[g_2(y)]\} = f^{-1}[k(x) \circ m(y)], \qquad (55)$$

$$H(x, y) = f_1[g_1(x)] \circ h_2(y) = k(x) \circ h(y), \qquad (56)$$

$$K(x, y) = h_2^{-1}\{f_1[g_2(x)] \circ h_2[k_2(y)]\} = h^{-1}[m(x) \circ g(y)], \qquad (57)$$

where $x \circ y$ is an arbitrary associative operation on Q.

Actually, $(Q; \circ)$ has to be a monoid as follows from:

Proof. Let $e = h_2[k_2(a)]$. Then

$$
\begin{aligned}
x \circ e &= x \circ h_2 k_2(a) \\
&= F(f_1^{-1}(x), k_2^{-1} h_2^{-1} h_2 k_2(a)) \\
&= F(f_1^{-1}(x), a) \\
&= f_1 f_1^{-1}(x) \\
&= x
\end{aligned}
$$

and

$$
\begin{aligned}
e \circ x &= h_2 k_2(a) \circ x = F(f_1^{-1} h_2 k_2(a), k_2^{-1} h_2^{-1}(x)) \\
&= F(f_1^{-1} h_2 K(a, a), k_2^{-1} h_2^{-1}(x)) \\
&= F(f_1^{-1} H(a, K(a, a)), k_2^{-1} h_2^{-1}(x)) \\
&= F(f_1^{-1} F(G(a, a), a), k_2^{-1} h_2^{-1}(x)) \\
&= F(f_1^{-1} f_1 G(a, a), k_2^{-1} h_2^{-1}(x)) \\
&= F(G(a, a), k_2^{-1} h_2^{-1}(x)) \\
&= H(a, K(a, k_2^{-1} h_2^{-1}(x))) \\
&= H(a, k_2 k_2^{-1} h_2^{-1}(x)) \\
&= h_2 h_2^{-1}(x) \\
&= x.
\end{aligned}
$$

$\square$

As Theorem 2 shows, rectangular quasigroups are very well structured groupoids, close to quasigroups, and the problem of solving generalized associativity on them comes naturally. We have the following result.

Definition 5. Rectangular quasigroups $(S; \cdot)$ and $(S; *)$ are *compatible* if there exist:

- a left zero semigroup $(L; \alpha)$
- two quasigroups $(Q; \beta)$ and $(Q; \gamma)$
- a right zero semigroup $(R; \delta)$
- a bijection $f : S \to L \times Q \times R$ which is an isomorphism of both $(S; \cdot)$ to $(L; \alpha) \times (Q; \beta) \times (R; \delta)$ and $(S; *)$ to $(L; \alpha) \times (Q; \gamma) \times (R; \delta)$.

Theorem 7. *If four compatible rectangular quasigroups A, B, C, D (on a set $S \neq \emptyset$) satisfy the generalized associativity equation (50) then they are all isotopic to the same rectangular group. The general solution of (50) (for compatible rectangular quasigroups) is given by (51)' where $\cdot$ is an arbitrary rectangular group on S and $A_1, A_2, B_1, B_2, C_1, C_2, D_1, D_2$ are arbitrary permutations of S satisfying (52).*

Proof. a) It is easy to check that rectangular quasigroups A, B, C, D defined by (51) indeed satisfy (50).

b) Let us prove the converse, i.e. that any particular solution of (50) is of the form (51).

Assume that four compatible rectangular quasigroups A, B, C, D on S satisfy (50). Compatibility implies that there is a bijection $f : S \to L \times Q \times R$ which is an isomorphism of:

- $(S; A)$ and $(L \times Q \times R; \Lambda) = (L; \cdot) \times (Q; \lambda) \times (R; \cdot)$
- $(S; B)$ and $(L \times Q \times R; \Pi) = (L; \cdot) \times (Q; \pi) \times (R; \cdot)$
- $(S; C)$ and $(L \times Q \times R; \Gamma) = (L; \cdot) \times (Q; \gamma) \times (R; \cdot)$
- $(S; D)$ and $(L \times Q \times R; \Delta) = (L; \cdot) \times (Q; \delta) \times (R; \cdot)$,

where:

- $(L; \cdot)$ is a left zero semigroup
- $\lambda, \pi, \gamma, \delta$ are quasigroup operations on Q
- $(R; \cdot)$ is a right zero semigroup.

Denote $f(x), f(y), f(z)$ by $(a, u, p), (b, v, q), (c, w, r)$ respectively and apply f to both sides of (50). We get

$$\Lambda((a, u, p), \Pi((b, v, q), (c, w, r))) = \Gamma(\Delta((a, u, p), (b, v, q)), (c, w, r)).$$
$$(58)$$

As $\Lambda, \Pi, \Gamma, \Delta$ are rectangular quasigroup operations on $L \times Q \times R$, (58) yields

$$(a, \lambda(u, \pi(v, w)), r) = (a, \gamma(\delta(u, v), w), r).$$

By Theorem 5, there is a group $+$ on Q and permutations $\lambda_1, \lambda_2, \pi_1, \pi_2,$ $\gamma_1, \gamma_2, \delta_1, \delta_2$ of Q, such that

$$\begin{cases} \lambda(x,y) = \lambda_1 x + \lambda_2 y, \\ \pi(x,y) = \lambda_2^{-1}(\lambda_2 \pi_1 x + \lambda_2 \pi_2 y), \\ \gamma(x,y) = \gamma_1 x + \gamma_2 y, \\ \delta(x,y) = \gamma_1^{-1}(\gamma_1 \delta_1 x + \gamma_1 \delta_2 y), \\ \lambda_1 = \gamma_1 \delta_1, \\ \lambda_2 \pi_1 = \gamma_1 \delta_2, \\ \lambda_2 \pi_2 = \gamma_2. \end{cases}$$

Therefore $(L \times Q \times R; *) = (L; \cdot) \times (Q; +) \times (R; \cdot)$ is a rectangular group.

Define an operation $\cdot$ on S thus: $x \cdot y = f^{-1}(fx * fy)$. As f is a bijection, $(S; \cdot)$ is a rectangular group isomorphic to $(L \times Q \times R; *)$. Moreover, if we define: $\Lambda_1 : (a, u, p) \longmapsto (a, \lambda_1 u, p)$, $\Lambda_2 : (b, v, q) \longmapsto (b, \lambda_2 v, q)$, $A_1 = f^{-1}\Lambda_1 f$ and $A_2 = f^{-1}\Lambda_2 f$, we get

$$\begin{aligned} A(x,y) &= f^{-1} f A(x,y) \\ &= f^{-1}\Lambda(fx, fy) \\ &= f^{-1}\Lambda((a, u, p), (b, v, q)) \\ &= f^{-1}((a, \lambda(u, v), q)) \\ &= f^{-1}((a, \lambda_1 u + \lambda_2 v, q)) \\ &= f^{-1}((a, \lambda_1 u, p) * (b, \lambda_2 v, q)) \\ &= f^{-1}(\Lambda_1(a, u, p) * \Lambda_2(b, v, q)) \\ &= f^{-1}(f f^{-1}\Lambda_1 f x * f f^{-1}\Lambda_2 f y) \\ &= f^{-1}\Lambda_1 f x \cdot f^{-1}\Lambda_2 f y \\ &= A_1 x \cdot A_2 y. \end{aligned}$$

Using

$$B_1 = f^{-1}(id, \pi_1, id)f, B_2 = f^{-1}(id, \pi_2, id)f, C_1 = f^{-1}(id, \gamma_1, id)f,$$

$$C_2 = f^{-1}(id, \gamma_2, id)f, D_1 = f^{-1}(id, \delta_1, id)f, D_2 = f^{-1}(id, \delta_2, id)f,$$

we can prove remaining equalities of (51) and (52).   □

Although the previous theorem generalizes the Four Quasigroups Theorem it is not comparable to Aczél's generalization. His solution depends on a monoid while the one from Theorem 7 depends on a rectangular group. Common groupoids from these two classes are groups, showing that the two generalizations lead in two different directions.

Rectangular quasigroups satisfying (50) need not be compatible, therefore Theorem 7 does not describe general solution of (50). One other type of solution (on S) is:

$$\begin{cases} A, B, C & - \quad \text{arbitrary right zero semigroups} \\ D & - \quad \text{arbitrary rectangular quasigroup.} \end{cases}$$

Therefore we pose:

Problem 2. *Find a general solution of (50) for rectangular quasigroups.*

References

[1] J. Aczél, *Lectures on Functional Equations and Their Applications*, Academic Press, New York–London, 1966.

[2] J. Aczél, V. D. Belousov, and M. Hosszú, *Generalized associativity and bisymmetry on quasigroups*, Acta Math. Acad. Sci. Hungar. **11** (1960), 127–136.

[3] T. Evans, *The word problem for abstract algebras*, J. London Math. Soc. **26** (1951), 64–71.

[4] G. Grätzer, *Universal Algebra*, D. Van Nostrand Co., Princeton–Toronto–London–Melbourne, 1968.

[5] R. A. Knoebel, *Product of independent algebras with finitely generated identities*, Algebra Universalis **3** (1973), 147–151.

[6] M. Petrich, *Structure of regular semigroups*, Université des sciences et techniques du Languedoc, Montpellier, 1977.

THE AGGREGATION EQUATION: SOLUTIONS WITH NON INTERSECTING PARTIAL FUNCTIONS

Mark Taylor

Department of Mathematics, Acadia University,
Wolfville, N.S. B0P 1X0
m.taylor@ns.sympatico.ca

Abstract   The Aggregation Equation, or functional equation of $m \times n$ generalized bisymmetry

$$G(F_1(x_{11},\ldots,x_{1n}),\ldots,F_m(x_{m1},\ldots,x_{mn}))$$
$$= F(G_1(x_{11},\ldots,x_{m1}),\ldots,G_n(x_{1n},\ldots,x_{mn}))$$

arose as a model linking micro-and macroeconomics.

It is shown that solutions to this equation from the class of functions with non intersecting partial functions are given by

$$F(z_1,\ldots,z_n) = \overline{F}(\gamma_1(z_1),\ldots,\gamma_n(z_n))$$
$$G(y_1,\ldots,y_m) = \overline{G}(\alpha_1(y_1),\ldots,\alpha_m(y_m))$$
$$F_i(x_{i1},\ldots,x_{in}) = \overline{F}_i(\beta_{i1}(x_{i1}),\ldots,\beta_{in}(x_{in}))$$
$$G_j(x_{1j},\ldots,x_{mj}) = \overline{G}_j(\delta_{1j}(x_{1j}),\ldots,\delta_{mj}(x_{mj})),$$

where $\overline{F},\overline{G},\overline{F}_i,\overline{G}_j$ are injective in all arguments and $\alpha_i,\gamma_j,\beta_{ij},\delta_{ij}$ ($i = 1,\ldots,m; j = 1,\ldots,n$) are surjections which satisfy

$$\overline{G}(\alpha_1\overline{F}_1(\beta_{11}(x_{11}),..,\beta_{1n}(x_{1n})),..,\alpha_m\overline{F}_m(\beta_{m1}(x_{m1}),..,\beta_{mn}(x_{mn})))$$
$$= \overline{F}(\gamma_1\overline{G}_1(\delta_{11}(x_{11}),..,\delta_{m1}(x_{m1})),..,\gamma_n\overline{G}_n(\delta_{1n}(x_{1n}),..,\delta_{mn}(x_{mn})))$$

This result is used to determine explicit solutions in the case $n = m = 2$.

Keywords: aggregation equation, generalized bisymmetry, partial function

Mathematics Subject Classification (2000): 39B52, 20M14, 90A99

351

Z. Daróczy and Z. Páles (eds.),
Functional Equations - Results and Advances, 351–360.
© 2002 *Kluwer Academic Publishers.*

1.   INTRODUCTION AND PRELIMINARY RESULTS

The Aggregation Equation, also known as the functional equation of $m \times n$ generalized bisymmetry is

$$G(F_1(x_{11}, \ldots, x_{1n}), \ldots, F_m(x_{m1}, \ldots, x_{mn}))$$
$$= F(G_1(x_{11}, \ldots, x_{m1}), \ldots, G_n(x_{1n}, \ldots, x_{mn})), \tag{1}$$

where m, n are integers greater than 1 and

$$F_i : X_{i1} \times \ldots \times X_{in} \to Y_i; \qquad G_j : X_{1j} \times \ldots \times X_{mj} \to Z_j;$$

$$F : Z_1 \times \ldots \times Z_n \to S; \qquad G : Y_1 \times \ldots \times Y_m \to S$$

$$(i = 1, \ldots, m; j = 1, \ldots, n).$$

This equation arose as a model linking micro- and macroeconomics. The model assumes m producers each of which has a single product. Each producer uses n inputs (raw material, labour, capital, research costs, etc.) and the ith producer has inputs $x_{ij}(j = 1, \ldots, n)$ from sets X_{ij}. The output of producer i is given by the production function $F_i(x_{i1}, \ldots, x_{in})$. The aggregate output of the m producers is given by

$$G(F_1(x_{11}, \ldots, x_{1n}), \ldots, F_m(x_{m1}, \ldots, x_{mn})).$$

The inputs $x_{1j}, x_{2j}, \ldots, x_{mj}$ are of the same "type" and are aggregated by the function $G_j(x_{ij}, \ldots, x_{mj})$. The *problem of consistent aggregation* is to determine conditions under which there exists a macroeconomic production function F which satisfies equation (1).

For a historical view of the equation, the reader is directed to [1] or [6]; more recent developments may be found in [2], [3], [4] and [5].

Despite the fact that the Aggregation Equation provided an important theoretical economic model, prior to the 1996 work of Aczel and Maksa [2] there was little hope of a thorough understanding of the equation without placing strong continuity constraints on the functions. Aczel and Maksa solved the equation for functions, defined on abstract sets by imposing surjective and injective conditions on their partial functions.

Throughout this paper functions

$$F, G, F_i, G_j \qquad (i = j, \ldots, m; j = 1, \ldots, n)$$

will be assumed to be defined on abstract non empty sets as given above. They will also be restricted to their images as they appear in the equation. Thus we assume F, G, F_i and G_j are all surjective.

Partial functions have played a major role in investigations of the equation without regularity assumptions. A qth *partial function* of a function

$$H : S_1 \times \ldots \times S_q \times \ldots \times S_p \to S \qquad (1 \le q \le p)$$

is any function obtained from H by keeping all but the qth variable constant.

Let $\mathcal{H}^q = \{H(a_1, \ldots, x, a_{q+1}, \ldots, a_p) | a_i \in S_i\}$ be the set of all q^{th} partial functions of H. Then H is said to have *non intersecting q^{th} partial functions* if for all $f, g \in \mathcal{H}^q$ and for all $a \in S_q$,

$$f(a) = g(a) \qquad \text{implies} \qquad f = g.$$

H is said to have *non intersecting partial functions* if it has non intersecting q^{th} partial functions for all $q, 1 \le q \le p$.

Examples of such functions are:

$$f(x_1, \ldots, x_p) = \sqrt{(x_1 - a_1)^2 + \ldots + (x_p - a_p)^2}, \qquad a_i \text{ constant.}$$

$$g(x_1, \ldots, x_p) = x_1^{b_1} x_2^{b_2} \ldots x_p^{b_p}, \qquad x_i > 0, b_i \text{ constant.}$$

Given a function $H : S_1 \times \ldots \times S_p \to S$, define the binary relation ${}^t H$ on $S_t (t = 1, \ldots, p)$ by; for all $a_t, b_t \in S_t$,

$$a_t \, {}^t H b_t \quad \text{if and only if for all } f \in \mathcal{H}^t, \quad f(a_t) = f(b_t).$$

Clearly ${}^t H$ is an equivalence relation on S_t. Denote by ${}^t H[a]$ the equivalence class of $a \in S_t$ under ${}^t H$, and let $S_t / {}^t H$ be the set of all such equivalence classes.

Lemma 1. *Let* $H : S_1 \times \ldots \times S_t \ldots \times S_p \to S$ *be a function,* $1 \le t \le p$, $p \ge 2$.
Then $\overline{H} : S_1 / {}^1 H \times \ldots \times S_t / {}^t H \times \ldots \times S_p / {}^p H \to S$ *given by*

$$\overline{H}({}^1 H[a_1], \ldots, {}^t H[a_t], \ldots, {}^p H[a_p]) = H(a_1, a_2, \ldots, a_p)$$

is a well defined function.

Proof. Suppose $a_t \, {}^t H b_t$; $a_t, b_t \in S_t$, $t = 1, \ldots p$. Then

$$H(a_1, x_2, \ldots, x_p) = H(b_1, x_2, \ldots, x_p) \qquad \text{for all} \quad x_i \in S_i, i = 2, \ldots p.$$

In particular,

$$H(a_1, a_2, a_3, \ldots, a_p) = H(b_1, a_2, a_3, \ldots, a_p).$$

However $a_2\,{}^2Hb_2$ implies

$$H(x_1, a_2, x_3, \ldots, x_p) = H(x_1, b_2, x_3, \ldots, x_p)$$
$$(x_k \in S_k, k = 1, 3, 4, \ldots, p)$$

with the consequence that

$$H(b_1, a_2, a_3, \ldots, a_p) = H(b_1, b_2, a_3, \ldots, a_p),$$

and therefore

$$H(b_1, a_2, a_3, \ldots, a_p) = H(b_1, b_2, a_3, \ldots, a_p).$$

Eventually,

$$H(a_1, a_2, \ldots, a_p) = H(b_1, b_2, \ldots, b_p),$$

and $\overline{H}$ is seen to be well defined. $\square$

The function $\overline{H}$ defined in Lemma 1 is called the *reduction of H*. If

$$\varphi_i : S_i \to S_i/{}^iH, \qquad (i = 1, \ldots, p) \quad \text{is the canonical surjection}$$

$$\varphi_i(a_i) = {}^iH[a_i], \quad \text{then} \quad \overline{H}(\varphi_1(a_1), \ldots, \varphi_p(a_p)) = H(a_1, \ldots a_p).$$

Lemma 2. *Let* $H : S_1 \times \ldots \times S_p \to S$ *have non intersecting partial functions. Then every partial function of* $\overline{H}$, *the reduction of H, is injective.*

Proof. We show $\overline{H}$ is injective in the ith variable, $1 \le i \le p$. Suppose

$$\overline{H}(\varphi_1(a_1), \ldots, \varphi_i(a_i), \ldots \varphi_p(a_p)) = \overline{H}(\varphi_1(a_1), \ldots, \varphi_i(b_i), \ldots, \varphi_p(a_p))$$
$$(a_k \in S_k, k = 1, \ldots p, b_i \in S_i).$$

Then, from the definition of $\overline{H}$,

$$H(a_1, \ldots, a_q, \ldots, a_i, \ldots, a_p) = H(a_1, \ldots, a_q, \ldots, b_i, \ldots, a_p)$$

and because H has non intersecting partial functions

$$H(a_1, \ldots, x_q, \ldots, a_i, \ldots, a_p) = H(a_1, \ldots, x_q, \ldots, b_i \ldots, a_p).$$

This leads to

$$H(x_1, \ldots, a_i, \ldots, x_p) = H(x_1, \ldots, b_i, \ldots, x_p)$$

and consequently,

$$\varphi_i(a_i) = \varphi_i(b_i).$$

□

Lemma 3. *The function* $H : S_1 \times \ldots \to S$ *has non intersecting partial functions if and only if there exist a function* $\overline{H} : \overline{S}_1 \times \ldots \times \overline{S}_p \to S$ *with all partial functions injective and non intersecting, together with surjections* $f_q : \overline{S}_q \to S_q$ *such that*

$$H(x_1, \ldots, x_p) = \overline{H}(f(x_1), \ldots, f(x_p)), \qquad (x_q \in S_q, q = 1, \ldots, p).$$

Proof. Lemma 1 ensures for a function H with non intersecting partial functions, the function $\overline{H}$ exists, together with the canonical surjections f_q. That $\overline{H}$ has all partial functions injective is guaranteed by Lemma 2. It remains to show that $\overline{H}$ also has non intersecting partial functions.

Suppose

$$\overline{H}(a_1, \ldots, a_q, \ldots, q_p) = \overline{H}(b_1, \ldots, a_q, \ldots, b_p)$$
$$(a_k, b_k \in \overline{S}_k, k = 1, \ldots, p).$$

Then there exist $s_k, t_k \in S_k$ such $f_k(x_k) = a_k, f_k(t_k) = b_k, k = 1, \ldots, p$ which leads to

$$H(s_1, \ldots, s_q, \ldots, s_p) = H(t_1, \ldots, s_q, \ldots, t_p).$$

H has non intersecting partial functions, therefore

$$H(s_1, \ldots, s, \ldots, s_p) = H(t_1, \ldots, s, \ldots, t_p) \qquad \text{for all} \quad s \in S_q.$$

Consequently,

$$\overline{H}(a_1, \ldots, f_q(s), \ldots, a_p) = \overline{H}(b_1, \ldots, f_q(s), \ldots, b_q), \qquad \text{for all} \quad s \in S_q.$$

The surjectivity of f_q now ensures $\overline{H}$ has non intersecting q^{th} partial functions and consequently all its partial functions are non intersecting. The converse has a similar simple proof. □

2.   SOLUTIONS OF EQUATION (1) WITH NON INTERSECTING PARTIAL FUNCTIONS

The preliminary results lead immediately to,

Theorem 1. *Equation* (1) *is satisfied by functions* F, G, F_i, G_j *with non intersecting partial functions if and only if there exist functions* $\overline{F}, \overline{G}, \overline{F}_i, \overline{G}_j$ *which have all partial functions non intersecting and injective, and surjections* $\alpha_i, \gamma_j, \beta_{ij}, \delta_{ij}$ $(i = 1, \ldots, m; j = 1, \ldots, n)$, *which* *together satisfy*

$$\overline{G}(\alpha_1 \overline{F}_1(\beta_{11}(x_{11}), \ldots, \beta_{1n}(x_{1n})), \ldots, \alpha_m \overline{F}_m(\beta_{m1}(x_{m1}), \ldots, \beta_{mn}(x_{mn})))$$
$$= \overline{F}(\gamma_1 \overline{G}_1(\delta_{11}(x_{11}), \ldots, \delta_{m1}(x_{m1})), \ldots, \gamma_n \overline{G}_n(\delta_{1n}(x_{1n}), \ldots, \delta_{mn}(x_{mn})))$$
$$(2)$$

and

$$F(z_1, \ldots, z_n) = \overline{F}(\gamma_1(z_1), \ldots, \gamma_n(z_n))$$
$$G(y_1, \ldots, y_m) = \overline{G}(\alpha_1(y_1), \ldots, \alpha_m(y_m))$$
$$F_i(x_{i1}, \ldots, x_{in}) = \overline{F}_i(\beta_{i1}(x_{i1}), \ldots, \beta_{in}(x_{in}))$$
$$G_j(x_{1j}, \ldots, x_{mj}) = \overline{G}_j(\delta_{1j}(x_{1j}), \ldots, \delta_{mj}(x_{mj})).$$

The solutions of the aggregation equation obtained in [2] and [4] depend on results obtained for the specific form of the equation when $m = n = 2$. We now apply Theorem 1 to this case.

First the definitions of quasigroup and loop.

A *quasigroup* $(Q, \cdot)$, consists of a non empty set Q together with a binary operation "$\cdot$" such that the binary operation, which is a function of two variables, has all partial functions bijective.

A *loop* $(L, \cdot)$ is a quasigroup with an identity element i.e. an element $e \in L$ such that $ex = xe = x$, for all $x \in L$.

Theorem 2. *Equation (1) is satisfied for $m = n = 2$ by functions $F, G, F_i, G_j (i, j = 1, 2)$ with non intersecting partial functions and further F_i, G_j have surjective partial functions if, and only if, there exist loops $(Y_i, +_i), (Z_j, \oplus_j)$ and an abelian group $(S, +)$ with epimorphisms $g_i : Y_i \to S, h_j : Z_j \to S$, surjections $H_{ij} : X_{ij} \to Y_i$, $K_{ij} : X_{ij} \to Z_j$ such that $g_i H_{ij} = h_i K_{ij}$ and*

$$F(z_1, z_2) = h_1(z_1) + h_2(z_2)$$
$$G(y_1, y_2) = g_1(y_1) + g_2(y_2)$$
$$F_i(x_{i1}, x_{i2}) = H_{i1}(x_{i1}) +_i H_{i2}(x_{i2})$$
$$G_j(x_{1j}, x_{2j}) = K_{1j}(x_{1j}) \oplus_j K_{2j}(x_{2j}) \qquad (i, j = 1, 2).$$

Proof. Under the supposition that $F, G, F_i, G_j (i, j = 1, 2)$ all have non intersecting partial functions, Theorem 1 allows us to use

$$\overline{G}(\alpha_1 \overline{F}_1(\beta_{11}(x_{11}), \beta_{12}(x_{12})), \alpha_2 \overline{F}_2(\beta_{21}(x_{21}), \beta_{22}(x_{22})))$$
$$= \overline{F}(\gamma_1 \overline{G}_1(\delta_{11}(x_{11}), \delta_{21}(x_{21})), \gamma_2 \overline{G}_2(\delta_{12}(x_{12}), \delta_{22}(x_{22})))$$
$$(3)$$

instead of the corresponding form of (1). The partial functions of $\overline{F}, \overline{G}$, $\overline{F}_i, \overline{G}_j$ are injective, and from the condition that F_i, G_j have surjective partial functions it follows that the partial functions of $\overline{F}_i$ and $\overline{G}_j$ are bijective. This is also true for the partial functions of $\overline{F}$ and $\overline{G}$; we will illustrate the proof:

Consider the partial function of $\overline{G}$ given by $\overline{G}(x, c)$ where c is constant.

The surjectivity of β_{2j} and α_2 ensure the existence of $a_{2j} \in X_{2j}(j = 1, 2)$ such that $c = \alpha_2 \overline{F}_2(\beta_{21}(a_{21}), \beta_{22}(a_{22}))$, and the variable x may be replaced by $\alpha_1 \overline{F}_1(\beta_{11}(x_{11}), \beta_{12}(x_{12}))$ where $x_{ij} \in X_{ij}$. Thus

$$\overline{G}(\alpha_1 \overline{F}_1(\beta_{11}(x_{11}), \beta_{12}(x_{12}), \alpha_2 \overline{F}_2(\beta_{21}(a_{21}), \beta_{22}(a_{22}))$$
$$= \overline{F}(\gamma_1 \overline{G}_1(\delta_{11}(x_{11}), \delta_{21}(a_{21})), \gamma_2 \overline{G}_2(\delta_{12}(x_{12}), \delta_{22}(a_{22})).$$

The functions given by $\gamma_j \overline{G}_j(\delta_{1j}(x_{1j}), \delta_{2j}(a_{2j})), (j = 1, 2)$ are surjective and $\overline{F}$ is onto S, consequently the given partial function of $\overline{G}$ is also onto S.

Now choose $a_{ij} \in X_{ij}$ and let $b_{ij} = \beta_{ij}(a_{ij}), \delta_{ij}(a_{ij}) = d_{ij}$,

$$\overline{F}_i(b_{i1}, b_{i2}) = b_i, \quad \overline{G}_j(d_{1j}, d_{2j}) = d_j, \alpha_i(b_i) = a_i \quad \text{and} \quad \gamma_j(d_j) = c_j,$$
$$(i, j = 1, 2).$$

Define bijections

$$\overline{G}^1(u_1) = \overline{G}(u_1, a_2); \qquad \overline{G}^2(u_2) = \overline{G}(a_1, u_2)$$
$$\overline{F}^1(v_1) = \overline{F}(v_1, c_2); \qquad \overline{F}^2(v_2) = \overline{F}(c_1, v_2).$$

Binary operations $+$ and $\oplus$ on S are defined by

$$\overline{F}(v_1, v_2) = \overline{F}^1(v_1) + \overline{F}^2(v_2)$$
$$\overline{G}(u_1, u_2) = \overline{G}^1(u_1) \oplus \overline{G}^2(u_2).$$

Then $(S, +)$ and $(S, \oplus)$ are loops with identity elements $\overline{F}(c_1, c_2)$ and $\overline{G}(a_1, a_2)$ respectively.

Equation (3) may then be written as

$$\overline{G}^1(\alpha_1 \overline{F}_1(\beta_{11}(x_{11}), \beta_{12}(x_{12})) \oplus \overline{G}^2(\alpha_2 \overline{F}_2(\beta_{21}(x_{21}), \beta_{22}(x_{22}))$$
$$= \overline{F}^1(\gamma_1 \overline{G}_1(\delta_{11}(x_{11}), \delta_{21}(x_{21})) + \overline{F}^2(\gamma_2 \overline{G}_2(\delta_{12}(x_{12}), \delta_{22}(x_{22})). \tag{4}$$

Substituting in equation (4), $x_{ij} = a_{ij}$ except for $i = j = 1$.

$$\overline{G}^1(\alpha_1 \overline{F}_1(\beta_{11}(x_{11}), b_{12})) = \overline{F}^1(\gamma_1 \overline{G}_1(\delta_{11}(x_{11}), d_{21})). \tag{5}$$

Introducing the notation

$$\overline{F}_i(x_1, b_{i2}) = \overline{F}_i^1(x_1); \qquad \overline{G}_j(y_1, d_{2j}) = \overline{G}_j^1(y_1)$$
$$\overline{F}_i(b_{i1}, x_2) = \overline{F}_1^2(x_2); \qquad \overline{G}_j(d_{1j}, y_2) = \overline{G}_j^2(y_2)$$

allows us to write equation (5) as

$$\overline{G}^1\alpha_1\overline{F}_1^1\beta_{11}(x_{11}) = \overline{F}^1\gamma_1\overline{G}^1\delta_{11}(x_{11}).$$

Further appropriate substitutions into equation (4) yield

$$\overline{G}^i\alpha_i\overline{F}_i^j\beta_{ij}(x_{ij}) = \overline{F}^j\gamma_j\overline{G}_j^i\delta_{ij}(x_{ij}), \tag{6}$$

where $i, j = 1, 2$.

We now let $x_{12} = a_{12}$ and $x_{21} = a_{21}$ in equation (4) and through the relationship in (6) it becomes clear that the operations $+$ and $\oplus$ are identical.

Turning our attention to the inner functions $\overline{F}_i$ and $\overline{G}_j$, we define operations $+_i$ on Y_i $(i = 1, 2)$ by

$$\overline{F}_i(x_1, x_2) = \overline{F}_i^1(x_1) +_i \overline{F}_i^2(x_2)$$

and operations $\oplus_j$ on $Z_j (j = 1, 2)$ by

$$\overline{G}_j(y_1, y_2) = \overline{G}_j^1(y_1) \oplus_j \overline{G}_j^2(y^2).$$

The bijectivity of all partial functions of $\overline{F}_i$ and $\overline{G}_j$ ensures that $(Y_i, +_i)$ and $(Z_j, \oplus_j)$ are quasigroups. Moreoever each quasigroup has an identity element and is, therefore, a loop.

Equation (4) may now be written as

$$\begin{aligned}
\overline{G}^1(\alpha_1(\overline{F}_1^1(\beta_{11}(x_{11})) &+_1 \overline{F}_1^2(\beta_{12}(x_{12}))) \\
+ \overline{G}^2(\alpha_2(\overline{F}_2^1(\beta_{21}(x_{21})) &+_2 \overline{F}_2^2(\beta_{22}(x_{22}))) \\
= \overline{F}^1(\gamma_1(\overline{G}_1^1(\delta_{11}(x_{11})) &\oplus_1 \overline{G}_1^2(\beta_{21}(x_{21}))) \\
+ \overline{F}^2(\gamma_2(\overline{G}_2^1(\delta_{12}(x_{12})) &\oplus_2 \overline{G}_2^2(\delta_{22}(x_{22}))).
\end{aligned} \tag{7}$$

Setting $x_{21} = a_{21}$ and $x_{22} = a_{22}$ in equation (7) gives

$$\begin{aligned}
\overline{G}^1(\alpha_1(\overline{F}_1^1(\beta_{11}(x_{11})) &+_1 \overline{F}_1^2(\beta_{12}(x_{12}))) \\
= \overline{F}^1\gamma_1\overline{G}_1^1\delta_{11}(x_{11}) &+ \overline{F}^2\gamma_2\overline{G}_1^2(\delta_{12}(x_{12}))
\end{aligned}$$

and because of equation (5) this becomes

$$\begin{aligned}
\overline{G}^1(\alpha_1(\overline{F}_1^1(\beta_{11}(x_{11})) &+_1 \overline{F}_1^2(\beta_{12}(x_{12}))) \\
= \overline{G}^1\alpha_1\overline{F}_1^1\beta_{11}(x_{11}) &+ \overline{G}^1\alpha_1\overline{F}_1^2(\beta_{12}(x_{12})).
\end{aligned} \tag{8}$$

The functions $\overline{F}_i^j\beta_{ij}(i = 1, 2)$ are all surjections which allows us to write

$$\overline{F}_1^1\beta_{11}(x_{11}) = y_1, \overline{F}_1^2\beta_{12}(x_{12}) = y_2$$

and transform equation (8) into

$$\overline{G}^1\alpha_1(y_1 +_1 y_2) = \overline{G}^1\alpha_1(y_1) + \overline{G}^1\alpha_1(y_2) \qquad (y_1, y_2 \in Y_1).$$

Thus $\overline{G}^1\alpha_1$ is an epimorphism from $(Y_1, +_1)$ to $(S, +)$.

In a similar manner it can be shown that $\overline{G}^2\alpha_2$ and $\overline{F}^i\gamma_i (i = 1, 2)$ are also epimorphisms onto $(S, +)$. This information together with equation (6) allows equation (7) to be written as

$$(u + v) + (s + t) = (u + s) + (v + t) \qquad (u, v, s, t \in S).$$

It quickly follows that $(S, +)$ is an abelian group.

Writing

$$\overline{G}^i\alpha_i = g_i, \qquad \overline{F}^j\gamma_j = h_j$$

$$H_{ij} = \overline{F}_i^j\beta_{ij} \quad \text{and} \quad K_{ij} = \overline{G}_j^i\delta_{ij} \qquad (i, j = 1, 2)$$

transforms equation (6) into

$$g_i H_{ij}(x_{ij}) = h_j K_{ij}(x_{ij}) \qquad (i, j = 1, 2)$$

Theorem 1 then leads us to

$$\begin{aligned}
F(z_1, z_2) &= \overline{F}(\gamma_1(z_1), \gamma_2(z_2)) \\
&= \overline{F}^1\gamma_1(z_1) + \overline{F}^2\gamma_2(z_2) \\
&= h_1(z_1) + h_2(z_2).
\end{aligned}$$

In a similar way,

$$G(y_1, y_2) = g_1(y_1) + g_2(y_2)$$
$$F_i(x_{i1}, x_{i2}) = H_{i1}(x_{i1}) +_i H_{i2}(x_{i2})$$
$$G_j(x_{ij}, x_{2j}) = K_{ij}(x_{ij}) \oplus_j K_{2j}(x_{2j}) \qquad (i, j = 1, 2).$$

This completes the first part of the proof.

For the converse, it is easily shown that the given F, G, F_i, G_j satisfy equation (1). It remains to demonstrate that the given functions have non intersecting partial functions. Let $H : S_1 \times S_2 \to S$, $f_i : S_i \to S$ $(i = 1, 2)$ be surjections and let $(S, *)$ be a loop such that

$$H(s_1, s_2) = f_1(s_1) * f_2(s_2).$$

Suppose $a_1, b_1 \in S_1$ and $a_2 \in S_2$ are such that

$$H(a_1, a_2) = H(b_1, a_2).$$

Then
$$f_1(a_1) * f_2(a_2) = f_1(b_1) * f_2(a_2).$$
The operation is cancellative, therefore $f_1(a_1) = f_1(b_1)$.

From this we see that

$$f_1(a_1) * f_2(x) = f_1(b_1) * f_2(x) \qquad \text{for all} \quad x \in S_2.$$

This shows
$$H(a_1, x) = H(b_1, x) \qquad \text{for all} \quad x \in S_2.$$

Similarly if
$$H(a_1, a_2) = H(a_1, b_2)$$

then
$$H(x, a_2) = H(x, b_2) \qquad \text{for all} \quad x \in S_1.$$

This completes the proof.   □

References

[1] J. Aczél, *Bisymmetry and Consistent Aggregation: Historical Review and Recent Results*, in *Choice, Decision, and Measurement* (ed. by A. A. J. Marley), Lawrence Erlbaum Assoc. Publ., Mahwah, NJ, 225–233, 1997.

[2] J. Aczél and Gy. Maksa, *Solution of the rectangular $m \times n$ generalized bisymmetry equation and of the problem of consistent aggregation*, J. Math. Anal. Appl. **203** (1996), 104–126.

[3] J. Aczél, Gy. Maksa, and M. A. Taylor, *Equations of generalized bisymmetry and of consistent aggregation: weakly surjective solutions which may be discontinuous at places*, J. Math. Anal. Appl. **214** (1997), 22–35.

[4] M. A. Taylor, *The generalized equation of bisymmetry: solutions based on cancellative abelian monoids*, Aequationes Math. **57** (1999), 288–302.

[5] M. A. Taylor, *On the equation of generalized bisymmetry with outer functions injective in each argument*, Aequationes Math. **60** (2000), 283–290.

[6] J. van Daal and A. Merkies, *The problem of aggregation of individual economic relations: consistency and representativity in a historical perspective*, in *Measurement in Economics* (ed. by W. Eichhorn), Physica, Heidelberg Verlag, 607–637, 1987.

Advances in Mathematics

1. A. Nagy: *Special Classes of Semigroups*. 2001   ISBN 0-7923-6890-8
2. P.A. Grillet: *Commutative Semigroups*. 2001   ISBN 0-7923-7067-8
3. Z. Daróczy and Z. Páles (eds.): *Functional Equations - Results and Advances*. 2002
                                                                      ISBN 1-4020-0485-0

KLUWER ACADEMIC PUBLISHERS – DORDRECHT / BOSTON / LONDON